CAN System Engineering

Wolfhard Lawrenz
Editor

CAN System Engineering

From Theory to Practical Applications

Editor
Wolfhard Lawrenz
C&S group GmbH
Wolfenbüttel
Germany

ISBN 978-1-4471-6802-7 ISBN 978-1-4471-5613-0 (eBook)
DOI 10.1007/978-1-4471-5613-0
Springer London Heidelberg New York Dordrecht

© Springer-Verlag London 1997, 2013
Softcover reprint of the hardcover 2nd edition 2013
This work is subject to copyright. All rights are reserved by the Publisher, whether the whole or part of the material is concerned, specifically the rights of translation, reprinting, reuse of illustrations, recitation, broadcasting, reproduction on microfilms or in any other physical way, and transmission or information storage and retrieval, electronic adaptation, computer software, or by similar or dissimilar methodology now known or hereafter developed. Exempted from this legal reservation are brief excerpts in connection with reviews or scholarly analysis or material supplied specifically for the purpose of being entered and executed on a computer system, for exclusive use by the purchaser of the work. Duplication of this publication or parts thereof is permitted only under the provisions of the Copyright Law of the Publisher's location, in its current version, and permission for use must always be obtained from Springer. Permissions for use may be obtained through RightsLink at the Copyright Clearance Center. Violations are liable to prosecution under the respective Copyright Law.
The use of general descriptive names, registered names, trademarks, service marks, etc. in this publication does not imply, even in the absence of a specific statement, that such names are exempt from the relevant protective laws and regulations and therefore free for general use.
While the advice and information in this book are believed to be true and accurate at the date of publication, neither the authors nor the editors nor the publisher can accept any legal responsibility for any errors or omissions that may be made. The publisher makes no warranty, express or implied, with respect to the material contained herein.

Printed on acid-free paper

Springer is part of Springer Science+Business Media (www.springer.com)

Preface

Controller Area Network—CAN—is a communication protocol, which had been developed by R. Bosch GmbH in the beginning of the 1980s.The design focus was to support robust applications in cars. The protocol then was introduced to the market in cooperation of Bosch and Intel. In the year 1990 Mercedes Benz was the first car manufacturer who applied CAN in a series application, in the S-class car for networking of body electronics. The first suppliers of CAN modules at that time were Intel and Motorola. Currently almost all minor and major semiconductor manufacturers have CAN products in their portfolio. In 1997 24 million CAN interfaces were produced in 1 year; 2 years later there were already more than three times as many. Currently there will probably be more than a billion per year.

In the CAN introduction phase—in the end of the 80s and the beginning of the 90s—mostly so called "stand-alone" solutions were offered, which could be easily connected to any type of micro-controller. This strategy fostered the broad application of CAN, because CAN application was not only limited to the micro-controller families of the pioneer semiconductor manufacturers. Since the 90s more and more integrated solutions—CAN together with a microcontroller on one single chip—were launched on the market. However the so called transceiver—the interface between the analogue signals on the bus lines and the digital signals of the CAN controller—is even still today typically implemented in a separate chip due to its different semiconductor technology. The integrated version of CAN and micro-controller saves die size and connecting pins, increases reliability and reduces cost. These factors and the CAN quality itself are some reasons for the overwhelming CAN market dissemination. Currently, CAN communication networking is applied widely in cars, but also in avionics, trains, military applications, industrial controls, etc.

The big interest in CAN is mirrored by the huge amount of publications related to CAN. This is another reason for this newly revised second edition of the English-language CAN book, which corresponds to the 5th edition of the German-language CAN book. This edition addresses the various issues and questions when applying CAN for communication networks. As such, there is a short introduction into the CAN basics. Furthermore, problems and solutions are discussed for the physical layout of networks including EMC issues and topology layout. Additionally, quality issues and, especially, test techniques are addressed. A special feature of this CAN book is that all the technical details have been contributed by different authors who are widely known specialists in their field of expertise.

Wolfhard Lawrenz, Wolfenbuettel
July 2013

Contents

1 CAN Basic Architectures 1
Wolfhard Lawrenz

2 Physical Layer 41
Wolfhard Lawrenz, Cornelius Butzkamm, Bernd Elend, Thorsten Gerke, Magnus-Maria Hell, Ursula Kelling, Bernd Koerber, Kurt Mueller, Christian Schmitz, Radoslaw Watroba and Rolf Weber

3 Data Link Layer Implementation 131
Wolfhard Lawrenz, Florian Hartwich, Ursula Kelling, Vamsi Krishna, Roland Lieder and Peter Riekert

4 Higher Level Protocols 173
Gangolf Feiter, Lars-Berno Fredriksson, Karsten Hoffmeister, Joakim Pauli and Holger Zeltwanger

5 Applications 255
Guenter Reichart, Gabriel Leen, Nathalie Courmont, Ralph Knüppel, Christian Schmid and Markus Brockmann

6 Testing 283
Wolfhard Lawrenz, Federico Cañas, Maria Fischer, Stefan Krauß, Lothar Kukla and Nils Obermoeller

Bibliography 345

Index 351

Contributors

Markus Brockmann WILO AG, Nortkirchenstrasse, Dortmund, Germany

Cornelius Butzkamm C&S group GmbH, Am Exer, Wolfenbuettel, Germany

Federico Cañas Quellweg 27, Berlin, Germany

Nathalie Courmont Airbus France S.A.S., 316 Route de Bayonne, Toulouse Cedex, France

Bernd Elend NXP Seminconductors Germany GmbH, Streesemannallee, Hamburg, Germany

Gangolf Feiter Concepts & Services Consulting, Alte Landstrasse, Heinsberg, Germany

Maria Fischer C&S group GmbH, Waldweg, Wolfenbuettel, Germany

Bernd Koerber Westsächsische Hochschule Zwickau, Dr.-Friedrichs-Ring, Zwickau, Germany

Lars-Berno Fredriksson Kvaser AB, Aminogatan, Mölndal, Sweden

Thorsten Gerke Synopsys GmbH, Karl-Hammerschmidt-Strasse, Aschheim-Dornach, Germany

Florian Hartwich Robert Bosch GmbH, Tuebinger Strasse, Reutlingen, Germany

Magnus-Maria Hell Infineon Technologies AG, Am Campeon, Neubiberg, Germany

Karsten Hoffmeister Elektrobit Automotive GmbH, Max-Stromeyer-Strasse, Konstanz, Germany

Ursula Kelling Infineon Technologies AG, Am Campeon, Neubiberg, Germany

Stefan Krauß Vector Informatik GmbH, Ingersheimer Strasse ,Stuttgart, Germany

Vamsi Krishna Xilinx India Technology Services Pvt. Ltd., Cyber Pearl, Hi-tec City, Madhapur, Hyderabad, India

Ralph Knüppel Airbus Deutschland GmbH, Hünefeldstr. Bremen, Germany

Lothar Kukla C&S group GmbH, Am Exer, Wolfenbüttel, Germany

Wolfhard Lawrenz C&S group GmbH, Waldweg, Wolfenbuettel, Germany

Gabriel Leen BMW AG, Petuelring, Munich, Germany

Roland Lieder Renesas Electronics Europe GmbH, Arcadiastrasse, Duesseldorf, Germany

Kurt Mueller Synopsys, Inc., 2025 NW Cornelius Pass Road, Hillsboro, OR, USA

Nils Obermoeller C&S group GmbH, Am Exer, Wolfenbüttel, Germany

Joakim Pauli Volvo Powertrain Corporation, Gropegårdsgatan, SE, Göteborg, Sweden

Peter Riekert Ingenieurbüro für IC-Technologie, Kleiner Weg, Wertheim, Germany

Christian Schmitz ELMOS Semiconductor AG, Heinrich Hertz Strasse, Dortmund, Germany

Christian Schmid Airbus Deutschland GmbH, Hünefeldstr. Bremen, Germany

Radoslaw Watroba STMicroelectronics Application GmbH, Bahnhofstraße, Aschheim-Dornach, Germany

Rolf Weber ELMOS Semiconductor AG, Heinrich Hertz Strasse, Dortmund, Germany

Holger Zeltwanger CAN in Automation (CiA) GmbH, Kontumazgarten, Nuremberg, Germany

Abbreviations

A

ABS	*Anti-lock Braking System*
ABT	*Automatic Block Transfer* for the transmission of multiple CAN messages without CPU interaction
AC	*Alternating Current*
ACC	*Adaptive Cruise Control*
ACK	*Acknowledge*, e.g. ACK-Bit in CAN-messages
ADC	*Analog Digital Converter*
AEE	*Architecture électronique électrique*—Denotes the electrionic architecture at PSA PEUGEOT CITROËN
AEEC	*Airlines Electronic Engineering Committee*—The AEEC developed standards and technical solutions for avionics, cabin systems and networks
AFCAN	*Advanced Full-CAN*—controller architecture microcontrollers from NEC with optional diagnostic functionality called "DAFCAN"
AFDX	*Avionics Full Duplex Switched Ethernet*—Ethernet-based protocol, supplemented by data rate control (QoS) and deterministic routing
AFIR	*(Xilinx) Acceptance Filter ID Register*
AFMR	*(Xilinx) Acceptance Filter Mask Register*
AFR	*(Xilinx) Acceptance Filter Register*
AM	*Amplitude Modulation*
AMP	*Arbitration on Message Priority*
API	*Application Programming Interface*
ARINC	*Aeronautical Radio Inc.*—various technical standards of the aerospace industry; create and maintained by ARINC
ARPANET	*Advanced Research Projects Agency Network*—precursor to the Internet
ASAP	*Asynchronous Service Access Protocol*—describes a standardized way to start, manage, and monitor long running services
ASC	*ASCII*—ASCII-encoded text file

ASCB-D	*Avionics Standard Communications Bus, rev. D*—bus structure for networking avionics modules from Honeywell
ASCII	*American Standard Code for Information Interchange*—known 7-bit character encoding with 128 characters, consisting of 95 printable and 33 non-printable characters
ASIC	*Application-Specific Integrated Circuit*—also: *Custom-Chip*, called an application-specific integrated circuit
ASP	*Abstract Service Primitive*
ATA	*Air Transport Association*
ATM	*"Anyone-to-Many"*—principle of communication
AUTOSAR	*AUTomotive Open System ARchitecture*

B

BACnet	*Building Automation and Control Networks*—a network protocol for building automation
BCC	*Block Check Character*—approval of a data block in the longitudinal parity checking
BCI	*Bulk Current Injection*
BIOS	*Basic Input Output System*—Software that allows the start of a PC
BRP	*Baud Rate Prescaler*—also: BRPR, Baud rate prescaler, directs the TQ-stroke from the oscillator clock
BSI	*Boîtier de Servitude Intelligent*—Acronym refers to the central control unit in PSA PEUGEOT CITROËN
BSP	*Bit Stream Processor*—Bit stream processor, responsible for serialization and deserialization of messages and the insertion and removal of transport information (Stuff-Bits, CRC etc.)
BSW	*Basic Software*
BTL	*1) Bit Timing Logic* *2) Backplane Transceiver Logic*
BTR	*Bit Timing Register*
BUSK	Bus Coupler for decentralized heating control system GENIAX

C

CAN	*Controller Area Network*
CAN_H	*CAN High*—also: CANH. One of the CAN bus lines, bus signal with dominant at a higher potential than CAN-Low (CAN_L)
CAN_L	*CAN Low*—also: CANL. One of the CAN bus lines, bus signal with dominant at a lower potential than CAN-High (CAN_H)
CAT	*Category* (of Twisted-Pair-Cables)
CCT	*CAN Conformance Tester*—test control software of the C&S group for conformance testing of CAN controllers
CD	*Compact Disc*
CEN	*(Xilinx) CAN Enable*—Bit in the software reset register (SRR), that activates the CAN controller

CFI	*Canonical Format Indicator*—Bit in the ethernet-frame-header
CiA	*CAN in Automation*—international users and manufacturers with stakeholders *CAN Application Layer* (IG CAL), *CANopen* (IG CANopen) and *SAE J1939* (IG J1939) with the aim of spreading and standardization of CAN in the industry
CIP	*Common Industrial Protocol*—an open protocol for industrial automation applications
CISPR	*Comité international spécial des perturbations radioélectriques*—official translation: International Special Committee on Radio Interference
CK	*CanKingdom*—short form for the CanKingdom Meta-High-Level-Protocol
CLK	*Clock*
CMC	*Common Mode Choke*
CO_2	*Carbon Dioxide*—chemical compound of carbon and oxygen, acidic, non-flammable, colorless and odorless gas
COAX	*Coaxial*—short form for "Coaxialcable", a cable with a central inner conductor, a dielectric surrounding the insulation and an outer conductor which also serves as the shield, enclosed by a protective jacket
COB-ID	*Communication Object Identifier*
CPU	*Central Processing Unit*
CRC	*Cyclic Redundancy Check*—Method for determining a check value for data in order to detect errors in transmission or storage may
CRI	*Certification Review Item*—Subject to an acceptance test
CS	*Chip Select*
CSMA/CA	*Carrier Sense Multiple Access/Collision Avoidance*
CSMA/CD	*Carrier Sense Multiple Access/Collision Detection*
CSMA/CR	*Carrier Sense Multiple Access/Collision Resolution*
CSR	*Command and Status Register*
CSV	*Comma Separated Values*

D

DAL	*Design Assurance Level*
DBC	CANdb-database-file, Channel definition for CAN bus in the format of Vector computer science
DC	*Direct Current*
DDM	*Driver Door Module*
DID	*Data Identifier*
DIN	formerly: German industrial standard; Today, the name is regarded as characteristic of DIN community work of the German Institute for Standardization. DIN standards are recommendations and can be applied
DI	*Digital I/O*

DIP-Schalter *DIP Dual In-line Package*—several typical single-pole switches in a design with two parallel rows of connection
DLC *Data Length Code*—length code for the amount of user data in a CAN message
DMA *Direct Memory Access*
DMIPS/MHz *Dhrystone MIPS/Megahertz*—A synthetic benchmark program to measure the integer performance; Loop cycles through the program are counted for performance evaluation, regardless of the number of pure instructions; Expressiveness by specifying better than pure MIPS
DN *Data New*—Flag that indicates the content of a receive buffer as updated
DOC *Data Object Code*
DP *Process Fieldbus for Decentralized Peripherals*—Profibus solution for Low-Cost-Sensor-/Actuator-networks specializing in brief messages at high speed
DPI *Direct Power Injection*—Method for direct interference coupling into the pins of the test object
DSC *Digital Signal Controller*
DSD *Debug Service Data*
DSO *Digital Storage Oscilloscope*
DSP *Digital Signal Processor*
DTC *Diagnostic Trouble Code*
DVD *Digital Versatile Disc*

E
EAN *European Article Number*—Product labeling for commercial items, today: International Article Number
EASA *European Aviation Safety Agency* URL: www.easa.europa.eu
EAST-ADL *Electronics Architecture and Software Technology—Architecture Description Language*
EBCDIC *Extended Binary Coded Decimals Interchange Code*
eBUS *Energy Bus*—Simple UART protocol for heating applications
ECC *Error Correction Code*—Procedures that can help detect and correct errors of redundancy in stored data
ECM *Engine Control Module*
ECU *Electronic Control Unit*
EDK *Embedded Development Kit*
EDS *Electronic Data Sheet*
EEC *Exception Event Channel*—Acronym for logical communication channel ARINC 825 for exception events
EED *Emergency Event Data*
EEPROM *Electrically Erasable Programmable Read-only Memory*

EIB	*European Installation Bus* – European installation of building automation, according to standard EN 50090, current name: KNX, standardized in ISO/IEC 14543-3
EMC	*Electromagnetic Compatibility*
EME	*Electromagnetic Emissions*
EMI	*Electromagnetic Interference*
EN	*European Standard*
EN, ENT	*Enable (Not), Enable Transmitter*—Mode selection control signal for a transceiver, and for activating the transmission stage
EOF	*End Of Frame*
ESD	*Electro-Static Discharge* –can destroy a semiconductorblock without adequate safeguards
ESPRIT	*European Strategic Program on Research in Information Technology* –
EU	*European Union*

F

FAA	*Federal Aviation Administration* URL: www.faa.gov
FCS	*Frame Check Sequence*—Check information generated by the CRC method and transmitted along with the protected message, in representations of data frames often called “CRC field” be lines, in an Ethernet frame, for example, a 32-bit CRC include field
FGID	*Function Group Identifier*
FID	*Function Code Identifier*
FIFO	*First In, First Out*—Operation of data buffer structures and items that were first stored, first out
FLASH	Flash-EEPROM, see EEPROM
FLDU	*Front Left Door Unit*
FM	*Frequency Modulation*
FMC	*CAN Base Frame Migration Channel* – Acronym for logical communication channel ARINC 825
FMS	*Process Fieldbus Message Specification*
FPGA	*Field Programmable Gate Array*—An integrated circuit (IC) in digital technology, in which a logic circuit can be programmed
FRDU	*Front Right Door Unit*
FSA	*Finite State Automation*
FTDMA	*Flexible Time Division Multiple Access*—Minislotting method

G

GA	*General Aviation*
GIFT	*Generalized Interoperable Fault-tolerant CAN Transceiver*—The GIFT-Group has developed the specification for low-speed CAN transceiver
GLARE	*Glass-fibre reinforced aluminium*

GND *Ground*-Zero potential, connection of devices, ECUs etc.
GNU *GNU is not Unix*—recursive acronym as the name of a project to create a fully free operating system
GPL General Public License—published by the Free Software Foundation License Copyleft (pun as opposed to copyright) for the licensing of free software
GPS *General Public License*—published by the Free Software Foundation License Copyleft (pun as opposed to copyright) for the licensing of free software
GPS *Global Positioning System*—officially NAVSTAR GPS is a global navigation satellite system for determining the position and timing
GPT *General Purpose Timer*
GTR *Global Technical Regulation*
GUI *Graphical User Interface*

H
HBM *Human Body Model*—An electrical model of the human body, in particular with respect to the capacitance and the resistance value of
HDV *Heavy Duty Vehicle*
HF *High Frequency*
HIL *Hardware In the Loop*—Method for testing and validating embedded systems by an existing hardware can be integrated into a simulated system to test the function
HIS *Herstellerinitiative Software*—Joint activities of the automobile manufacturer Audi, BMW, Daimler, Porsche and Volkswagen to develop uniform standards: standard software modules for networks that process maturity determination, software testing, software tools and programming of control units
HLP *High Level Protocol*
HPB *(Xilinx) High Priority Buffer*—Transmission buffer having a high priority in order to avoid the need to transmit FIFO

I
I/M *Inspection/Maintenance*
I/O *Input/Output*
IC *Integrated Circuit*
ICSP *In-Circuit Serial Programming*—In circuit serial programming (programming of a semiconductor in installed)
ICT *International Transceiver Conformance Test*—The ICT project has developed conformance testing for CAN transceiver after the date specified in POISON standard components and standardized

ID	*Identifier*—The identifier of CAN messages (CAN-ID), also used as an example, data identifier, connection identification, equipment identification
IDE	*Identifier Extension*—Bit in the header of the CAN frame: high—followed by 18 additional ID bits (extended frame), low—standard frame
IE	*Interrupt Enable*—Flag for switching of interrupts
IEC	*International Electrotechnical Commission*—The International Electrotechnical Commission developed and published international standards URL: http://www.iec.ch
IEEE	*Institute of Electrical and Electronics Engineers*—Worldwide professional association of engineers in the fields of electrical engineering and computer science, originally acting as a standards body engineering association in the United States
IFS	*Interframe Space*—Minimum interval between two CAN frames, three recessive bits
IIC	*Inter-Integrated Circuit*—serial data bus for connecting different circuit parts, mostly inside the device
IMA	*Integrated Modular Avionics*—a computer-/network- architecture in aircraft
INH	*Inhibit*—Output signal of a transceiver, for example, serves to control of connected voltage regulator to switch off the control unit in sleep mode until an alarm is received via the CAN bus
IP	*Internet Protocol*
IP65	*International Protection (Class) 65*– For protection against contact and foreign bodies and water protection (1. digit: 6– Protected against access with a wire and dustproof, 2. digit: 5– Protection against water jets (nozzle) from any angle)
IP-Core	*Intellectual Property Core*—reusable part of a chip designs, including the intellectual property of the developer and is passed against royalties
IPIEF	*(Xilinx) Intellectual Property Inter-Connect*—OPB IPIF signals for connecting the processor and peripheral on-chip
IPIF	*(Xilinx) Intellectual Property Interface*—Used in connection with the *on-chip peripheral bus (OPB)*,
IRT	*Isochronous Real Time*—Simultaneous real-time method for e.g. Industrial Ethernet for real-time systems with time-synchronized participants
ISA/SP50	*International Society of Automation, Signal Compatibility of Electrical Instruments*—Objective: Define common interfaces between components of electrical measurement and/or control systems URL: http://www.isa.org
ISO	*International Organization for Standardization*
ISO/CD	*ISO/Committee Draft*
ISO/PAS	*ISO/Public Available Standard*

ISO/WD *ISO/Working Draft*
ISPF *Interoperable System Project Foundation*—Predecessor of the Fieldbus Foundation, which has emerged from the ISPF and the North American offshoot of WorldFIP
ISR *1) Interrupt Service Routine*—routine which is executed at the interrupt event occurs
2) (Xilinx) Interrupt Status Register—Register for querying the status of the status of interrupt sources, eg, if no interrupts are used (polling)
ITID *InfoType Identifier*
IUT *Implementation Under Test*

J, K
JSAE *Japanese Society of Automotive Engineers*

L
LAN *Local Area Network*
LBA Federal Aviation
LCC *Logical Communication Channel*
LCL *Local Bus Only*—ARINC 825 bit in the identifier
LDV *Light Duty Vehicle*
LED *Light Emitting Diode*
LEN *Length-Field*
LIN *Local Interconnect Network*—especially for low-cost applications
LLC *Logical Link Control*—Logical Link Control, the extension of the layer 2 Ethernet protocol provides the connection to the layer 3 with various transmission methods ago
LON *Local Operating Network*—mainly inserted in building automation fieldbus
LRC *Longitudinal Redundancy Check*
LRU *Line Replaceable Unit*—By defined hardware interfaces, network layer, security mechanisms, data types and coordinate system definitions replaceable network nodes in CANaerospace
LSB *Least Significant Bit*
LSS *Layer Setting Services*—CANopen services for Node ID configuration
LT *Lower Tester*—Part of the test system, which uses the interface of the IUT to deeper layers
LWL Fiber

M
MAC *Media Access Control*
MAC ID *Media Access Control Identifier*

MAST A hardware description language (HDL) from Synopsys and a de facto industry standard, first published in 1986 and applied for analog and mixed-signal applications
MBRB *Multi Buffer Receive Block*—Block of a plurality of receiving buffers for receiving messages complying with the same filtering conditions
MCAL *Micro-Controller Abstraction Layer*
MCNet *Mobile Communication Network*
MCU *Micro-Controller Unit*
MDI *Medium Dependent Interface*—Functional interface to the transmission medium
MDV *Medium Duty Vehicle*
MI *Malfunction Indicator*
MID *Monitor Identifier*
MIL *Malfunction Indicator Light*
MIPS *Mega Instructions Per Second*
MLT-3 *Multilevel Transmission Encoding, 3 Levels*
MMU *Memory Management Unit*
MOST *Media Oriented Systems Transport*—Bus system, especially for the connection of multimedia components in the vehicle
MPDO *Multiplexed Process Data Object*
MPU *Micro Processor Unit*
MSB *Most significant Bit*
MSC *Message Status Counter*
MSR *(Xilinx) Mode Select Register*—Register to select the operating mode

N
NACK *Negative Acknowledge*
NASA *National Aeronautics and Space Administration*
NBT *Nominal Bit Time*—Nominal (ideal) bit time without consideration of oscillator tolerances or other deviations from the nominal bit time
NEFZ New European Driving Cycle
NERR *(Not) Error*—Output signal of a transceiver is used to indicate detected bus errors or e.g. overtemperature
NIC *Network Interface Controller*
NID *Node Identifier*
NM/NMT *Network Management*
NOC *Normal Operation Channel*—Acronym for logical communication channel ARINC 825 for normal operation (user data)
NOD *Normal Operation Data*—Gewöhnliche Nutzdaten
NOx *Nitrogen Oxide*—Stickstoffoxid, oft auch kurz Stickoxid

NRZ *Non-Return-to-Zero*—Method of line coding for binary signals without neutral symbol (zero-symbol), the binary coded one on one with two defined levels

NRZI *Non-Return-to-Zero Invert*—A method for encoding binary signals without line neutral symbol (zero), in contrast to the NRZ coding, the coded binary signal not level but by a level change (for example, 1' = switch to the beginning of the stroke and '0'=no exchange)

NSC *Node Service Channel*—Acronym for logical communication channel ARINC 825 for node services

NSH/NSL *High/Low Priority Node Service Data*

NSTB *(Not) Stand-by*—Mode selection control signal for a transceiver

NTU *Network Time Unit*—network-wide measure of the time at TTCAN

NVM *Non-Volatile Memory*

O

OBD *On-Board Diagnostics*

ODX *Open Diagnostic Data Exchange*—Formal description language for essential information of vehicle or ECU diagnostics (request and documentation), standardized in ISO/DIS 22901-1

OEM *Original Equipment Manufacturer*

OPB *(Xilinx) On-chip Peripheral Bus*—Bus between the processor and the peripherals on a microcontroller chip

OSEK/VDX *Offene Systeme und deren Schnittstellen für die Elektronik im Kraftfahrzeug/Vehicle Distributed Executive*—Panel of automotive manufacturers, suppliers and software companies to - and the French initiative VDX

OSI *Open Systems Interconnect*

P

PA *Process Automation*—Profibus solution for Chemical and Petrochemical industry in hazardous areas

PAD *Padding-Field*—Padding field in the Ethernet-frame, if this is not achieved with the minimum length of the payload transported by 64 bytes

PC *Personal Computer*

PCB *Printed Circuit Board*

PCO *Point of Control and Observation*

PDIP *Plastic Dual In-line Package*—A housing for electronic devices having double-row-shedding plastic, wherein there are two rows of pins (pins) for through-hole mounting on opposite sides of the housing

PDM *Passenger Door Module*

PDO *Process Data Object*

PDU *Protocol Data Unit*
PEC *Peripheral Event Controller*—A kind of "hidden", infiltrates realized in hardware interrupt, the transfer commands in the normal program flow without new commands must be fetched from the program memory
PGN *Parameter Group Number*
PHY *Physical Layer (common abbr.)*—Identifies the module that is responsible for mediating between purely digital and analog bus signals
PID *Parameter Identifier*
PLB *(Xilinx) Processor Local Bus*—A fast interface between the processor and high-performance peripherals
PLL *Phase-Locked Loop*—Phase locked loop electronic circuit to minimize the phase error of a clock signal from a reference signal, whereby the resulting frequency is stabilized
PLS *Physical Layer Signalling*—includes Bit-Encoding/-Decoding, bit timing and synchronization
PMA *Physical Medium Attachment*
PN *Partial Networking*—Partial network operation by switching off of ECUs by the partial networking transceiver that is awakened only by dedicated wake-up frames again
PoE *Power over Ethernet*—Method for supplying power to Ethernet network devices on the bus
POF *Polymeric Optical Fiber/Plastic Optical Fibre*
PPDT *Peer-to-Peer Data Transport*
PRNG *Pseudorandom number generator*
PTP *"Peer-to-Peer"*—Communication with participants addressing
PVT *Private*—ARINC 825 bit in the identifier
PWM *Pulse Width Modulation*

R

RAM *Random Access Memory*
RCI *Redundancy Channel Identifier*—ARINC 825 supports redundant transmission: The RCI defines which redundancy channel is assigned to a message
RCP *Rapid Control Prototyping*—Computer-aided design method for regulating and controlling development (rapid prototype design)
REC *Receive Error Counter*
RF *Radio Frequency*—Denotes oscillation frequencies in the range from about 30 kHz to 300 GHz
RHL *Receive History List*—List of reception history of CAN messages
RID *Routine Identifier*
RLDU *Rear Left Door Unit*
RM *Receive-only Mode*
ROM *Read Only Memory*

RPDO	*Receive Process Data Object*
RRDU	*Rear Right Door Unit*
RS	*Recommended Standard*—Recommended Standard EIA (Electronic Industries Alliance), former importance: Radio Sector of the U.S. standardization committees, a well-known representative is such as RS 232
RSD/SMT	*Reserved/Service Message Type* ARINC 825 bit in the identifier
RT	*Real Time*
RTC	*Real Time Clock*
RTE	*Run-Time Environment*
RTH	*Resistor Termination High* In low-speed transceiver CAN_H can be connected to the stabilization of the bus level (idle) on the external termination resistor to ground
RTL	*Resistor Termination Low*—In low-speed transceiver CAN_L can be connected to the stabilization of the bus level (idle) on the external termination resistor to V_{CC}
RTR	*Remote Transmission Request*
RTT	*Real-Time Testing*
RX	*Receive*
RxD	*Receive Data*
RZ	*Return to Zero*—Line coding, wherein the transmitter is returned to zero within each transmitted symbol (usually + 1 and − 1)

S

S/N	*Signal/Noise*
SAE	*Society of Automotive Engineers*
SAM	*System of Aviation Modules*—A completely based CANaerospace avionics system for small transport aircraft
SAP	*Service Access Point*
SAPI	*Service Access Point Identifier*
SBP	*Serial Bus Protocol*—Basis for the PPDT with FireWire
SCI	*Serial Communication Interface*—Generic term for serial bus systems, often confused with e.g. RS232
SCU	*System Control Unit*—Microcontroller unit that allows, for example, by reducing or stopping the clock frequency of peripheral elements for saving energy
SDCS	*Safety Decision Control System* – Control system, which automatically makes security decisions
SDL	*System Description Language*
SDO	*Service Data Object*
SERCOS	Efficient, deterministic real-time communication based on Ethernet protocol is used by CNC machines via motion control to general automation URL: http://www.sercos.com

SFD	*Start Frame Delimiter*—Ends the preamble in an Ethernet frame
SFID	*Sub-Function Identifier*
SFR	*Special Function Register*
SGN	*Signal Group Number*
SID	*Service Identifier/Server Identifier*
SIL	*Safety Integrity Level*—A method for determining the potential risk to persons, systems, devices and processes in the case of a malfunction
SJW	*Synchronization Jump Width*—Maximum extent to which the phase segments may be changed in the post-synchronization during receiving in order to adjust the sampling
SLIO	*Serial Linked IO*—Input-/Output port, which operates without local software and configured via the bus can be accessed and controlled
SOF	*Start Of Frame*—Bit, marks the beginning of a CAN frame
SOFIA	*Stratospheric Observatory For Infrared Astronomy*—A Boeing 747SP with a 2.7-m telescope for infrared astronomy
SOIC	*Small-Outline Integrated Circuit*—Housing for integrated circuits, SO-ICs are 30 to 50% smaller than the corresponding DIL-ICs
SOP	*Start Of Production*
SOVS	*System Operation Vector Space*
SPI	*Serial Peripheral Interface*
SPICE	*Simulation Program with Integrated Circuit Emphasis*—a software for simulation of analog, digital, and mixed electronic circuits PSPICE—PC-Version of SPICE HSPICE—UNIX-Version of SPICE
SPLIT	Output pin of a transceiver, provides 2.5 V voltage to stabilize the bus level ready to be connected via split termination resistors for CAN_H and CAN_L
SPN	*Suspect Parameter Number*—Signal and parameter identification, such as J1939-71
SR	*(Xilinx) Status Register*
SRDO	*Safety-related Data Object*
SRR	*1) Substitute Remote Request*—Bit in the header of a CAN frame, replaces the RTR bit *2) (Xilinx) Software Reset Register*—The control register to perform a software reset
STP	*Shielded Twisted Pair*
SUT	*System Under Test*
SWC	*Software Component*
SWITCH	*Selective Wake-up Interoperable Transceiver in CAN High-Speed*

T

TADL	*Timing Augmented Description Language*—Language for system-wide time behavior description of automotive functions and their interactions
TC	*Test Case*
TCM	*Transmission Control Module*
TCP	*Test Coordination Procedure*
TCP/IP	*Transmission Control Protocol/Internet Protocol*—Network protocol suite, also called Internet protocol family
TDMA	*Time Division Multiple Access*
TEC	*Transmit Error Counter*
THL	*Transmit History List*—List of broadcasting history of CAN messages
TID	*Test Identifier*
TIMMO	*Timing Model*—A ITEA 2 project developed a standardized infrastructure for development-related handling of time information in real-time embedded systems in the automotive industry
TMC	*Test and Maintenance Channel*—Acronym for logical communication channel ARINC 825 for test and maintenance purposes
TPDO	*Transmit Process Data Object*
TQ	*Time Quantum*—Time step, a sampling designation for the bus sampling of the CAN controller, the length is derived from the oscillator clock, divided by the value of the baud rate prescaler
TRQ	*Transmit Request*
TSEG1/2	*Timing Segment 1/2*
TTCAN	*Time-Triggered CAN*
TTL	*Transistor-Transistor-Logic*—Transistor-transistor logic family of logic, logic circuits are used in the planar npn bipolar transistors as active components
TTP	*Time-Triggered Protocol*—Transmission protocol that allows timed and fault-tolerant communication and satisfies hard real-time requirements
TUR	*Time Unit Ratio*—Part relationship (node-specific) between the local clock source and the network time unit (NTU) in TTCAN
TV	*Television*
TX	*Transmit*
TxD	*Transmit Data*
TxEN	*Transmit Enable (Not)*—Control signal to the transceiver of the CAN controller that activates the transmitter
TxRQ	*Transmit Request*—also: TXREQ or TRQ

U

UAF	*(Xilinx) Use Acceptance Filter*—Bit n AFR-register, on the use of an acceptance filter pair free

UART *Universal Asynchronous Receiver Transmitter*—Universal asynchronous receiver [and] transmitter electronic circuit for the realization of digital, serial interfaces and byte-oriented protocols
UDC *User-defined Channel*—Acronym for logical communication channel ARINC 825 for custom data
UDH/UDL *High/Low Priority User-Defined Data*
UDP/IP *User Datagram Protocol*—Protocol for data transfer on the Internet
UDS *Unified Diagnostic Services*
UML *Universal Modelling Language*
USB *Universal Serial Bus*
USIC *Universal Serial Interface Controller*
UT *Upper Tester*—Part of the test system, which serves the IUT interface to higher layers
UTP *Unshielded Twisted Pair*—Unshielded, twisted-pair cable, often used for CAN

V
VDA Association of the Automotive Industry
VHDL-AMS *Very High Speed Integrated Circuit Hardware Description Language*
VFB *Virtual Function Bus*
VIN *Vehicle Identification Number*
VLAN *Virtual Local Area Network*—Virtual local area network, The method for displaying a plurality of logical networks on a single physical network, such as Ethernet
VRC *Vertical Redundancy Check*

W
WK *Wake*—Local Wake-pin with which the transceiver can be awakened, for example by means of switches from sleep mode
WUF *Wake-up Frame*—CAN message with specific useful data that exist only to wake from partial network transceivers on the busWWH-OBD *Wold-wide harmonized OBD*—globally harmonized OBD

X
XLP *eXtreme Low Power*
XML Extensible *Markup Language*—free "extensible markup language" for representing hierarchically structured data in the form of text data
XOR *eXclusive OR*

Chapter 1
CAN Basic Architectures

Wolfhard Lawrenz

1.1 CAN History

The driving factor in the automotive industry for the development of controller area network (CAN) and other so-called bus protocols was the need to search for new solutions for the problem field, which resulted with the increasing need for communication in vehicles—with the boundary condition to reduce the associated impending increase in the complexity of the wiring harness in the car. The wiring in cars may include several miles of cable length and may be heavier than 100 kg. It is understandable that such wiring harnesses are difficult to install.

In addition, for example, a family car requires for covering the variants of only one single model hundreds of different types of wiring harnesses. This shows clearly which problem exists with the cabling, also in conjunction with the just-in-time-logistic and, of course, with the question of cost reduction. Since the early 1980s, almost all automobile manufacturers had started intensive efforts to find or develop suitable communication protocols. One application—the so-called Class-C applications—aimed at the communication link between real-time critical control units for engine management, transmission, brakes, vehicle stabilization, etc. Any other application—so-called Class-A applications—addresses the area of body electronics, e.g. for mirror control, power windows, door locks, seat adjustment, climate control, lighting, etc. The latter area is less real-time critical, therefore, it requires a lower communication bandwidth, but is, however, extremely cost-sensitive. In response to the needs of different classes of applications, different bus protocols have emerged, such as CAN, Local Interconnect Network (LIN), Media Oriented Systems Transport (MOST) and FlexRay.

Since 1994–1995, CAN is the most common protocol for automotive applications, and is described in the International Organization for Standardization (ISO) standards 11898-1 to 11898-5. In Part 1 of ISO 11898 (ISO 11898-1), the

W. Lawrenz (✉)
Waldweg 1,
38302, Wolfenbuettel, Germany
e-mail: W.Lawrenz@gmx.net

W. Lawrenz (ed.), *CAN System Engineering,* DOI 10.1007/978-1-4471-5613-0_1,
© Springer-Verlag London 2013

data link layer and physical layer are set according to the ISO reference model ISO/IEC 7498-1. The "High Speed" CAN bus access (up to 1 Mbit/s) is specified in ISO 11898-2 and mainly used in the propulsion of a vehicle. The "Low Speed" CAN (40…125 kbit/s) for the comfort section is described in ISO 11898-3. ISO 11898-4 allows a time-triggered communication to ensure a smooth data transfer with high communication traffic. Available since 2007, the expansion of the ISO 11898-2 describes the behaviour of a CAN high-speed node in the power saving mode. Since the early 1990s, CAN was at first used in series-production cars and also in industrial control, usually in conjunction with the higher CANopen protocol. CANopen (see Sect. 6.1) is specified in the European standard EN 50325-4 and is maintained by the user organization CiA (CAN in Automation). Reasons for the series use are certainly the high performance standard, low costs and the wide range of semiconductor manufacturers, which offer CAN modules. CAN was used first in the Mercedes S-Class in 1992 and served there as a high-speed network for communication between engine control, transmission control and dashboard. A second CAN network was—at low speed—used for distributed climate control. BMW, Porsche, Jaguar, etc., put CAN into series-production cars shortly after.

Typical applications of CAN are:

- Transportation with CAN applications: passenger cars, trucks, planes, trains, ships, agricultural machinery, construction machinery, etc.
- Industrial control automation technology with CAN communications such as: programmable controllers, automatic handling equipment, robots, intelligent motor controllers, intelligent sensors/actuators, hydraulic systems, intelligent meters for different applications such as water consumption, energy consumption, etc., textile applications such as spinning, weaving, etc., medical technology, laboratory automation, building systems, elevators, automatic teller machines (ATMs), toys, mechanical tools and much more.

Nowadays, it is increasingly important, also through the tightening of legal regulations, to reduce CO_2emissions and fuel consumption of new vehicles. At the same time, with shorter development times, the driving performance and the comfort of a vehicle are to be increased, costs reduced and quality improved. Thus, there are always new challenges to the development of electronic systems in vehicles; some of them are only possible by a high degree of networking. Many of today's standard techniques would not be feasible without the use and dissemination of field buses.

1.1.1 Standardizations of CAN

Robert Bosch GmbH began as an electronic control unit (ECU) manufacturer in 1983 with the development of a communications protocol for bus systems, which was initially only intended for the use in the automotive sector. The result was CAN. The first CAN chip was offered to car manufacturers in the late 1980s by Intel. CAN was finally standardized in November 1993 as ISO 11898 *Road vehicles—*

Interchange of digital information—Controller area network (CAN) for high-speed communication.

1.1.1.1 The ISO 11898 family

The standard describes the architecture of CAN in terms of the layers of the International Organization for Standardization/*Open Systems Interconnection* (ISO/OSI) model[1]. It specifies both parts of the physical layer and parts of the link layer for transmission rates up to 1 Mbit/s.

Part 1 and Part 2 correspond together to the first version of the standard:

- **ISO 11898-1** *Road vehicles—Controller area network (CAN) —Part 1: Data link layer and physical signalling*
- **ISO 11898-2** *Road vehicles—Controller area network (CAN) —Part 2: High-speed medium access unit*

The following parts were developed later:

- **ISO 11898-3** *Road vehicles—Controller area network (CAN)—Part 3: Low-speed, fault-tolerant, medium-dependent interface*
- **ISO 11898-4** *Road vehicles—Controller area network (CAN)—Part 4: Time-triggered communication*
- **ISO 11898-5** *Road vehicles—Controller area network (CAN)—Part 5: High-speed medium access unit with low-power mode*

Besides the ISO 11898 family, there are other standards for CAN in other organizations of standardization. Examples of this are the Society of Automotive Engineers (SAE) Standards J2284-1 bis-3, which specify the physical layer for CAN networks with different bit rates of 125 kbit/s (J2284-1), 250 kbit/s (J2284-2) and 500 kbit/s (J2284-3).

1.2 CAN Specifications

The first publication of CAN was a paper at the SAE conference in Detroit (SAE 860391), in February 1986. The first published version of the whole CAN protocol was the specification version 1.1. The protocol was further developed with version 1.2, followed by version 2.0A and version 2.0B. The versions are compatible; CAN modules of different protocol versions may be, with few limitations, operated in a single network. CAN 1.2 allows a higher oscillator tolerance than CAN 1.1.

---

[1] As a basis for a standardization of the various protocols, the Open Systems Interconnection (OSI) reference model was designed. The protocol hierarchy—also known as the ISO/OSI model—is available as standard ISO/IEC 7498–1 *Information Technology— Open Systems Interconnection—Basic Reference Model: The Basic Model.*

CAN 2.0A is identical to 1.2, while the CAN 2.0B specification introduces—in addition to those messages with *11-bit identifiers* defined in CAN 2.0A—frames with *29-bit identifiers* called "*Extended Frame*".

The CAN protocol has been standardized by the ISO, in the standards *ISO 11519* for bit rates up to 125 kbit/s and ISO 11898 for bit rates up to 1,000 kbit/s. Later, these standards have been merged into *ISO 11898*. *ISO 11898-1* describes the higher *ISO/OSI* layer of the protocol up to the Data Link Layer, *ISO 11898-4* describes the time triggered option on CAN (TTCAN, see Sect. 3.2), and the other parts of the standard describe different variants of the lower *ISO/OSI* layers, the CAN transceivers and the CAN bus line.

The following sections describe the CAN frame types and the associated system features.

1.2.1 Bit Coding

The bit stream on the CAN bus line is coded according to the Non Return to Zero (NRZ) method with *bit-stuffing*. This has the advantage to require only a minimum bandwidth for signal transmission. On the other hand, the NRZ encoding, where the bit level is constant during the bit time, contains no information about the bit clock. For long data lengths, this may cause problems with synchronization and thus lead to erroneous bit detection since each node in the network has its own clock generator.

Therefore, additional stuff bits are added to the main part of the bit stream. In that part of the bit stream where the stuffing method applies, the transmitter will, after having sent five consecutive bits of identical value, insert ("stuff") an additional bit of inverse value into the bit stream. The receiver will recognize a sequence of five consecutive bits of identical value and discard the following stuff bit.

Due to the bit-stuffing mechanism, the distance between edges in the bit stream is at most five bits. The edges are used to synchronize the local bit clocks of all nodes in the network. The chosen stuff-width is a compromise between the lengthening of the frames and the tolerance range of the clock generators.

1.2.2 CAN Frames

There are four types of frames in the CAN protocol: *Data Frame, Error Frame, Remote Frame* and *Overload Frame*. Only the *Data Frame* transports message data; the other frames are for fault containment, triggering and synchronization.

Each CAN message is divided into different fields of specific length. These fields are, for example, the End of Frame, the *CyclicRedundancyCheck (CRC) Field*, the *Data Field* and the *Arbitration Field*. The *Arbitration Field* combines, in the identifier, the priority of the message with the logical address of the information.

Data Frame CAN 2.0A (11-Bit-Identifier)

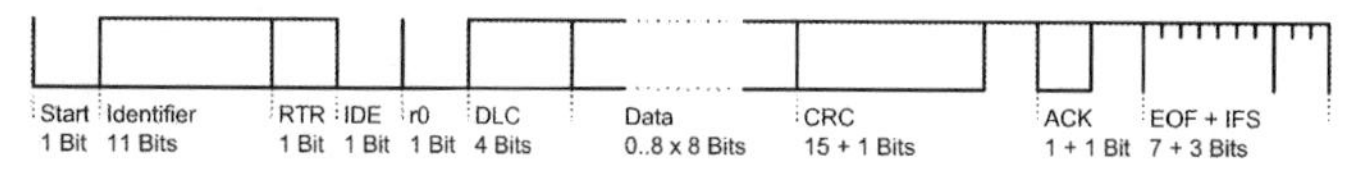

Data Frame CAN 2.0B (29-Bit-Identifier)

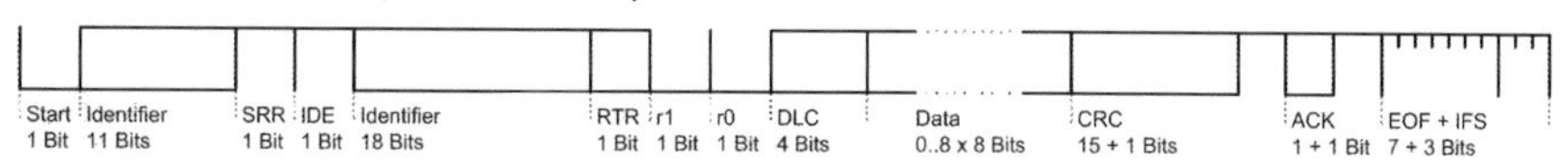

Fig. 1.1 Structure of a CAN data frame

Bit stuffing applies to the fields from the *Start of Frame* to the *CRC Field*, in *Data Frames* and *Remote Frames*. There is no bit stuffing in the other fields and also not in *Error Frames* or *Overload Frames*.

There are two identifier formats in CAN. The first is 11 bits long, specified in CAN 2.0A, allowing 2,048 different logical addresses (0–2,047). This means that a maximum of 2,048 different messages can be sent in a network if no additional distinction is performed, for example, in the data bytes.

In some applications, this number may be too low. Therefore, the specification has been extended so that messages with a total of 29 identifier bits are possible. This allows 2^{29}(=536.870.912) different messages, called *Extended Frames* according to CAN specification 2.0B.

The exact structure of the frames is shown in Fig. 1.1.

The individual bit fields, initially for frames with 11-bit identifiers (also called *Standard Frames*) are described below:

- *Start Of Frame*=1 bit=low (*dominant*):

Marks the start of a frame. After an idle time on the bus, the falling edge is used for phase synchronization of all network nodes.

- *Arbitration Field*=12 bits:

The Arbitration Field contains the 11-bits-long Identifier, which is the logical address and priority of the message. The lower the numerical value of the Identifier is, the higher is the priority.

- The Identifier is followed by the ***R**emote-**T**ransmission-**R**equest* (RTR) bit. The *RTR* bit identifies this message as one that either contains the data itself which belong to the logical address (RTR=low=*dominant*) or as a frame which itself contains no data but only triggers the actual transmitter of this information to send a frame with the current data. The *RTR* bit is the last bit of the *Arbitration Field*. The request (which is called *Remote Frame*) and the response of the corresponding data transmitter are two separate messages. This means that one or more higher priority messages may be transmitted on the CAN bus between request and response.
- *Control Field*=6 bits:

The first bit in the Control Field is the Identifier Extension Flag (IDE) bit. In 11-bit identifier frames it is low (*dominant*), which indicates that the identifier is completed. The following bit *r0* is reserved. The last four bits of the *Control Field* contain the *Data Length Code* (*DLC*) for the following *Data Field.*

- *Data Field*=0–8 bytes of data:

Contains the actual data of the message.

- *CRC Field*=16 bits:

Contains the checksum for the preceding bits of the frame. The 15-bits-long CRC checksum (*CRC Sequence*) is only used for fault detection, not for error correction. The CRC polynomial of CAN results in a Hamming distance of 6, meaning that up to five single-bit errors in the message can be detected. Furthermore, so-called burst-errors—directly consecutive bit errors—can be recognized up to the length of the *CRC Sequence.* The checksum is followed by one bit, the *CRC Delimiter* (high=*recessive*).

- *Acknowledge Field*=2 bits:

All nodes that have seen a syntactically correct message on the CAN bus acknowledge by sending a *dominant* level in the so-called *ACK Slot.* A Transmitter sends a *recessive* level in the ACK Slot and expects that this *recessive* level is overwritten by a *dominant* level. The transmitter of the message considers a missing *dominant* level in the *ACK Slot* to be an acknowledgement error. Detection of this *dominant* level does *not* mean that the frame was received by all intended recipients of this information! It only states that the frame was recognized by at least one active node to be a correct CAN frame. The *ACK Slot* is followed by one bit, the *ACK Delimiter* (high=*recessive*).

- *End Of Frame (EOF)*=7 bits (high=*recessive*):

The *End Of Frame (EOF*) is characterized by a deliberate breach of coding. According to the bit-stuffing mechanism, a *Stuff-Bit* would be inserted after the fifth *recessive* bit. This does not happen in *EOF;* the *EOF* indicates the end of the data frame.

- *Inter Frame Space (IFS)*=3 bits (high=*recessive*):

The *Inter Frame Space (IFS*) separates the current frame from the following frame. Inside the CAN node, this period is also used for transferring a correctly received message from protocol controller into the appropriate space of the receive buffer, or for transferring a message from the transmit buffer to the protocol controller.

- *Idle*≥0 bit (high=*recessive*):

The bus is unused. Each CAN node may, beginning with a *Start of Frame* bit, send a message to the bus.

Frames with 29-bit identifiers (also called *Extended Frames*) differ from frames with 11-bit identifiers in the *Arbitration Field* and in the first bit of the *Control Field*:

- *Arbitration Field*=32 bits:

The *Arbitration Field* of frames with 29-bit identifiers consists of three parts. It starts with the 11 most significant bits (MSBs) of the identifier (*Base Identifier*), followed by two *recessive* (high) bits: the ***S**ubstitute **R**emote **R**equest bit (SRR*) and the ***ID**entifier **E**xtension Flag (IDE*) bit. Then follow the 18 least significant bits (LSBs) of the identifier (*Identifier Extension*). The concatenated values of *Base Identifier* and *Identifier Extension* are the logical address and priority of the message.

The last bit of the *Arbitration Field* is the *RTR* bit.

- *Control Field*=6 bits:

In contrast to frames with 11-bit identifiers, the *Control Field* of frames with 29-bit identifiers does not start with the *IDE* flag and one reserved bit, but with two reserved bits, *r1* and *r0*. The last four bits of the *Control Field* contain the *DLC* for the following *Data Field.*

Valid values for the *DLC* are 0…15. The *DLC* values 8…15 are treated as *DLC* 8, i.e. the maximum length of a frame's *Data Field* is 8 bytes. Reserved bits (*r1* and *r0*) are transmitted *dominant* (low); receivers accept both *dominant* and *recessive* values for *r1* and *r0*.

Old CAN controllers, designed prior to the release of CAN 2.0B, would detect errors in *Extended Frames* and would destroy them with *Error Frames*, but these CAN controllers are no longer in production. All CAN controllers that can process *Extended Frames* are also able to process frames with 11-bit identifiers. Therefore, it is possible to use both frame formats in the same CAN network. Limited CAN implementations that can send and receive frames with 11-bit identifiers but ignore *Extended Frames* may be used for specific applications.

The IDE bit distinguishes between the frame formats. This bit is *recessive* in *Extended Frames;* hence, an *Extended Frame* will lose arbitration to another frame with an 11-bit identifier that is identical to the *Extended Frame's Base Identifier*.

The *Data Frames* shown in Fig. 1.1 are the regular messages that distribute network data. Under normal circumstances, such a message is sent on the initiative of the application program in the transmitting node. Alternatively, CAN also provides the option that a recipient of a specific piece of information may prompt the sender to transmit the actual data values. This is done via a *Remote Frame*. Such a frame is characterized by its *recessive RTR* bit and its Data Field—independent of the DLC—contains no data. A Remote Frame will prompt a sender to transmit its corresponding Data Frame. This response can be done automatically by the CAN controller (with "Full CAN" controller) or under the control of the application program ("Basic CAN" controller). It should be noted that several higher priority messages may be sent on the CAN bus between the query, i.e. the Remote Frame, and the answer, i.e. the Data Frame.

Remote Frame may, e.g. be used when a receiver was temporarily inactive, that is, not participating in bus communication, and needs the current data value before their sender reaches its next scheduled transmit time.

The structure of a *Remote Frame* is shown in Fig. 1.2. The frame structure is identical in CAN 2.0A and CAN 2.0B except for the length of the *Arbitration Field*.

A *Remote Frame* must always be sent with the same *DLC* as the corresponding Data Frame. If more than one CAN node would simultaneously start *Remote Frames* with the same identifier but different *DLCs*, these frames would destroy each other.

Remote Frame CAN 2.0A (11-Bit-Identifier)

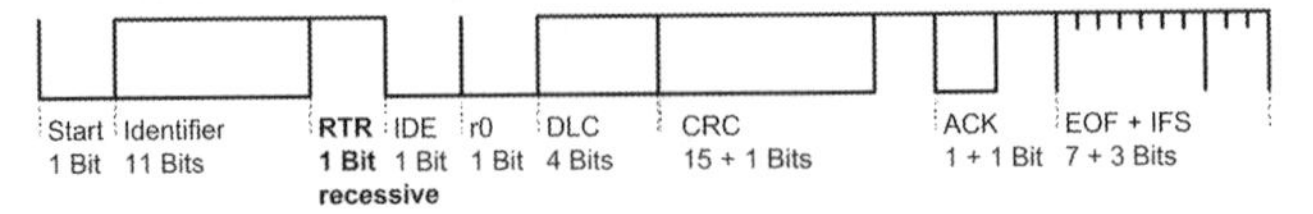

Remote Frame CAN 2.0B (29-Bit-Identifier)

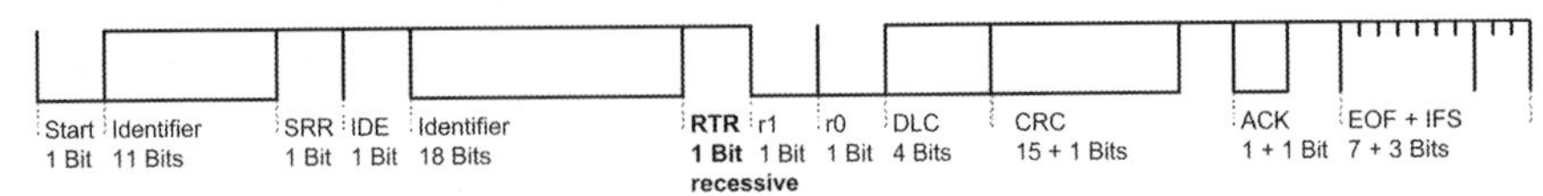

Fig. 1.2 Structure of a CAN remote frame

1.2.3 Arbitration

The CAN bus access is regulated by the method of nondestructive bitwise arbitration. Nondestructive means that the frame that is winner of the arbitration—i.e. the higher priority message—is not disturbed and does not need to be restarted. This mechanism requires the relevant physical drivers to be implemented in a certain way: The two logical levels on the CAN bus must be *dominant* and *recessive*, meaning that one node, sending a *dominant* level, overwrites all other nodes that send a *recessive* level. In the CAN protocol, a logical one is sent *recessive* and a logical zero is sent *dominant*.

In multimaster networks, the "philosophy" of the access right allocation on the bus is a crucial factor that characterizes the throughput, the transmission delay and thus the real-time capability of a networked system. CAN shows a very good performance in this regard: When multiple nodes simultaneously try to access the CAN bus, the most important competitor is automatically selected. When a CAN node with a pending transmission request detects that the bus is already occupied, that request is delayed until the bus returns to the idle state.

An example illustrates arbitration between two competing CAN nodes (see Fig. 1.3).

Two CAN nodes A and B start a transmission at the same time. According to the Carrier Sense Multiple Access with Collision Detection and Arbitration on Message Priority (CSMA/CD+AMP) access method, both nodes had to wait until the bus is free (Carrier Sense). When this is detected, they both send their *dominant Start Of Frame* bit (Multiple Access). Throughout the frame, each CAN node will, via its transceiver, read back the logical value which occurs on the CAN bus and compare it with the transmitted logical value (Collision Detection). The two *dominant Start Of Frame* bits are superimposed on the bus and a *dominant* bus level is read back by both nodes. Next, the MSBof the identifier is sent. In our example, the MSB is *recessive* in both identifiers. Here, too, the *recessive* level appears on the bus and is again recognized by both nodes. Therefore, no node notices the competing node until the first difference in the identifier.

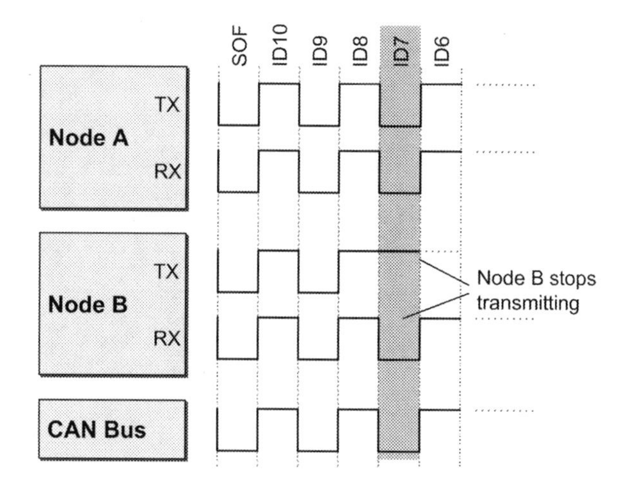

Fig. 1.3 CAN arbitration

In the example, the first difference between the two frames is at identifier bit 7. Node A sends a *dominant* level, node B sends a *recessive* level. According to the protocol specification, a *dominant* level appears on the bus. Node A reads back the level that it had sent; hence, it does not see a collision. Node B also reads back the *dominant* level and compares it with the *recessive* level that it had sent; therefore, it sees a *Bit-Error*. At this point, node B recognizes that it has lost arbitration (Arbitration on Message Priority) and immediately stops the transmission of its own frame. Furthermore, node B becomes receiver of node A's frame, because the frame that has won the arbitration may contain data that need to be processed by node B.

A *Bit-Error* is detected when the level read back from the CAN bus differs from the level that is sent to the CAN bus. If a *Bit-Error* is detected outside the *Arbitration Field* (Identifier and *RTR* bit) or *ACK Slot*, or if there is a *Bit-Error* on a *dominant* bit inside the *Arbitration Field*, that is an error that is handled by the Error Management Logic. More about that follows in the next section.

1.2.4 Error Management

Besides the above-mentioned *Bit-Errors*, other errors are also detected by CAN and treated accordingly. The following description refers to the error management in the *Data Link Layer*. This error management is also part of the standardization in *ISO 11898*, which means that each CAN integrated circuit (IC) has, implemented in silicon, the same error handling.

It is a principle of CAN that as many errors as possible should be detected and treated inside the CAN IC. Any error that is detected by a network node is immediately communicated to all other nodes. After this error notification, all the participants in the network discard the message currently in progress. The error correction is done by automatic retransmission, which is a function of the *Data Link Layer*.

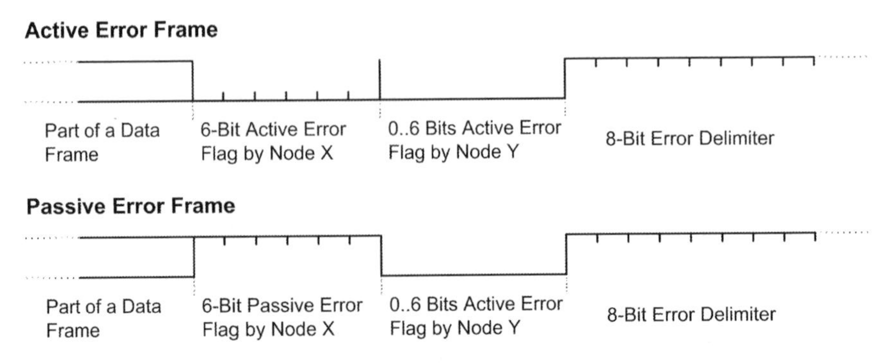

Fig. 1.4 CAN error frame

The notification of a detected fault to all other participants is done by an *Error Frame*, which is characterized—like the *End Of Frame*—by a deliberate breach of the bit-stuffing rules. The structure of such an *Error Frame* is shown in Fig. 1.4.

The *Active Error Flag* overwrites, with a series of six or more consecutive *dominant* bits, all other current bus levels. This pattern is—by violating the bit-stuffing rules—recognized by all other network nodes.

To prevent a persistent local disturbance of a CAN node or a group of CAN nodes from permanently paralyzing the CAN bus with *dominant Active Error Flags*, the affected CAN nodes—according to a specific algorithm—gradually withdraw from CAN bus activity. After the first stage of the withdrawal from the CAN bus, a CAN node may send only so-called *Passive Error Flags* (see Fig. 1.4). This algorithm is described in detail below.

The fault confinement consists of three parts: error detection, error handling and fault isolation.

1.2.4.1 Error Detection

The error management is able to identify five different types of errors:

- *Bit-Error*

A transmitted bit is not received with the same logical value with which it was sent. Excluded are the *Arbitration Field* and the *ACK Slot*.

- *Stuff-Error*

More than five consecutive bits of the same level have been detected. Excluded are *End Of Frame* and *Interframe Space (IFS)*.

- *CRC-Error*

The calculated CRC checksum does not match with the received CRC checksum.

- Form-*Error*

There has been a violation of the frame format, e.g. *CRC Delimiter* or *ACK Delimiter* was not recognized as a *recessive* bit, or *End Of Frame* was disturbed.

- *Acknowledgement-Error*

A transmitter did not detect a dominant bit in the *ACK Slot*, which means that the message was not recognized as fault-free by any other CAN node.

1.2.4.2 Error Handling

After one of the errors described above is recognized, each CAN node is notified by an *Error Frame*. The *Error Frame* overwrites other frames and is detected by the violation of the bit-stuffing rules. As a result of an *Active Error Flag*, which consists of actually only six *dominant* bits, other CAN nodes see a *Stuff-Error* and therefore also start to send an *Error Frame*. This effect occurs when the error is not a global error (detected simultaneously by all CAN nodes), but a local error (detected only by one CAN node or by a group of CAN nodes) and leads to a possible superposition of *Error Flags* up to a length of 12 *dominant* bits. The *Error Frame* is terminated by an *Error Delimiter* (a series of eight *recessive* bits).
The error handling is done in the following order:

a. An error is detected.
b. An *Error Frame* is sent by any CAN node that has detected the error.
c. The message currently in progress is rejected by all CAN nodes.
d. The error counters (see fault isolation) in each CAN node are affected according to the fault confinement rules.
e. The disturbed message is repeated.

1.2.4.3 Fault Isolation

The CAN protocol defines—with the aim of fault isolation—specific methods to prevent local disturbances from disturbing the CAN bus with multiple *dominant Active Error Flags*. For this purpose, three states are defined for the CAN nodes:

- *Error Active*

This is the normal state of a CAN node, in which messages can be sent and received. In the event of a fault, an *Active Error Flag*, consisting of *dominant* bits, is transmitted.

- *Error Passive*

This state is reached after several errors have been detected on the CAN bus. In this state, the CAN node may continue to send and receive messages, but in the case of an error, only a *Passive Error Flag* is sent, consisting of *recessive* bits. Consequently, a CAN node that is *Error Passive* cannot impede the other CAN traffic. It will mark only errors in its own transmitted messages. This may happen when the node sees a local disturbance. Such a CAN node that terminated its own transmitted frame switches—after the *Error Delimiter*—for 8 bits into a *Suspend* state where it cannot start a frame, but can receive every message.

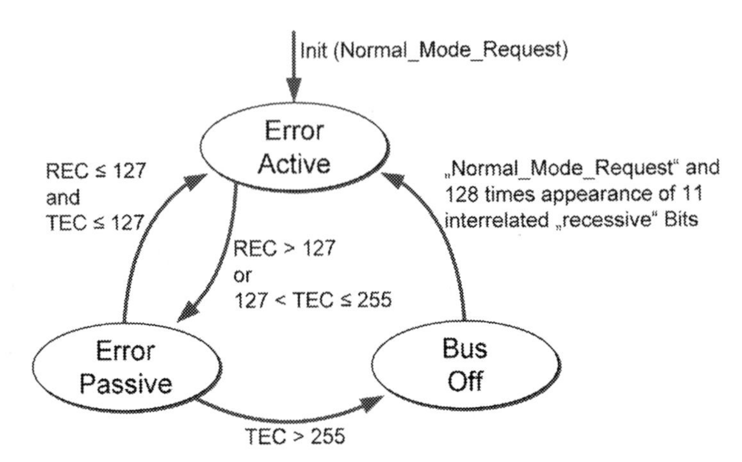

Fig. 1.5 CAN error states

- *Bus Off*

In this state, the CAN node is completely disconnected from the CAN bus. It therefore cannot send or receive messages or *Error Flags*. This state is reached when there is a long disturbance on the CAN bus or a frequently repeated disturbance.

The transition from *Error Active* to *Error Passive* and vice versa is done automatically by the CAN IC. The *Bus Off* state can be exited only by appropriate actions of the host controller (software or hardware reset). The transitions between the states are shown in Fig. 1.5.

The error state transitions are controlled by so-called error counters. Each CAN node has an error counter for receive errors (*Receive Error Counter—REC*) and an error counter for transmit errors (*Transmit Error Counter—TEC*).

The states are characterized by the following count values:

- *Error Active*:

The counts of the *Receive Error Counter* and *Transmit Error Counter* are both less than or equal to 127.

- *Error Passive*:

At least one of the two error counts is greater than 127 and the *Transmit Error Counter* is less than 256.

- *Bus Off*:

The *Transmit Error Counter* is greater than 255.

The error counters are incremented on detected errors or decremented on successful messages, according to the fault confinement rules. There are algorithms in the error management that allow CAN nodes to distinguish between local and global errors. Nodes seeing more local errors will reach the *Error Passive* state earlier than the other CAN nodes.

The fault confinement rules are explained below (*Receive Error Counter—REC*; *Transmit Error Counter—TEC*):

a. A receiver detects an error → $REC = REC + 1$
except the error is a *Bit-Error* that is detected while the receiver sends an *Active Error Flag* or an *Overload Flag*.
b. A receiver detects a *dominant* bit after sending an *Error Flag* → $REC = REC + 8$.
c. A transmitter starts an *Error Flag* → $TEC = TEC + 8$
Exception 1: When a transmitter is in *Error Passive* state and detects an *Acknowledgment-Error* (no *dominant* bit in the *ACK Slot*) and it does not detect a *dominant* bit while sending its *Passive Error Flag*, the *TEC* is not changed.
Exception 2: If a transmitter detects a *Bit-Error* on a *recessive* stuff-bit in the *Arbitration Field* before the *RTR* bit, the *TEC* is not changed.
d. A transmitter detects a *Bit-Error* while sending an *Active Error Flag* or an *Overload Flag* → $TEC = TEC + 8$.
e. A receiver detects a *Bit-Error* while sending an *Active Error Flag* or an *Overload Flag* → $REC = REC + 8$.
f. Any CAN node tolerates up to seven consecutive *dominant* bits after sending an *Active Error Flag*, *Passive Error Flag*, or *Overload Flag*. After detecting the 14th consecutive *dominant* bit (in case of an Active *Error Flag* or an *Overload Flag*) or after detecting the eighth consecutive *dominant* bit following a *Passive Error Flag*, and after each sequence of additional eight consecutive *dominant* bits, every transmitter increases → $TEC = TEC + 8$ and every receiver increases → $REC = REC + 8$.
g. A transmitter with $TEC > 0$ will decrement after each successful transmission of a *Data Frame* or a *Remote Frame* → $TEC = TEC - 1$.
h. A receiver with $REC > 0$ will decrement after each reception of a fault-free *Data Frame* or *Remote Frame* → $REC = REC - 1$.

Exception 1 of rule c) was introduced to avoid complicating the initialization phase of a network. After a system is started, the initialization times of the individual CAN nodes in a distributed control network will—under normal circumstances—be different. If we suppose now that the first CAN node that has completed its initialization sends a frame, it will always get an *Acknowledgment-Error* since there is no other CAN node ready to give an acknowledge, with the transmission attempt automatically repeated. This would mean that the *Transmit Error Counter* is incremented by 8 at each attempt, until it finally reaches the *Bus Off* threshold and the CAN operation is stopped. One can easily imagine that the initialization of a network would be very difficult under these circumstances. Therefore, this *Acknowledgment-Error* is treated differently.

After reaching the *Error Passive* state, the *Transmit Error Counter* is no longer incremented in the case of an *Acknowledgment-Error* —if there is no other node sending an *Active Error Flag*. This means that the first transmission attempt will be repeated until a second node finishes its initialization and gives an acknowledge.

One single error may cause the application of more than one rule, e.g. under the following circumstances:

If CAN node A, that is currently receiving a frame, detects a local error and then sends an *Active Error Flag*, this *Error Flag* will not completely overlap with the

Error Flags of the other CAN nodes. The other CAN nodes did not see the local error, but the *Active Error Flag* causes them to see a *Stuff-Error* (or a *Form-Error*), and therefore they start their *Error Flags* later. This causes node A to detect more *dominant* bits after sending its sixth *dominant* bit of the *Active Error Flag*. Rule a) applies to all receivers and their *Receive Error Counters* are incremented by 1. Rule b) additionally applies in CAN Node A, that had seen the local error; therefore, its *Receive Error Counter* is increased by a total of 9. As a result, CAN Node A will reach the *Error Passive* state earlier. Once a *REC* has exceeded the *Error Passive*-threshold, it is no longer incremented.

The process of error counting and the changing error states are shown in Fig. 1.6. The *Error Warning* flag in the CAN IC's status register is set—and if enabled, an interrupt is triggered—upon reaching the warning threshold of 96 (*Receive Error Counter* or *Transmit Error Counter*), just before reaching the *Error Passive* state.

When the *Transmit Error Counter* reaches the *Bus Off* threshold of 256, the *Bus Off* flag is set in the CAN IC's status register and that may in turn trigger an interrupt. The CAN node is now in the *Bus Off* state and will not take part in CAN bus activity before being restarted by the host controller.

The two above-mentioned *Error Warning* and *Bus Off* flags are the minimum required means for the host controller to detect the presence of noise or other disturbances. Their advantage is that the host controller does not need to attend to infrequent faults.

Most variants of CAN ICs also allow the host controller direct read access to the error counters as well as information on the type of the detected error; hence, early action can be taken to troubleshoot the application.

The measures described above are entirely functions of the *Data Link Layer*, so they are implemented in the hardware of the CAN IC.

The impact of the error states (*Error Passive* or *Bus Off*) on the application and what steps need to be taken are highly dependent on the application area. This can, for example, mean that even a single faulty CAN node may require a complete shutdown of the overall system, or it may require the system to be set into a safe state.

In less safety-relevant systems, suitable measures may be implemented that at least enable an emergency operation. There are, e.g. applications in which the measurements of a certain failed sensor can be replicated by other sensors and thus the overall function is still guaranteed. Likewise, the function of a failed brake light in a vehicle could be taken over by the rear fog light. All these measures are only made possible by the use of a network.

Faults in a CAN network may have several causes, such as electromagnetic interference, short circuits, defective contacts, or a malfunction in the CAN IC itself. One error-handling measure in the application therefore is to completely turn off a CAN node that—in the considered period—switches too often into the *Bus Off* state. If the root cause was a local error in the disconnected CAN node, this measure will allow the rest of the network—and the host controller—to continue operation, at least in an emergency mode.

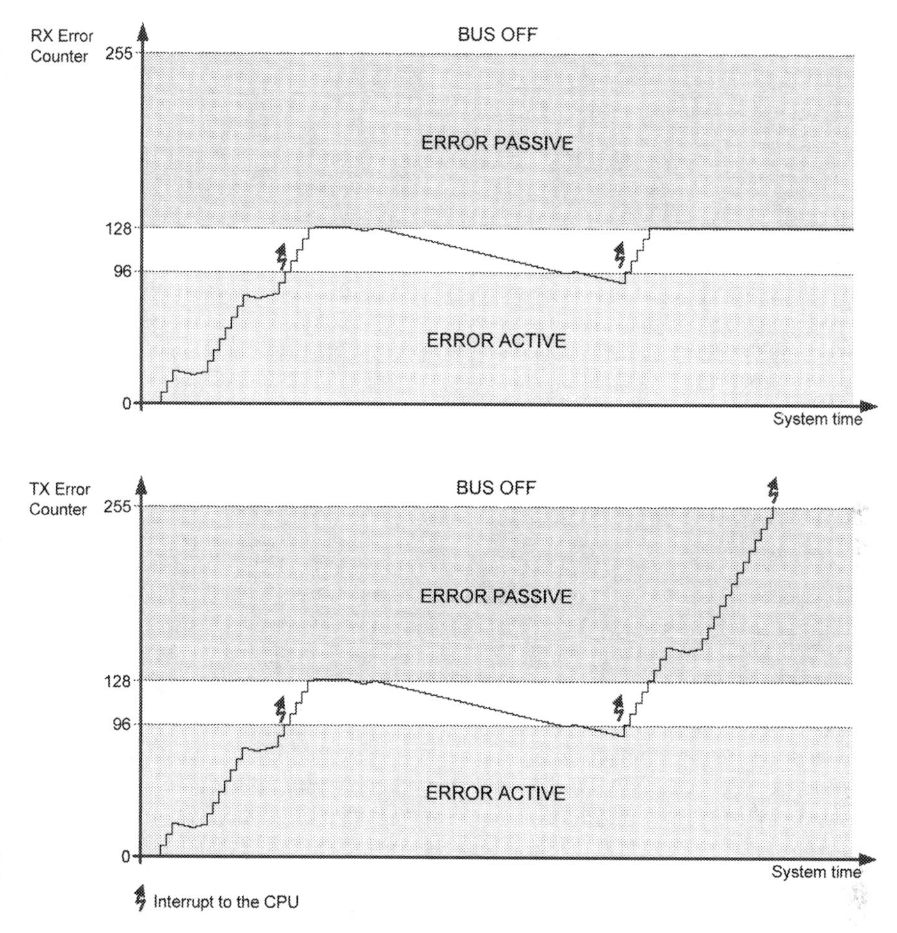

Fig. 1.6 CAN error counting

1.2.4.4 Overload Handling

Another means of exception handling is the so-called *Overload Frame*. The structure of an *Overload Frame* is exactly the same as that of an Error Frame. The only difference is that an *Error Frame* overwrites and destroys normal messages while the *Overload Frame* is exclusively started in the *IFS* (see Fig. 1.7).

There are two conditions for sending an *Overload Frame*. First, an *Overload Frame* is sent when a *dominant* bit is seen in the first two bits of the *IFS*. Here, the *Overload Frame* is used to synchronize all CAN nodes. On the other hand, a CAN controller may use an *Overload Frame* to notify all other CAN nodes that it is overloaded (not able to handle a new message immediately), caused by (IC) internal delay times. The other CAN nodes will see the first *dominant* bit of the *Overload*

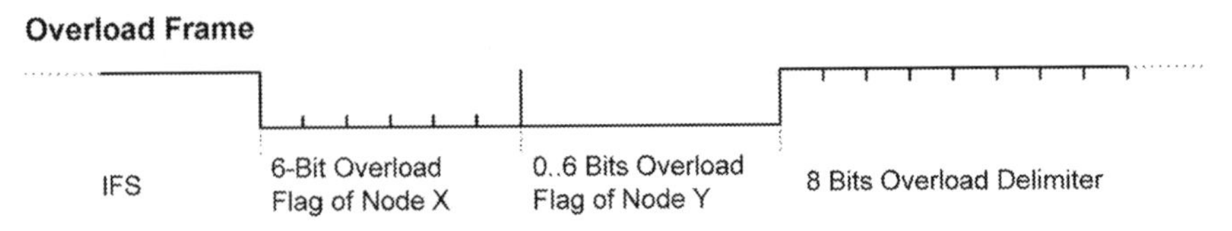

Fig. 1.7 CAN overload frame

Flag inside the *IFS* and will respond with their own *Overload Frames;* the *Overload Flags* will overlap each other. In this case, the start of a next frame is delayed by the length of an *Overload Frame*. In both cases, the overload condition has no effect on the error counters. Today's CAN controllers are fast enough; they are able to handle all bus traffic without the additional delay of *Overload Frames*.

1.2.5 *Timing Considerations*

As described in Sect. 1.2.2, there are two different frame types, those with 11-bit identifiers (also called *Standard Frame*) and those with 29-bit identifiers (also called *Extended Frame*). Furthermore, each message may contain from 0 to 8 bytes of data. Due to this, there are differences in effective data rates and minimum latency times for high priority messages.

Under normal circumstances (in undisturbed operation), a minimum latency time can be calculated only for the message with the highest priority in the network. Delay times for lower priority messages are not deterministic; they can only be determined statistically, using appropriate measuring equipment such as simulators, emulators, or dedicated network analyzers. Alternatively, the time-triggered option of CAN (TTCAN—time-triggered CAN) may be used, in which all messages are sent in a predefined time schedule. TTCAN is described in Sect. 1.3.

The maximum delay time for the message with the highest priority depends on the length of the longest possible message and on the bit rate. The length of the longest message (with a *Data Field* length of 8 bytes) is shown in Table 1.1; the result on the left is for networks where only frames with 11-bit (short) identifiers are used, while the result on the right is for networks where frames with 29-bit (long) identifiers are used.

This means that in networks where only frames with 11-bit identifiers are used, the message with highest priority has to wait at most 135-bit times frames for bus access, 135 μs at a bit rate of 1 Mbit/s. If frames with 29-bit identifiers are used, the maximum waiting time is 160 bit times, 160 μs at 1 Mbit/s.

The effective data rate of CAN frames is calculated from the ratio of data bits to the length of the frame. The data rate depends on the length of the *Data Field*, the length of the identifier and on the transmission rate. Table 1.2 gives an indication of

Table 1.1 Maximum frame length for data frames

Short identifier	Elements of a CAN frame	Long identifier
1	Start of frame	1
11	(Base) Identifier	11
	Substitute remote request (SRS)	1
	Identifier extension flag (IDE)	1
	Identifier extension	18
1	Remote transmission request (RTR)	1
6	Control field (IDE, r0, DLC or r1, r0, DLC)	6
64	Data field (8 × 8 bit)	64
16	CRC field (CRC sequence and CRC delimiter)	16
24	(Maximum number of) stuff-bits	29
2	Acknowledge field (ACK slot and ACK delimiter)	2
7	End of frame	7
3	Interframe space	3
Total 135	Maximum frame length	Total 160

the effective data rate for different lengths of the Data Field at a bit rate of 1 Mbit/s. At these values, the number of stuff bits is not considered.

Of course, no data rate can be calculated for a *Data Field* with a length of zero. However, such messages still have information content. The mere fact that a message with a certain identifier appears on the bus can be used for a so-called "Life Guarding" mechanism, i.e. key CAN nodes must—in a certain time schedule—send such a frame on the network, allowing other nodes to keep track of their operation state. Furthermore, such frames can be used, e.g. for synchronization purposes in a distributed application.

Considering the previous results for the calculation of the host controller's time budget needed to handle the CAN messages, the following values have to be taken into consideration:

The worst-case conditions regarding the processing time of the interrupt service routine for received frames occur at the maximum transfer rate of 1 Mbit/s, 100 % bus load and a message length of zero data bytes.

This means that with these parameters a message can be received every 47 µs. If the interrupt service routine in this example needed exactly 47 µs, the central processing unit (CPU) would be loaded to 100 % with this task alone and would have no capacity left for other tasks. For networks using only frames with 29-bit identifiers, this time would be—under the same conditions—67 µs.

The tasks of the received interrupt service routine depend on the type of CAN module; they are different for modules of the type "Full CAN" or the type "Basic CAN", see Sect. 1.2.7.

Table 1.2 Effective data rate at 1 Mbit/s

Data field in bytes	Effective data rate for frames with 11-bit identifier	29-bit identifier
0	—	—
1	72.1 kbit/s	61.1 kbit/s
2	144.1 kbit/s	122.1 kbit/s
3	216.2 kbit/s	183.2 kbit/s
4	288.3 kbit/s	244.3 kbit/s
5	360.4 kbit/s	305.3 kbit/s
6	432.4 kbit/s	366.4 kbit/s
7	504.5 kbit/s	427.5 kbit/s
8	576.6 kbit/s	488.5 kbit/s

1.2.6 Bit-Timing and Synchronization

CAN supports data rates of less than 1 kbit/s and also up to 1,000 kbit/s. Each CAN node in a CAN network is clocked by an individual clock generator (usually a quartz oscillator). The parameters of the bit time (i.e. the inverse of the bit rate) are individually adjusted in each CAN node, to achieve a uniform bit rate even from different oscillator frequencies (fosc).

The parameters are written into the configuration registers of the bit timing logic (*BTL*). The Baud Rate Prescaler (*BRP*) determines the length of the *time quantum* (tq), which is the basic unit of the bit time, while the timing segments determine the number of *time quanta* in the bit time. The clock frequencies of these oscillators are not absolutely stable, caused by temperature or voltage fluctuations and the aging of components. As long as the deviations remain within a certain oscillator tolerance range (*df*), the CAN nodes are able to compensate for the differences by synchronizing to the edges in CAN frames.

According to the CAN specification, a CAN bit time is divided into four segments (see Fig. 1.8): the *Synchronization Segment*, the *Propagation Time Segment*, the *Phase Buffer Segment 1* and the *Phase Buffer Segment 2*. Each segment consists of a certain (programmable) number of *time quanta*. The duration of a *time quantum* (t_q) is determined by the CAN controller's system clock f_{sys} and the *BRP*: $t_q = BRP/f_{sys}$. Common clocks are $f_{sys} = f_{osc}$ or $f_{sys} = f_{osc}/2$.

The synchronization segment, *Sync_Seg*, is that part of the bit time where edges of the CAN bus level are expected to occur. The Propagation Time Segment, *Prop_Seg,* is intended to compensate for the physical delay times within the CAN network. The Phase Buffer Segments, *Phase_Seg1* and *Phase_Seg2,* surround the Sample Point. The (re-) synchronization jump width (*SJW*) defines how far a resynchronization may move the Sample Point inside the limits, defined by the Phase Buffer Segments.

The individual parameters are configurable in the following ranges:

- *Sync_Seg*: 1
- *Prop_Seg*: [1…8]

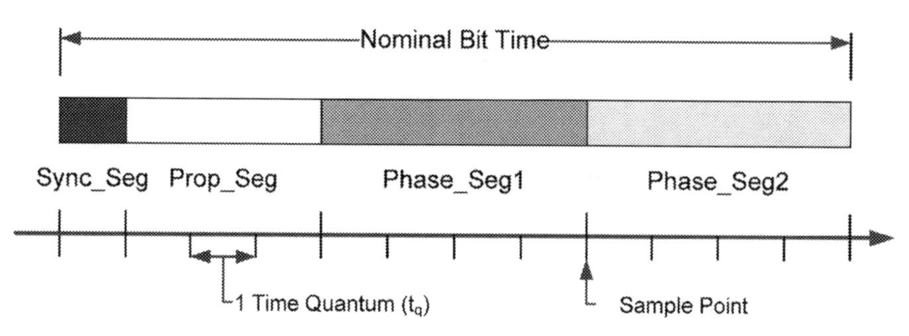

Fig. 1.8 Bit timing

- *Phase_Seg1*: [1…8]
- *Phase_Seg2*: [1…8]
- *SJW*: [1…4]
- *BRP*: [1…32]

In most CAN implementations, the sum of (*Prop_Seg+Phase_Seg1)* is collected (as *TSEG1*) together with *Phase_Seg2* (as *TSEG2*) in a first configuration register, while *SJW* and *BRP* are collected in a second register. It should be noted that the values programmed into these bit timing registers are—for each of *TSEG1*, *TSEG2*, *SJW* and *BRP*—the formal values reduced by one; hence, the values are written [0…n − 1] instead of [1…n]. This allows, for example, to represent SJW (actual range [1…4]) by only two bits. When *Prop_Seg* and *Phase_Seg1* are configured as a sum, *Prop_Seg* may be even larger than 8 while *Phase_Seg1* will be correspondingly shorter.

Therefore, the length of a bit time is (programmed values) [*TSEG1*+*TSEG2*+3] tq or (formal values) [*Sync_Seg*+*Prop_Seg*+*Phase_Seg1*+*Phase_Seg2*] tq.

The sequential flow of the bit time and possible synchronizations are controlled by the CAN protocol controller's *BTL*, a state machine that is evaluated once per *time quantum*. The *BTL* has the task to evaluate the CAN bus level and to determine the position of the *Sample Point*. The remaining part of the CAN protocol controller, the Bit Stream Processor (BSP) state machine, is evaluated only once per bit time, at the *Sample Point*, with the CAN bus level evaluated at the *Sample Point* taken as the sampled bit value.

1.2.6.1 "Hard" and "Soft" Synchronization

At each time quantum, the CAN controller's *BTL* compares the actual level of the CAN bus with the stored value of the last *Sample Point*, to detect edges for synchronization.

There are two types of synchronization:

- "Hard" synchronization

is performed on the falling edge (from *recessive* to *dominant*) of a *Start Of Frame* bit. This edge defines the beginning of a CAN frame; the *BTL* restarts at *Sync_Seg*.
- “Soft” (re-)synchronization is performed within a CAN frame at edges from *recessive* to *dominant*; it lengthens or shortens single bits of that frame.

If the CAN bus level changes outside of *Sync_Seg*, the distance between *Sync_Seg* and that edge is called the *Phase_Error*. When a *Phase_Error* is detected, it is compensated by synchronization. In case of a resynchronization, *Phase_Seg1* is extended (when the edge was between *Sync_Seg* and the *Sample Point*) or *Phase_Seg2* is shortened (when the edge was between the *Sample Point* and the next *Sync_Seg*). This is intended to keep the distance between the edge of the input signal and the following *Sample Point* at the configured value. In a hard synchronization, the *Phase_Error* will be fully eliminated, while a single resynchronization will reduce the *Phase_Error* only by an amount of up to the value of the (Re-)Synchronization Jump Width (*SJW*). A residual error (*Phase_Error* – *SJW*) may remain until the next synchronization.

There may be at most one synchronization between two *Sample Points*. The distance between the edge and *Sample Point*, which is maintained by synchronizations, allows the CAN bus level time to stabilize and filters out spikes that are shorter than (*Prop_Seg*+*Phase_Seg1*). Internal delay in the CAN transceiver, from the transceiver’s transmit input to its receive output, may cause transmitters to see all their transmitted edges “late”, this happens especially at high bit rates. Therefore, to avoid lengthening their own transmitted *dominant* bits, transmitters do not synchronize on “late” edges. The bit-stuffing mechanism guarantees a maximum distance of ten bits between two edges for synchronization inside a frame. *Phase_Errors* caused by clock tolerances will accumulate between synchronizations, so a receiver’s actual *Sample Point* position (relative to the transmitter’s *Sample Point*) may need to be corrected. The size of the Phase Buffer around the *Sample Point* defines how large a *Phase_Error* can be tolerated and therefore limits the clock tolerance.

Figure 1.9 shows how the phase buffer segments are used to compensate for *Phase_Errors*. Two successive bit times are presented: At the top with a synchronization to a “late” edge, which is detected between *Sync_Seg* and the *Sample Point*, in the centre without synchronization (as reference) and at the bottom with a synchronization to an “early” edge, which is seen after the *Sample Point*.

The examples in Fig. 1.10 show how short dominant spikes on the CAN bus are filtered out by the *BTL*. In both examples, the spike starts at the end of *Prop_Seg* and has a length of less than (*Prop_Seg*+*Phase_Seg1*).

In the first example, *SJW* is at least as large as the *Phase_Error* of the “late” edge which starts the spike. Therefore, the *Sample Point* can be moved past the end of the spike. At the *Sample Point*, the CAN bus is returned to *recessive* and the spike is suppressed.

In the second example, *SJW* is smaller than the *Phase_Error* of the edge which starts the spike. The *Sample Point* cannot be shifted far enough; the *dominant* spike at the *Sample Point* is taken as actual bus level.

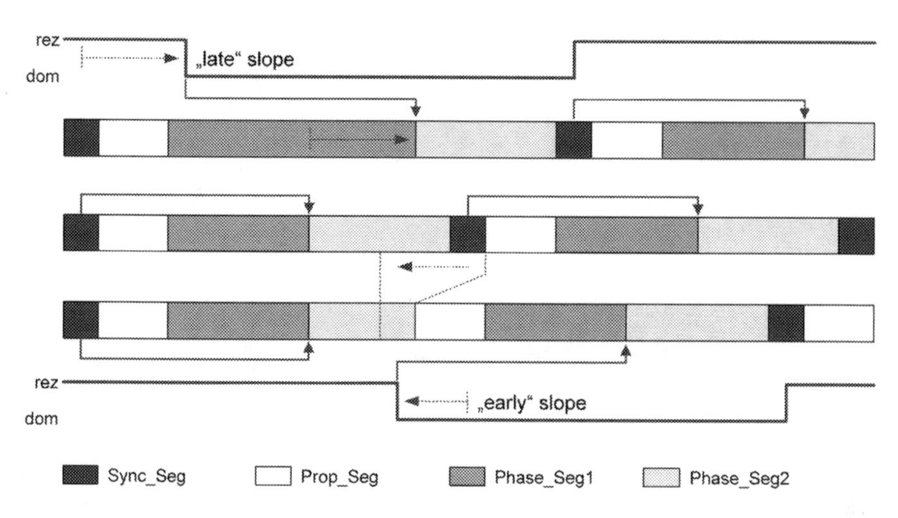

Fig. 1.9 Synchronizations on edge before and after the Sync_Seg

The advancement of the CAN protocol from version 1.1 to 1.2 increased the oscillator tolerance and the option to synchronize on edges from *dominant* to *recessive* became obsolete. Only the edges from *recessive* to *dominant* are used for synchronization. CAN protocol implementations according to version 1.1 are no longer in production.

Most synchronizations occur during arbitration. All CAN nodes synchronize "hard" on the first node that sends a *Start Of Frame* bit, but their phases will be shifted to each other, caused by differences in the signal propagation delays. During arbitration, there may be several transmitters (transmitters do not synchronize on "late" edges) and the transmitter, which triggers the hard synchronization, does not necessarily win the arbitration. Therefore, the receivers must synchronize themselves successively to different transmitters, whose edges arrive delayed by different propagation delays.

1.2.6.2 Propagation Delays

The maximum signal propagation time in the CAN network (between the most distant nodes, here called A and B) becomes relevant when both of them start CAN frames at the same time. Supposing that A causes the hard synchronization of B, then B will operate with a phase shift of *Delay A → B* relative to A. *Delay A → B* is the sum of the delay of A's bus driver, the propagation time on the CAN bus line between A and B and the delay of B's bus coupling circuit. The critical point is reached when A sends two consecutive *recessive* bits while B sends at the same time a *recessive* and a *dominant* bit. A loses the arbitration at the *dominant* bit. However, A sees that bit at the earliest 2 • *Delay A → B* (*Delay B → A* assumed equal to

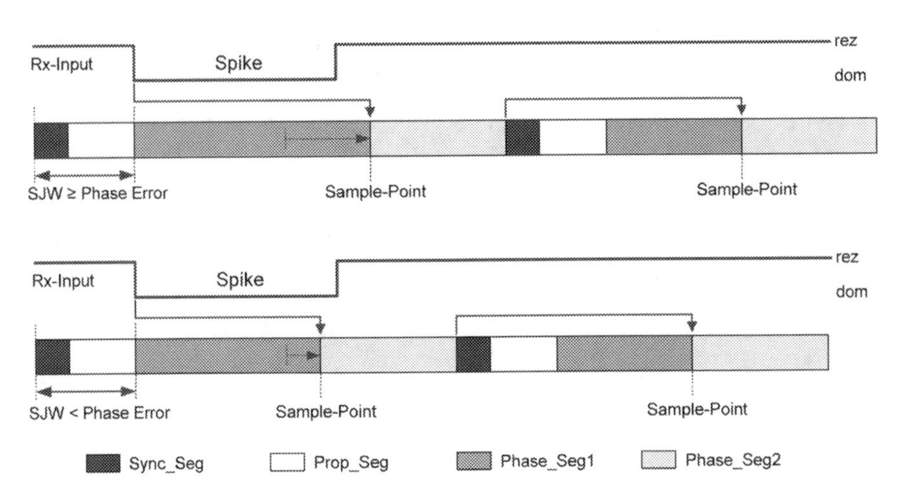

Fig. 1.10 Filtering of short dominant spikes

Delay A → B) after its *Sync_Seg*. Its *Phase_Seg1* may not yet have started at that time, since the beginning of *Phase_Seg1* is the earliest position of A's *Sample Point* if that *Sample Point* is virtually shifted by clock tolerance. *Prop_Seg* must therefore be programmed to a value of 2 • *Delay A → B* (rounded up to the nearest integer multiple of tq), as shown in Fig. 1.11.

The *Propagation Time Segment* must be programmed to a length that ensures edges to be seen before the onset of *Phase_Seg1*, even when the edges are affected by the maximum signal propagation delay. The *Phase Buffer Segments* before and after *Sample Point* have to be left for compensating oscillator tolerances. The impact of the bit time parameters on the oscillator tolerance *df* is shown by the formulae [1.1] and [1.2].

Some CAN implementations provide an optional so-called *Three_Sample_Mode*. In this mode, a digital low-pass filter is set before the CAN bus input of the *BTL*. Three *Sample Points* are not present in one bit time, but the bit value at the *Sample Point* is generated from the last three values of the CAN bus input (strobed each tq), evaluated by majority voting. The majority voting filters out spikes, but the CAN bus input signal is delayed by at least one tq and the Propagation Time Segment must be extended accordingly.

1.2.6.3 Oscillator Tolerance

The formulae [1.1] and [1.2] are used to calculate the oscillator tolerance:

$$df = \frac{\min(Phase\_Seg1,\ Phase\_Seg2)}{2 \cdot (13 \cdot Bitperiod - Phase\_Seg2)} \tag{1.1}$$

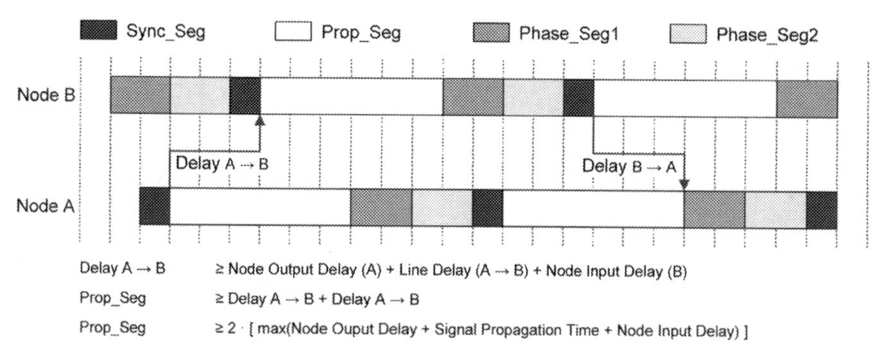

Fig. 1.11 Phase shift between the nodes in the CAN network

and

$$df = \frac{SJW}{20 \cdot Bitperiod}. \tag{1.2}$$

Both inequalities must hold, with *df* indicating how much a CAN node's oscillator frequency may deviate from its reference value without causing disturbances on the CAN bus.

The oscillator tolerance *df* (given in per cent) thus depends on the length of the *Phase Buffer Segments* as well as the length of the *Resynchronization Jump Width*, compared to the entire bit time. It must be noted that the *Resynchronization Jump Width* may not be longer than the shorter of the *Phase Buffer Segments* and that the partition of the bit time available for the *Phase Buffer Segments* is limited by the *Propagation Time Segment*.

1.2.6.4 Setting the Bit Timing Registers

Two cases have to be distinguished when programming a CAN node's bit time configuration. When the CAN node is integrated into an existing network, it needs to adopt the bit time configuration that was specified for the network. Here, the only variable is the *BRP*. Depending on the node's fsys frequency, *BRP* defines the length of the tq, with the requirement that an integer number of tq needs to be placed in one bit time, before and after the *Sample Point*. In CANopen-based networks, for example, the *Sample Point* should be set to 85 % of the bit time. The standardized configuration allows the easy integration of modules into a common CAN network, even when the modules are supplied by different vendors.

If no default configuration is available, the parameters of the bit time are largely determined by the desired bit rate and by the signal propagation delay. The components of the bit time are calculated as follows: First, the delay times in the network need to be determined. These consist of the delay on the CAN bus line (typically 5 ns/m), the transit times through the node's transceivers and, for example, in industrial applications, through additional optocouplers for galvanic isolation. The determined signal propagation delay gives an upper limit for the possible bit rate, since the CAN bit time (reciprocal of the bit rate) must be significantly longer than twice this value.

Then, based the desired bit rate and the available system clock frequency fsys, *BRP* values are selected, that can produce the desired bit time from an integer number (in the range of [5…25]) of tq, with tq = *BRP*/fsys. The length of *Prop_Seg* needs to be twice the signal propagation delay, rounded up to the next integer multiple of tq. Hence, the *BRP* value must allow *Prop_Seg* to be in the range of [1…8] (if needbe, in the range of [1…15] when *Prop_Seg* and *Phase_Seg1* are combined in *TSEG1*).

The desired bit time must be at least 3 tq longer than the time required for *Prop_Seg*. *Sync_Seg* is always 1 tq long; the rest is left for the two *Phase Buffer Segments*. *Phase_Seg1* and *Phase_Seg2* should have the same length, but if the subtraction (bit time – [*Sync_Seg* + *Prop_Seg*]) results in an odd number of tq, *Phase_Seg2* is set longer: *Phase_Seg2* = *Phase_Seg1* + 1. In order to optimize the oscillator tolerance, *SJW* needs to be set to the highest possible value, but not longer than one *Phase Buffer Segment*. The oscillator tolerance is mainly determined by the relation between the length of the *Phase Buffer Segments* and the length of the bit time.

The combination of *Prop_Seg* = 1 and *Phase_Seg1* = *Phase_Seg2* = *SJW* = 4 yields has an oscillator tolerance of 1.58 %, the largest value possible in the CAN protocol. This combination, with a *Propagation Time Segment* of only 10 % of the bit time, is not suitable for short bit times; at a 40-m bus length, it can be used for bit rates up to 125 kbit/s (with a bit time of at least 8 μs).

The bit timing concept of the CAN protocol has sufficient reserves, so that small deviations from the nominal values (temperature changes or aging of the components may cause, e.g. drift of the oscillators or longer signal delays) do not directly cause disturbances in the communication; but the deviations may make the network less resilient with regard to external sources of error (e.g. EMI). If a bit is disturbed at its *Sample Point*, the faulty bit is intercepted by higher protocol layers (CRC code, bit-stuffing, etc.) and the message is invalidated.

1.2.7 Characteristics of CAN Controllers

In principle, there are three types of CAN controllers: "Full CAN", "Basic CAN" and serial linked input/output (SLIO). CAN controllers are internally partitioned in CAN protocol controller and CAN message handler. While the function of the CAN protocol controller is defined by the CAN specification, the function of the CAN message handler is application specific. Different concepts have been developed for the CAN message handling.

A "Full CAN" manages a set of dedicated transmit and receive buffers. It is characterized by the fact that it sorts—based on the acceptance filtering—a received message into a receive buffer of the CAN controller's message memory that was specifically configured for that message. This means that the host controller can read the received data directly from their dedicated address range inside the CAN controller's data random access memory (RAM). Furthermore, a "Full CAN" will notify the host controller only of those received messages that successfully pass the acceptance filtering, meaning that the identifier of the received message matches with the identifier of at least one of the configured receive buffers. The host controller is not burdened with message filtering. The dedicated transmit buffers are also configured for specific identifiers. During operation, the host controller only needs to update the data content of the transmit buffer and request the transmission. This transmission may also be triggered by the reception of the matching *Remote Frame;* hence, the acceptance filtering enables the "Full CAN" to automatically reply to a *Remote Frame* with a *Data Frame*, not needing any further action of the host controller.

The "Basic CAN" variant has in principle only one receive buffer, which is designed as a first in, first out (FIFO). Therefore, the FIFO needs to be checked regularly and each received identifier needs to be compared with a list of relevant identifiers. If this comparison finds the relevant received messages, then the data are transferred into the corresponding address range of the host controller's RAM. The transmit buffer is managed dynamically; it is reconfigured for each transmission. To answer a *Remote Frame*, the host controller needs to prepare the corresponding *Data Frame*, load the transmit buffer and then request its transmission.

A "SLIO" is a CAN controller that requires no local software. It has several digital port pins and some analog/digital (A/D) and digital/analog (D/A) converters. These input and output functions are polled and controlled via CAN messages. Another feature is the automatic clock calibration. For this purpose, the SLIO observes the CAN communication and seeks dedicated calibration messages. There is a specific bit pattern in these messages that provides reference points for the measurement of the network's bit rate. The SLIOs calibrate and adjust their clock frequency to the determined bit rate. The SLIOs' transmission path remains switched off until the calibration is complete, to prevent disturbances on the CAN network. The calibration messages have to be repeated regularly.

The advantage of the "Full CAN" concept is the reduced load on the host controller; a disadvantage is the limited number of messages that can be managed. The advantage of the "Basic CAN" concept is that the number of different messages that can be treated is not limited by the number of receive (FIFO-) buffers; the disadvantage is the higher load on the host controller, for acceptance filtering and FIFO management.

In both concepts, received messages are lost when the host controller picks them up slower than they are loaded into the buffers of the CAN controller. In the "Basic CAN", an unexpectedly large number of messages ("babbling idiot" problem), which are not rejected by the acceptance filtering, may cause other, unread messages to be pushed out of the FIFO.

There are mixed concepts of "Full CAN" and "Basic CAN". A "Full CAN" may, for example, use mask registers to except selected bits of the identifier from acceptance filtering. This allows groups of messages to be stored in the same receive buffer. Further, some transmit buffers in a "Full CAN" may be managed dynamically, when more buffers are needed than the CAN controller implementation provides. Typical "Full CAN" controllers have 15–128 message buffers, which can be configured to receive buffers or transmit buffers.

"Basic CANs" may have the characteristics of "Full CANs" if, for example, more than one FIFO is provided, or if an extensive acceptance filtering loads only those messages into the FIFO that are relevant for the application.

CAN controllers may be implemented as dedicated CAN ICs ("stand alone") or they may be integrated as a peripheral module of a microcontroller. There are also microcontrollers available that are provided with several CAN modules. These modules may be connected to the same CAN network, operating as one CAN module with a large number of message buffers. Mostly they are connected to different CAN networks and may also be used, for example, to transfer messages between these networks, operating as *gateway* (switch).

In general, CAN controllers cannot be directly connected to a CAN bus; CAN transceivers are required for the connection. CAN controllers, especially the modules in a microcontroller, are manufactured in a semiconductor process that is optimized for high packing density. CAN transceivers are manufactured in a different process, which provides much more electrical driving strength and is also more robust against high voltage, e.g. caused by electrostatic discharges on the CAN bus line. The partitioning in CAN controller and CAN transceiver has also the advantage that the same CAN controller can be used with different types of CAN transceivers, such as for conventional two-wire differential bus lines (*ISO 11898-2*, *-3*, or *-5*) and also for single-wire bus lines (*SAE J2411*).

CAN transceivers mediate between the bi-directional CAN bus on the one side, that knows only dominant and recessive levels, and the CAN controller on the other side, which is connected by two unidirectional digital lines, one each for reception and for transmission.

1.3 Time-Triggered CAN

1.3.1 Motivation to Advance CAN Protocol Towards TTCAN

Safety-relevant systems in car electronics require a timely deterministic communication protocol with predefined latency time. This demand is a sine qua non prerequisite when designing distributed control systems consisting of a multitude of individual controllers fulfilling their special requirements on

- Safety
- Schedulability of timely behaviour
- Verifiability
- Fault tolerance
- Synchronous behaviour

CAN communication protocol implements a purely priority-driven bus-access technology which per se does not guarantee any specific latency time. CAN message transfer is thus characterized by a typically big jitter in best-case and worst-case latency times. Therefore, different methodologies had been considered over the years in order to achieve a message transfer characterized by predefined latency times.

When inspecting and revising the CAN standard ISO 11898-1 periodically, the experts checked whether the event-driven CAN protocol could be modified towards a time-triggered version; and it turned out that this attempt could be achieved. Hence, the time-triggered version of the CAN protocol was developed by a small team of experts of Robert Bosch GmbH. The resulting TTCAN protocol then became the ISO 11898-4 standard.

1.3.2 Constraints

- When developing TTCAN, one of the basic prerequisites was the compatibility with the existing CAN standard features such as data frames, the proven error detection and error handling mechanism. Therefore, CAN knowledge, experience and CAN development tools could easily be reused.
- Only in case of automatic retransmission on error detection of the CAN protocol there is a small restriction: In exclusive time windows—refer to Sect. 1.3.4—erroneous messages are not retransmitted immediately; otherwise, the cyclic communication would be corrupted. This message, anyhow, will be retransmitted within the next cycle.
- In time-triggered systems, a condition that is absolutely needed is a time stable basis, being available all over the whole system. A failure of an individual module in a system must not lead to chaos nor cause a breakdown of the communication process of the remaining system.
- The communication process must be synchronized by external events in order to synchronize further sub-networks, redundant channels, or events depending on engine revolution speed.
- The system must be configurable over the bus. There, easy synchronous switching between the event-driven and time-triggered mode is required.

1.3.3 Time Triggered Basics

A common synchronized clock (a global clock) in all nodes participating in a communication system is the basic principle in time-triggered systems .

Basically, there are multiple methods to synchronize the clocks of the individual communication participants to a global time in a system. Either all participants communicate their local clock to all other participants, and thus individually can calculate a common global clock, or there is one single global clock known to everybody, and all participants can synchronize themselves on this single global clock. There will be no further evaluation of the pros and cons of these methods following.

When developing a TTCAN protocol, the choice of the time base was driven by the characteristics of the already existing CAN applications: Many of the messages are communicated in non-periodic repetition rate in the range of some milliseconds. Furthermore, there are other messages which are transferred spontaneously, based on an event or based on long repetition rates.

In a TTCAN protocol, one of the participants is appointed to be the Time Master with its responsibility to provide the clock signal to all other participants. By applying an individual node-specific clock divider ratio, called Time Unit Ratio, TUR, to the Time Master clock, each node derives internally the Network Time Unit—NTU. The NTU therefore is identical within all participants, representing the time granularity for any communication in the system. The calculation of the local time within each node is shown in Fig. 1.12.

1.3.4 Communication Architecture

Communication is controlled by the time progress after being initialized by a reference message from the Time Master. This reference message is repeated periodically with a predefined cycle time. The reference message is followed by so called Time Windows which are used by the network participants to insert their information for the communication process.

The start of a time window is designated by time tags (*Tx_Ref_Trigger*, *Tx_Trigger*) resulting from the cycle time counter value (*Cycle_Time*). Cycle time will be described in more detail below.

There are various kinds of time windows: exclusive, arbitrating and free time windows. Furthermore, there are merged arbitrating windows, which are a special kind of arbitrating time windows.

An exclusive time window is used for periodic transfer of solely one message of a participant. Arbitrating time windows are available for any participant sharing bus access. Arbitrating time windows typically communicate sporadic messages or periodic messages with long cycle times.

Any number of consecutive arbitrating time windows can be combined to a so-called merged arbitrating window. This feature typically is applied to transfer messages which previously had lost arbitration or were a little late. The transfer is then performed immediately after the currently transferred message without waiting until the next start of a time window. As such, bus efficiency is increased and message latency times are reduced. In an extreme case, all time windows could be defined as merged arbitrating windows, thus resulting in an almost mere event-driven CAN system.

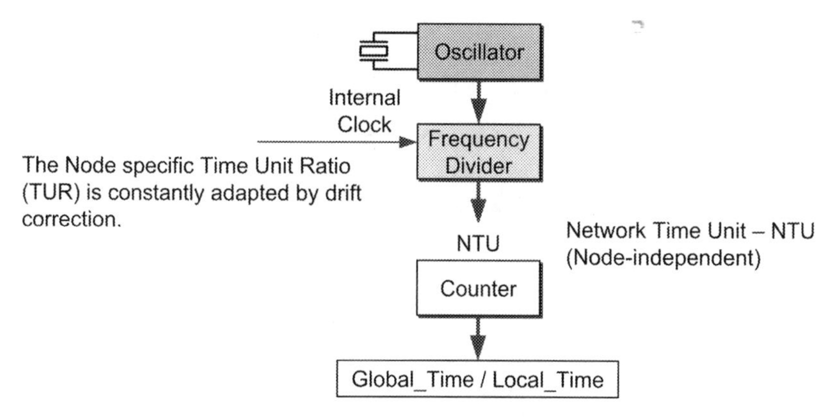

Fig. 1.12 Calculation of local time within a node

"Free" time windows are planned for further extension or they may result with the design of the system.

In order to optimize the required bus bandwidth lengths of time windows, it can be adjusted to the lengths of the messages to be transferred. Therefore, time windows have different lengths in a system. However, the design must make sure that the longest messages to be transferred fit into their designated time windows. (This constraint must be met especially when doing the layout of the arbitrating time windows and the system matrix.)

The sequences of time windows between two reference messages are referred to as "Basic Cycle".

The sequence of time windows in a basic cycle as shown in Fig. 1.13 is not always the same. Distribution and content of time windows may vary from one basic cycle to the next one. The entirety of all basic cycles is called "System Matrix". The columns of the system matrix represent the so-called Transmission Columns containing the time windows, while its rows represent the basic cycles. Lengths of time windows in the columns are restricted maximally to the width of the corresponding column. The only exception to this constraint are the merged arbitrating windows, which may comprise several columns but must start and end at column borders.

There is a restriction to the numbers of possible basic cycles due to system constraints. The maximum number of basic cycles in one single-system matrix (Fig. 1.14) is 64, or any lower number to the power of 2.

When specifying exclusive Time Windows only, access is uniquely time controlled. In that case, the identifier part of the data frame could be abandoned. However, for compatibility reasons and as learned from experience, it is required and advantageous to keep this arbitration feature.

All participants of a network know their related time windows in which they are allowed to transmit messages or in which they may receive messages from other participants. The start of each time window respectively is given by a time tag relative to the reference message. Therefore, each participant must be able to recognize

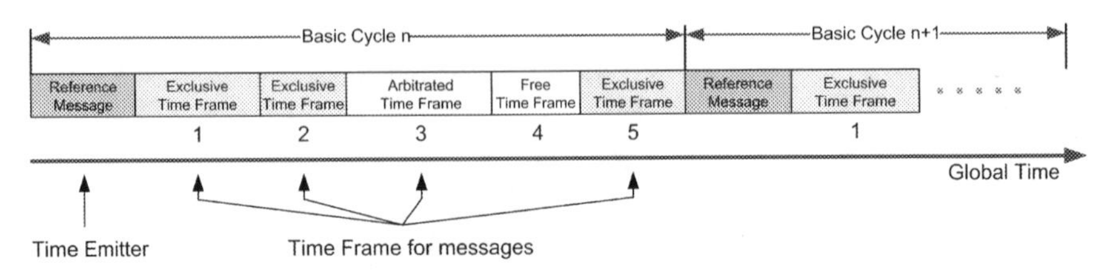

Fig. 1.13 Structure of a TTCAN "Basic Cycle"

the reference message. It is designated by a specific identifier (specified by the system designer) which is communicated to all network participants.

The reference message itself consists—in the simple case Level I implementation—of one data byte containing the Cycle Counter (number of the actual basic cycle) and the control bit, next_is_gap. In Level II implementation, the reference message contains further information for the interpretation of NTU (refer to Sect. 1.2.3), an additional Disc_Bit and a 16-bit information giving the actual time of the time master. The Disc_Bit signals a disruption in the time continuity by an external synchronization.

While in Level I implementation NTU typically is related to a bit time, in Level II implementation NTU may be related to the physical time entity "second".

1.3.5 Communication Sequence

Whenever the cycle time counter of the node, designated as the system clock, reaches the reference transmission threshold (Tx_Ref_Trigger), this node transmits its reference message, which is recognized as such by all other nodes due to the designated identifier and which causes synchronization of their local time counters or global time counters, in case of Level II implementation (see Sect. 1.2.6). If the internal local or global time counter of a participant reaches its transmit threshold (*Tx_Trigger*) signalling, the start of its designated transmission window and if the bus is not busy, this participant transmits its message which is assigned to this time window. In case this is an exclusive time window, a receive tag (*Rx_Trigger*) is set in this window in all those participants which are to receive this message. This causes error checks upon reception of this message, in order to initiate an error-handling process in case of a problem detected.

The transmit tags consist of a time tag (column number of the system matrix), a basic tag (designating the basic cycle in which the message is to be transmitted for the first time) and a repetition rate (specifying the number of basic cycles after which the message is to be retransmitted).

The receive tags contain the same information, but only the time tag is deferred by the window width.

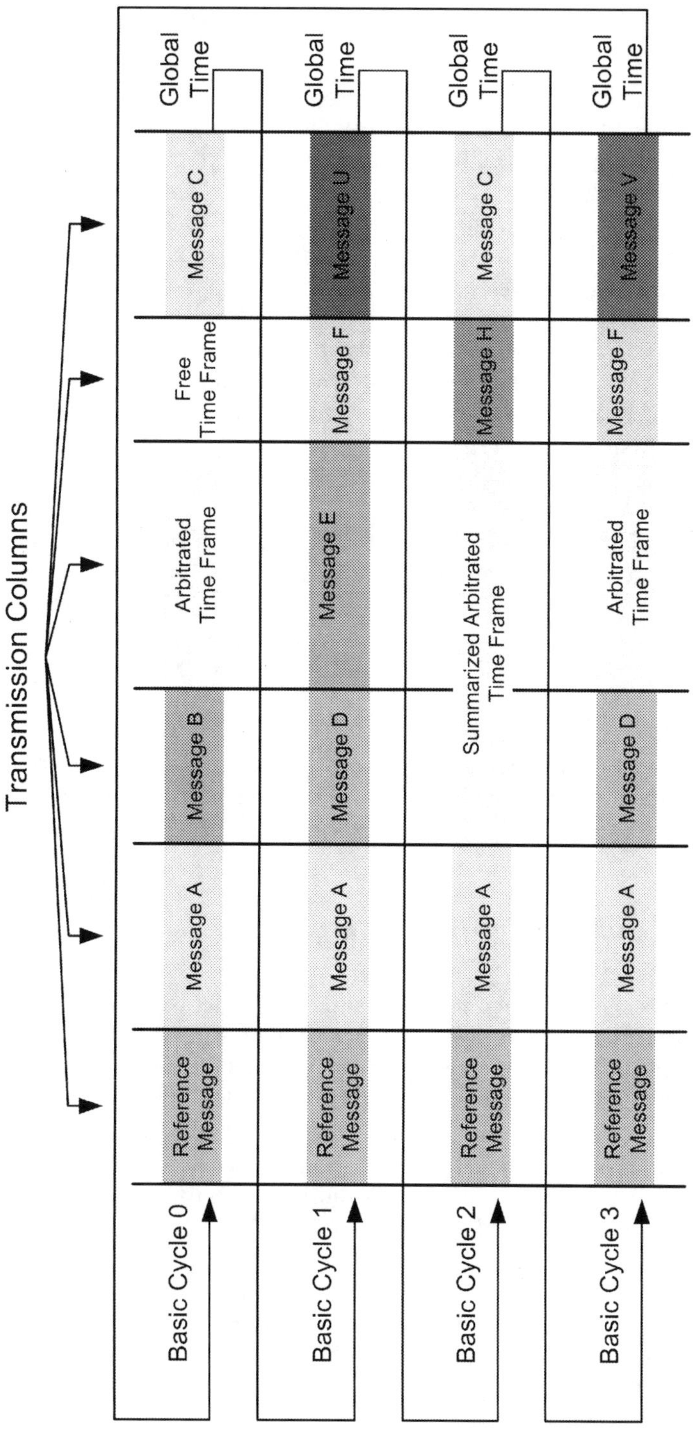

Transmission Columns

Basic Cycle 0	Reference Message	Message A	Message B	Arbitrated Time Frame	Free Time Frame	Message C	Global Time
Basic Cycle 1	Reference Message	Message A	Message D	Message E	Message F	Message U	Global Time
Basic Cycle 2	Reference Message	Message A	Summarized Arbitrated Time Frame		Message H	Message C	Global Time
Basic Cycle 3	Reference Message	Message A	Message D	Arbitrated Time Frame	Message F	Message V	Global Time

Fig. 1.14 Example of a system matrix

1.3.6 Synchronization, Local and Global Time

As already explained, all time windows respectively the start of time windows are related to the reference message. All participants receive the reference message at the same time—except the differences in their signal propagation delays—but they must also have a common reference point within the message. There is a distinctive item recognizable uniquely by all participants: the Start of Frame (SOF) of CAN messages. However, at this point of time it is not known whether or not the following message is a reference message and whether or not it will be received non-corrupted.

Therefore, on detection of an SOF all participants write the Local_Time value of their cycle time counter into a so-called Sync_Mark register.

If hereafter a reference message is recognized, the Sync_Mark register is copied into a Ref_Mark register. Then, the difference between this register and the Cycle_Time counter is calculated, thus achieving the time count being set in relation to the start of the CAN frame (see Fig. 1.15).

In Level II implementations, even absolute time references can be achieved relative to a predefined time given by the reference sender. In order to do so when detecting the sample point of SOF, the reference sender writes its local time into a Master_Ref_Mark, which is part of the data field of the currently to be transmitted reference message. All other participants can calculate from there their Ref_Mark as described above.

The difference between the Master_Ref_Mark and the node-specific Ref_Mark corresponds to the Offset (called the Local_Offset) of each node relatively to the time transmitter. Adding the offset to the local time of a node results in a Global_Time, which is common to all participants (Fig. 1.16).

1.3.7 Redundancy of Time Transmitters

TTCAN systems with only one single-time transmitter had to be considered very weak from a safety-critical systems-design point of view. Therefore, suitable provisions are required in order to meet safety and availability requirements.

One of these provisions required is the installation of redundant so-called potential time transmitters. Up to eight participants can be defined as potential time transmitters in one network. Therefore, specific reference messages are defined in these participants, all with identical identifiers except for their three LSBs. The participant with the lowest identifier (typically all 3 LSB = 0) becomes as such the standard time transmitter (Time Master), while all the others are the potential time transmitters.

As long as the standard time transmitter is functional, all the others will receive its reference message thus suppressing transmission of their own references messages. If this one fails, all potential time transmitters will try to send their

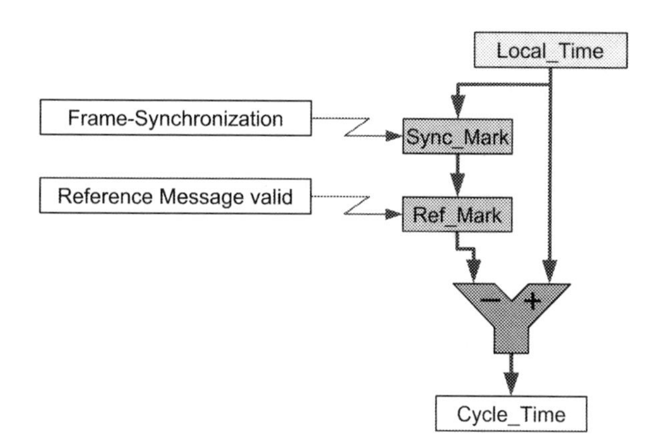

Fig. 1.15 Synchronization of local time by reference message (Level I)

reference messages, while in case of simultaneous transmissions the one with the highest priority will win. When the standard time transmitter recovers, this one will try to take over again the time transmitter function when the next basic cycle is to start.

This technique guaranties a fast and well-organized system start up: The first one of the potential time transmitters being operational will start the basic cycles. The other higher priority potential time transmitters becoming operational over time then will take over, while at the end of this start-up phase the standard time transmitter will finally control the communication process.

1.3.8 Event-Synchronized Cycle

In non-event-synchronized TTCAN systems, the reference message is transmitted periodically in equidistant steps.

However, TTCAN systems also allow us to interrupt this periodic process and to start basic cycles on an event in a time transmitter. The event may be derived from a software trigger or from a timer elapsed or it may come from an external signal source.

For synchronization, the currently running cycle will be halted at the end of the basic cycle while the next_is_gap bit is set in the reference message. This halts the next reference message until the predefined event will start a new cycle by transmitting the reference message.

This technique enables synchronization of communication on, e.g. crank shaft synchronous tasks, synchronization of multiple subnets, or smoothing cycle drifts caused by corrupted reference messages.

Figure 1.17 shows pure cyclic and event synchronized cyclic message communication.

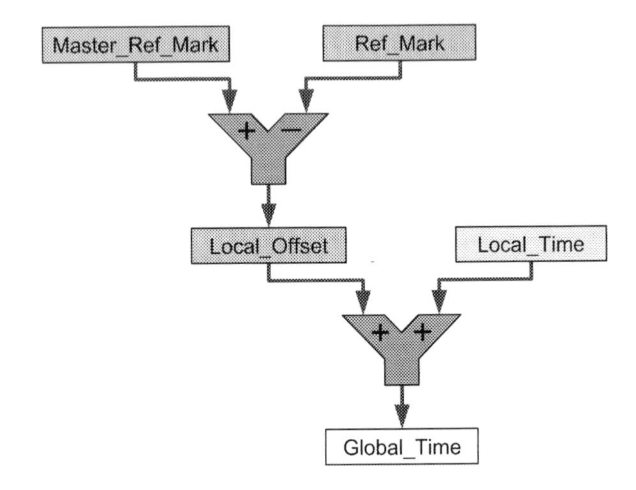

Fig. 1.16 Calculation of global time (Level II)

1.3.9 Drift Correction

At long basic cycles, slight differences in clock frequencies between individual nodes may cause nodes to fall out of synch. Level II implementations provide a drift-correction feature for compensation (Fig. 1.18).

Therefore, each node compares the length of the basic cycle, transmitted by the time transmitter with its own. The difference between the two results in a correction factor, which adjusts the frequency divider in each node respectively through TUR (see Sect. 1.2.3).

1.3.10 Extended Error Detection and Error Handling

Due to the cyclic character of TTCAN, there are additional diagnostic features on top of the well-known error-handling techniques of CAN:

- Based on the reception tags mentioned above, a loss of an expected message in exclusive time windows can be detected. The number of expected messages per system cycle is known. The application will be signalled, if a specific number of messages is lost.
- Further, the number of transmitted messages is counted. If for a specific participant the number of exclusive transmission windows exceeds the defined amount, a further transmission of messages is abandoned until the start of the next system cycle.
- Various watchdogs prevent the system from going into an uncontrolled state:
  - A synchronization watchdog monitors the periodic transmission of the reference message. In the event synchronous operation mode, another watchdog restarts communication in case of a loss of trigger.

Cyclic Communication

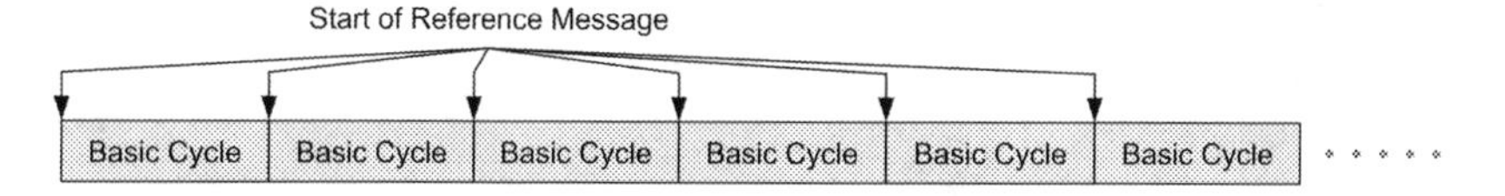

Event-synchronised Cyclic Commucication

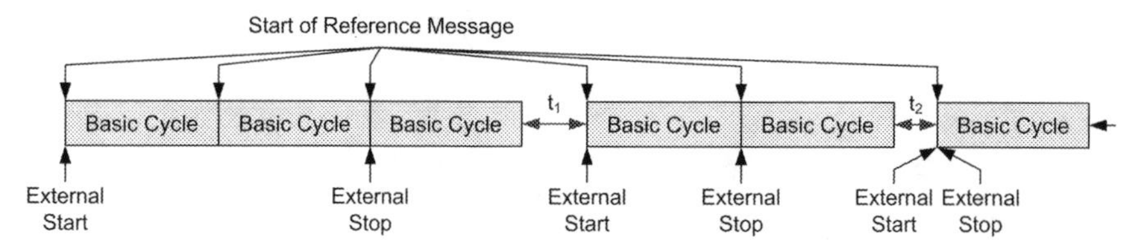

Fig. 1.17 Pure cyclic and event synchronized cyclic message communication

- An application watchdog checks the periodic application signal "alive". If this signal is not detected, the watchdog abandons further transmission of messages. This prevents any TTCAN controller from behaving unexpectedly and thus disturbing communication of other participants.

- The time windows are protected by so-called "Tx_Enable-Windows". A message is only allowed to be transmitted, if this can be done within the Tx_Enable-Window. The Tx_Enable-Window is a small fraction in the first part of a time window. This feature prevents corruption of the entire communication cycle when the bus is still busy due to overlapping messages or any sporadic time deviation. There are four error levels specified, related to the types of error:

  - S0 = No error, unrestricted communication
  - S1 = Warning, application may decide what to do
  - S2 = Error, neither date frames nor remote frames are allowed to be transmitted (ACK frames and error frames may be transmitted)
  - S3 = Severe error, no bus communication (dominant bits) allowed; reconfiguration required (corresponds to the Bus-Off state in CAN)

1.3.11 Summary

TTCAN provides deterministic communication at a bus load which may even exceed 90 %, theoretically. Synchronization of multiple busses in the area of an NTU can be done as well as synchronization on external events and the implementation of redundant safety critical systems.

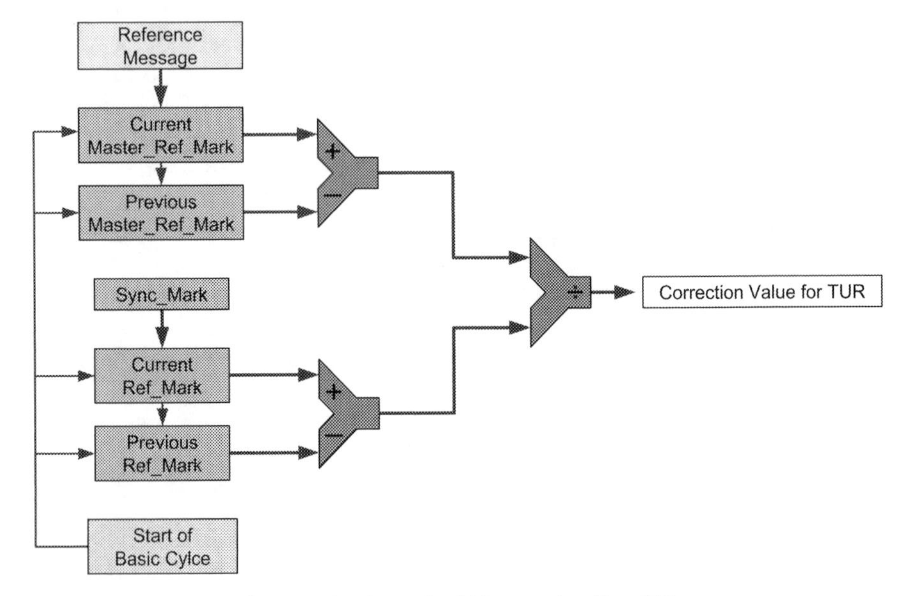

Fig. 1.18 Calculation of a correction factor for drift correction (Level II)

For further details and amendments, refer to ISO 11898-4, *Road vehicles—Controller area network (CAN)—Part 4: Time-triggered communication* as well as to www.semiconductors.bosch.de/en/20/can/3-literature.asp.

1.4 CAN FD—CAN with Flexible Data Rate

CAN with flexible data rate (CAN FD) is a new protocol that combines CAN's core features with a higher data rate. For automotive applications, CAN FD targets an average data rate of 2.5 MBit/s with existing CAN transceivers, resulting in the same effective payload as a low-speed FlexRay network. In other applications, where long bus lines limit the bit rate in the *Arbitration-Phase* to, e.g. 125 kbit/s, a higher acceleration factor is possible without exceeding the specification range of existing CAN transceivers in the *Data-Phase*. New specification ranges for the CAN transceivers are currently evaluated, while new transceiver designs may further increase the acceleration factor.

There is an easy migration path from CAN systems to CAN FD systems since CAN application software can be left (apart from configuration) unchanged. CAN FD controllers can operate in CAN systems. The Bosch CAN IP - CAN Intellectual Property - modules are currently being adapted to optionally support the CAN FD protocol. The first engineering samples of microcontrollers with integrated CAN FD modules are expected in 2013. International standardization of CAN FD is in preparation.

1.4.1 Further Development of the CAN Protocol

Increasing system complexity can fill a CAN network's communication bandwidth to its limit. Solving this problem by using multiple CAN buses or by switching to another protocol requires high effort in system design as well as replacing hardware and software.

Over the years, several concepts have been proposed regarding how to replace CAN with new bus systems that have a higher bandwidth and a similar controller host interface. The similarity avoids the need for major software modifications.

CAN's bandwidth limit is closely linked to one of its greatest advantages, its nondestructive arbitration mechanism for media access control. This mechanism requires that the signal propagation delay between any two nodes is less than half of one bit time and so defines an upper boundary for the bit rate as well as for the bus length.

Therefore, new concepts to increase the CAN bit rate avoid this limit mainly by two alternatives. Firstly, some concepts change CAN's multimaster bus line to a star (or tree) topology where arbitration (or even message routing) is performed inside an active star. Other concepts use two alternate bit rates and switch—after the arbitration—from the lower to the higher bit rate.

CAN FD has been developed with the goal to increase the bandwidth of a CAN network while keeping most of the software and hardware—especially the physical layer—unchanged. Consequently, only the CAN protocol controllers need to be enhanced with the CAN FD option. The new frame format makes use of CAN's reserved bits. Via these bits, a node can distinguish between the frame formats during reception. CAN FD protocol controllers can take part in standard CAN communication. This allows a gradual introduction of CAN FD nodes into standard CAN systems. CAN FD may be restricted to specific operation modes, e.g. software-download at end-of-line programming, while other controllers that do not support CAN FD are kept on standby.

1.4.2 CAN FD Concept

The development of CAN FD was based on the standard CAN protocol and had the requirement to accelerate the serial communication while keeping the physical layer of CAN unchanged. CAN FD started with the approach to increase the bandwidth by modification of the frame format. Two changes suggest themselves: firstly, improving the header to payload ratio by allowing longer data fields and secondly, speeding up the frames by shortening the bit time.

However, these steps are only the groundwork; some additional measures are needed, e.g. to keep the Hamming distance of the longer frames at the same level as in standard CAN and to account for the CAN transceiver's loop delay time.

The CRC polynomial of CAN is suited for patterns of up to 127 bits in length including the *CRC Sequence* itself. Increasing the CAN frame's payload makes longer polynomials necessary.

CAN nodes synchronize on received edges from *recessive* to *dominant* on the CAN bus line. The phases of their *Sample Points* are shifted relative to the phase of the transmitter's *Sample Point*. A node's specific phase shift depends on the signal delay time from the transmitter to that specific node.

The signal delay time between the nodes needs to be considered when more than one node may transmit a *dominant* bit. This is the case in the *Arbitration Field* or in the *ACK Slot*. The configuration of the CAN bit time, especially the *Propagation Time Segment's* length and the Sample Point's position, must ensure that twice the maximum phase shift fits between the *Synchronization Segment* and the *Sample Point*. Once the arbitration is decided, until the end of the *CRC Field*, only one node transmits *dominant* bits, all other nodes synchronize themselves to this single transmitter. Therefore, it is possible to switch to a predefined (shorter) bit time in this part of a CAN frame, here called the *Data-Phase*, see Fig. 1.19. The rest of the frame, outside the *Data-Phase*, is called the *Arbitration Phase*. Different coding of the *DLC* allows making the *Data Field* longer than the 8 bytes of standard CAN. The advantage of the improved header to payload ratio rises with the acceleration factor between the *Arbitration-Phase* and *Data-Phase*.

All nodes in the network must switch to this shorter bit time synchronously at the start of the *Data-Phase* and back to the standard bit time at the end of the *Data-Phase*. The factor between the short bit time in the *Data-Phase* and the standard bit time in the *Arbitration-Phase* decides how much the frames are speeded up. This factor has two limits. The first is the speed of the transceivers: Bits that are too short cannot be decoded. The second is the time resolution of the CAN synchronization mechanism: after switching to the short bit time, a *Phase_Error* of one *time quantum* in the standard bit time needs to be compensated for.

At the last bit of the *Data-Phase*, the *CRC Delimiter*, all nodes switch back to the standard bit time before the receivers send their acknowledge bit, followed by a *recessive Acknowledge Delimiter* and *End of Frame*.

CAN's fault confinement strategy, where a node that detects an error in an ongoing frame immediately notifies all other nodes by destroying that frame with an *Error Flag*, requires that all nodes monitor their own transmitted bits to check for *Bit-Errors*. Current CAN transceivers may have, according to *ISO 11898-5*, a loop delay (CAN_Tx pin to CAN_Rx pin) of up to 255 ns. That means, to detect a *Bit-Error* inside a bit time of the *Data-Phase*, this bit time has to be significantly longer than the loop delay. To make the length of a short bit time independent of the transceiver's loop delay, CAN FD provides the *Transceiver Delay Compensation* option, where the check for *Bit-Errors* and the responding *Error Flag* are delayed. All nodes that detect an error switch back to the standard bit time before sending an *Error Flag*.

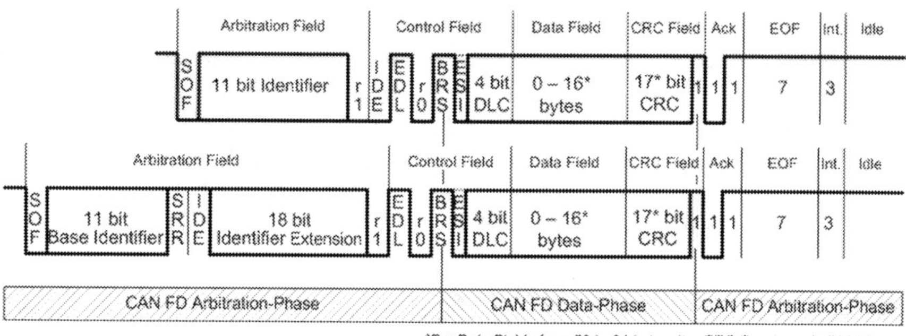

Fig. 1.19 CAN FD data frame format with 11-bit identifier and with 29-bit identifier

1.4.3 CAN FD Frame Format

The *Control Field* in standard CAN frames contains reserved bits which are specified to be transmitted *dominant*. In a CAN FD frame, the reserved bit after the *IDE* bit (11-bit Identifier) or after the *RTR* bit (29-bit Identifier) is redefined as *Extended Data Length* (*EDL*) bit and is transmitted *recessive*. This sets the receiving *BSP* and *BTL* FSMs - Finite State Machines - into a CAN FD decoding mode.

The following bits are new in CAN FD:

- *EDL Extended Data Length*
- *r1, r0* reserved, transmitted *dominant*
- *BRS Bit Rate Switch*
- *ESI Error State Indicator*

The *DLC* values from 0000b to 1000b still code a *Data Field* length from 0 to 8 bytes, while the (in standard CAN redundant) *DLC* values from 1001b to 1111b are redefined in CAN FD to code *Data Fields* with a length of up to 64 bytes:

DLC	1001	1010	1011	1100	1101	1110	1111
Byte	12	16	20	24	32	48	64

EDL distinguishes between the standard CAN frame format and the CAN FD frame format.

The value of *BRS* decides whether the bit rate in the *Data-Phase* is the same as in the *Arbitration-Phase* (*BRS dominant*) or whether the predefined faster bit rate is used in the *Data-Phase* (*BRS recessive*).

In CAN FD frames, *EDL* is always *recessive* and followed by the *dominant* bit, *r0*. This provides an edge for resynchronization before an optional bit rate switch. The edge is also used to measure the transceiver's loop delay for the optional *Transceiver Delay Compensation*.

In CAN FD frames, the transmitter's error state is indicated by *ESI*, *dominant* for *Error Active* and *recessive* for *Error Passive*. This simplifies network management.

There are no CAN FD *Remote Frames*; the bit at the position of the *RTR* bit in standard CAN frames is replaced by the *dominant* reserved bit, *r1*. However, standard CAN *Remote Frames* may optionally be used in CAN FD systems.

Receivers ignore the actual values of the bits r1 and *r0* in CAN FD frames; they are reserved for future expansion of the protocol, e.g. using *r1* as additional identifier bit.

Chapter 2
Physical Layer

Wolfhard Lawrenz, Cornelius Butzkamm, Bernd Elend, Thorsten Gerke, Magnus-Maria Hell, Ursula Kelling, Bernd Koerber, Kurt Mueller, Christian Schmitz, Radoslaw Watroba and Rolf Weber

The transfers of data which have been processed in the data link layer are done in the physical layer. The physical layer consists of:

C. Butzkamm (✉)
C&S group GmbH, Am Exer 19b, 38302, Wolfenbuettel, Germany
e-mail: C.Butzkamm@cs-group.de

W. Lawrenz
Waldweg 1, 38302, Wolfenbuettel, Germany
e-mail: W.Lawrenz@gmx.net

B. Elend
NXP Seminconductors Germany GmbH, Streesemannallee 101, 22529, Hamburg, Germany
e-mail: bernd.elend@nxp.com

T. Gerke
Synopsys GmbH, Karl-Hammerschmidt-Strasse 34, 85609, Aschheim-Dornach, Germany
e-mail: gerke.thorsten@yahoo.de

M.-M. Hell · U. Kelling
Infineon Technologies AG, Am Campeon 1-12, 85579, Neubiberg, Germany
e-mail: magnus-maria.hell@infineon.com

U. Kelling
e-mail: ursula.kelling@infineon.com

B. Koerber
Westsächsische Hochschule Zwickau, Dr.-Friedrichs-Ring 2A, 08056, Zwickau, Germany
e-mail: Bernd.Koerber@fh-zwickau.de

K. Mueller
Synopsys, Inc., 2025 NW Cornelius Pass Road, 97124, Hillsboro, OR, USA
e-mail: kurt.mueller@synopsys.com

C. Schmitz · R. Weber
ELMOS Semiconductor AG, Heinrich Hertz Strasse 1, 44227, Dortmund, Germany
e-mail: christian.schmitz@elmos.eu

R. Watroba
STMicroelectronics Application GmbH, Bahnhofstraße 18, 85609 Aschheim-Dornach, Germany

W. Lawrenz (ed.), *CAN System Engineering,* DOI 10.1007/978-1-4471-5613-0_2,
© Springer-Verlag London 2013

- Transceiver
- Controller area network (CAN) choke (optional)
- Electro magnetic compatibility (EMC) and electrostatic discharge (ESD) protection devices (optional)
- Connector
- Network or data bus, e.g. a two-wire cable

There are two different concepts for the physical layer, which is, on the one hand, the so-called *Fault-Tolerant Low-Speed CAN Physical Layer* and the *High-Speed CAN Physical Layer* on the other. They are different in the maximum data rate (1 Mbit/s for high-speed CAN and 125 kbit/s for low-speed CAN) and the concepts for bus termination which enable the fault-tolerant low-speed CAN transceiver to continue the communication in all error conditions. This is not feasible for high-speed CAN communication if there is, for instance, a short circuit condition between the bus lines. Nevertheless, high-speed CAN physical layer meanwhile is applied for the communication between all modules in cars. The area of application of low-speed CAN is typically limited to the vehicle interior applications—the so called "body". The fault-tolerant low-speed CAN physical layer is preferably applied by European car manufacturers. However, American and Japanese automotive manufacturers solely apply the more simple high-speed CAN physical layer. Nevertheless, European car manufacturers have dropped low-speed CAN in the meantime. In newer vehicle generations, they almost only apply the newer version of the high-speed CAN transceivers. The basic characteristics of the transceivers and the related physical layers are specified in the ISO 11898 standard.

2.1 Basic Elements

2.1.1 Transceiver

The transceiver transmits and receives the physical data to and from the bus. The basic concept of the transceiver is identical for high-speed CAN and low-speed CAN transceiver. For this basic concept, a transmitter and a receiver are needed. The transmitter is a buffer, which transforms the logical signal on pin TxD (transmit data) into a slew rate-controlled analog signal on the pins CAN_H and CAN_L. Both are open drain outputs. CAN_H is a high-side driver and CAN_L is a low-side driver. Both have reverse polarity diodes to protect these outputs against reverse operation. The minimum voltage range for both outputs is required to be between −27 and +40 V. The maximum output current is controlled to protect this output stage and the CAN coils (if implemented) against shorts to ground and supply…. The current will normally be limited between 40 and 200 mA, but in newer products the maximum value is reduced to 100 mA to protect the 100-µH CAN coils too. The receiver is a differential comparator and converts the differential signal into a logical signal on pin RxD (receive data). A logical high on RxD corresponds with a recessive differential level on the bus, and a logical low on pin RxD corresponds with a dominant signal

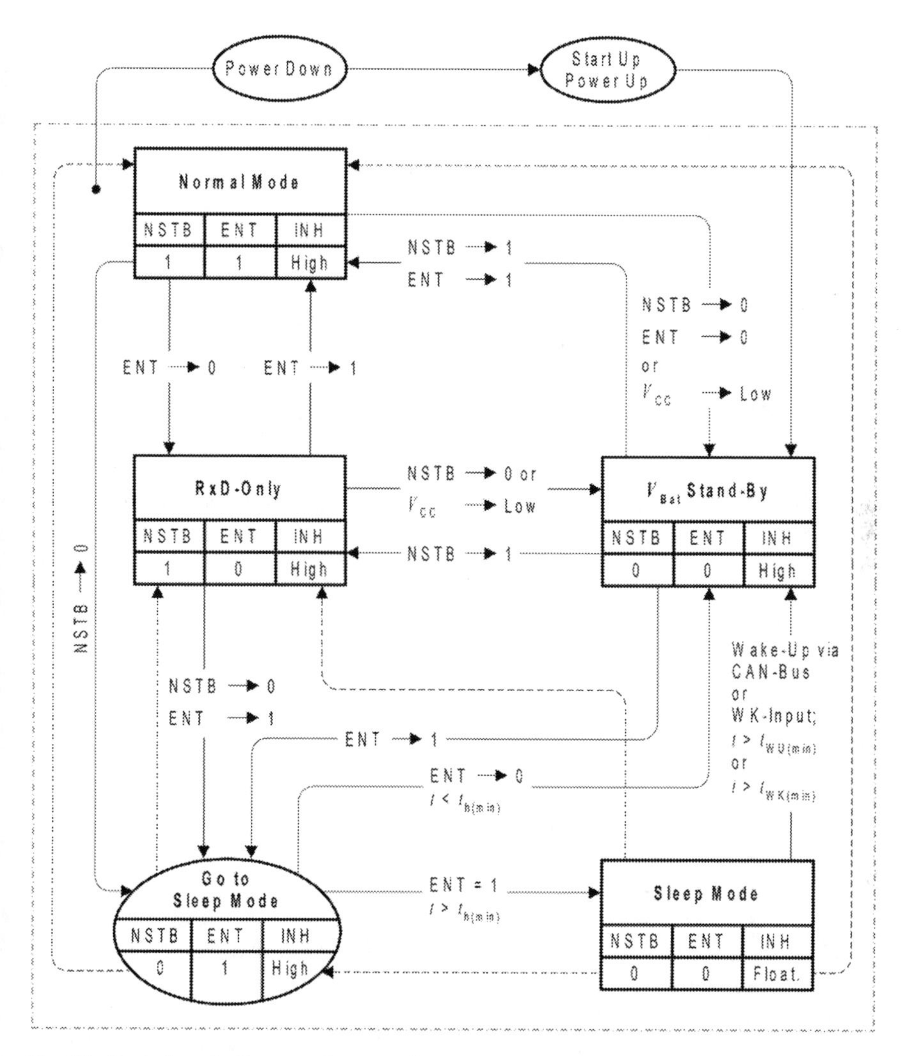

Fig. 2.1 Typical mode diagram of a transceiver

on the bus. The receiver in normal mode is always active and receives the signal on the bus, independent of whether these signals are transmitted from him or other participants on the bus. The differences between fault-tolerant low-speed and high-speed CAN transceivers will be discussed in the following chapters.

2.1.1.1 Transceiver Mode

In ISO 11898-3- (fault-tolerant low-speed CAN) and ISO 11898-5 transceiver (high-speed CAN with remote wake-up), special modes are implemented. A typical mode diagram is shown in Fig. 2.1 and will be described in more detail.

The most important transceiver modes are:

- Sleep mode
- Stand-by mode
- Normal mode
- Go-to-sleep mode

Sleep Mode

The fault-tolerant low-speed CAN transceiver (according to ISO 11898-3) and the high-speed CAN transceiver (according to ISO 11898-5) with remote wake function can be permanently supplied and have a sleep mode to reduce the current consumption below 30 μA. In this mode, the pin Inhibit (INH) is switched off. In the transceiver itself, the receiver is active with a very low current consumption to monitor CAN communication on the bus. A valid differential signal on the bus wakes the transceiver up, sets the transceiver in standby mode and switches on the INH pin. If a local wake pin is implemented, the transceiver can be woken up with level changes on this pin too.

Standby Mode

The standby mode is an intermediate mode after a transceiver wake up. The reasons for a wake-up can be:

- Power-up
- Level change on local wake pin
- Remote bus wake-up

In standby mode, the transmitter is blocked and set to recessive to guarantee no disturbance on the bus during microcontroller ramp-up. Depending on the implementation, the wake-up source is flagged. Normally, a remote wake-up is flagged on the pin RxD with a permanent dominant signal. A mode change in normal mode resets this flag. In standby mode, the INH pin (if available) is set to high. An INH-controlled voltage regulator will be switched on.

Normal Mode

In normal mode, the transceiver transmits and receives data. The transceiver can be set in normal mode via the mode pin EN. This mode pin is controlled from the microcontroller. In some transceiver products, a TxD-dominant time-out protection is implemented too. In case of a permanent dominant signal on the pin TxD, the bus is blocked and the communication on the bus is corrupted. After a defined time, the TxD-dominant time-out protection will set the transmitter to recessive and the communication on the bus can be continued. The release of this TxD-dominant time-out latch depends on the implementation of the product. Most of the time, a logical high on pin TxD releases the latch. An over-temperature to protect the output stages CAN_H and CAN_L is normally implemented in a standard transceiver and switches off the output stages CAN_H and CAN_L. A recessive level on pin TxD will reset the over-temperature protection.

Go-to-Sleep Mode

If an INH pin is available, a voltage regulator, connected to this INH pin, can be controlled from the transceiver. In sleep mode, the INH pin is switched off and the voltage regulator is switched off as well. The go-to-sleep mode is implemented to delay the host command sleep mode until the INH is switched off. During this time, the microcontroller can finalize the activities. After remote wake-up, the transceiver changes into standby mode and switches on the INH pin and the controlled voltage regulator.

2.1.2 CAN Coil

The advantages of a CAN coil include:

- Reduction of electromagnetic emission
- Improvement of immunity
- Sometimes improving the ESD performance

The typical used values for CAN coils in CAN networks are 22, 51 or 100 μH.

The CAN coil reduces the emission from the transceiver and increases the immunity robustness against disturbances.

2.1.3 Network Concepts

Two different concepts of physical layer implementations are available now:

- The high-speed CAN physical layer concept with baud rates up to 1 Mbaud
- The fault-tolerant low-speed CAN concept with baud rates up to 125 kbaud

These concepts are different from each other and will be described in the next chapter.

2.1.4 Fault-Tolerant Low-Speed CAN Physical Layer

The application areas for the fault-tolerant low-speed CAN physical layer are body application in cars and applications where the fault tolerance is needed and the lower baud rate can be accepted.

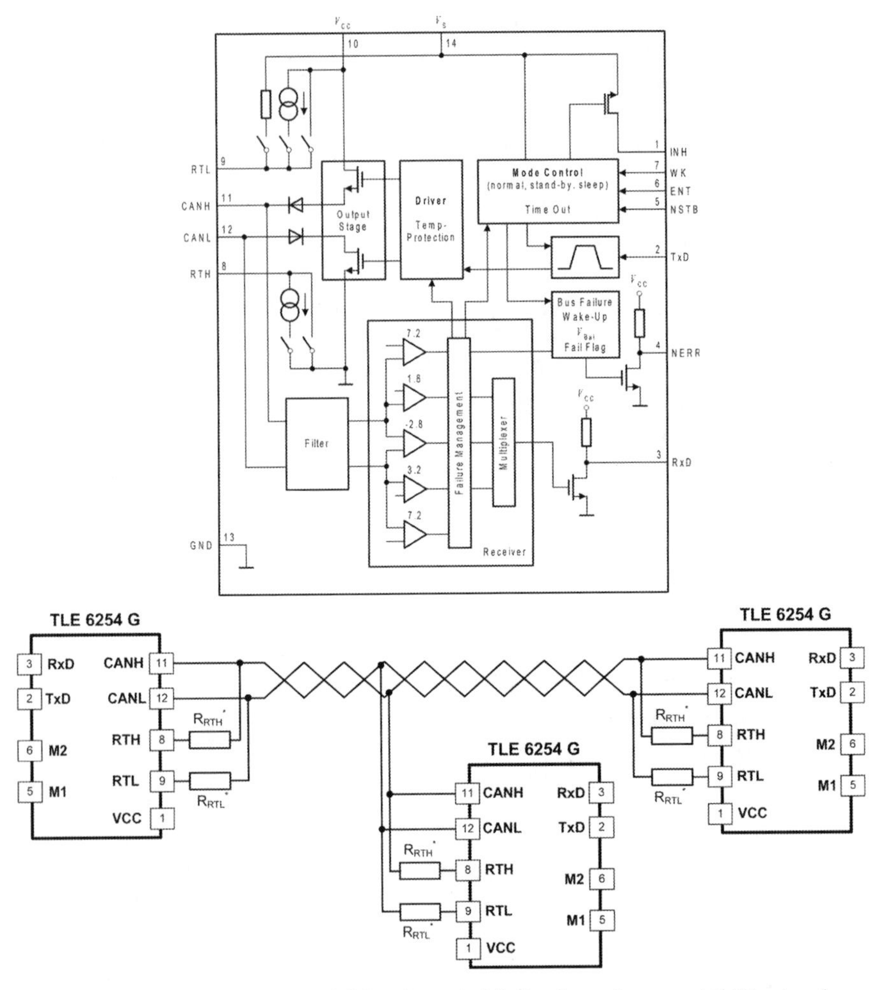

Fig. 2.2 Block diagram of TLE6254-3G and a typical fault-tolerant low-speed CAN network

This concept is tolerant against:

- CAN_H short to ground
- CAN_L short to ground
- CAN_H short to supply or battery voltage
- CAN_L short to supply or battery voltage
- CAN_H short to VCC (5 V supply for microcontroller and transceiver supply)
- CAN_L short to VCC (5V supply for microcontroller and transceiver supply)
- CAN_L short to CAN_H
- CAN_H open wire
- CAN_L open wire

In addition, the combination of all of these failures can be detected as double failure. In total, 120 combinations are possible and can be handled. How does this

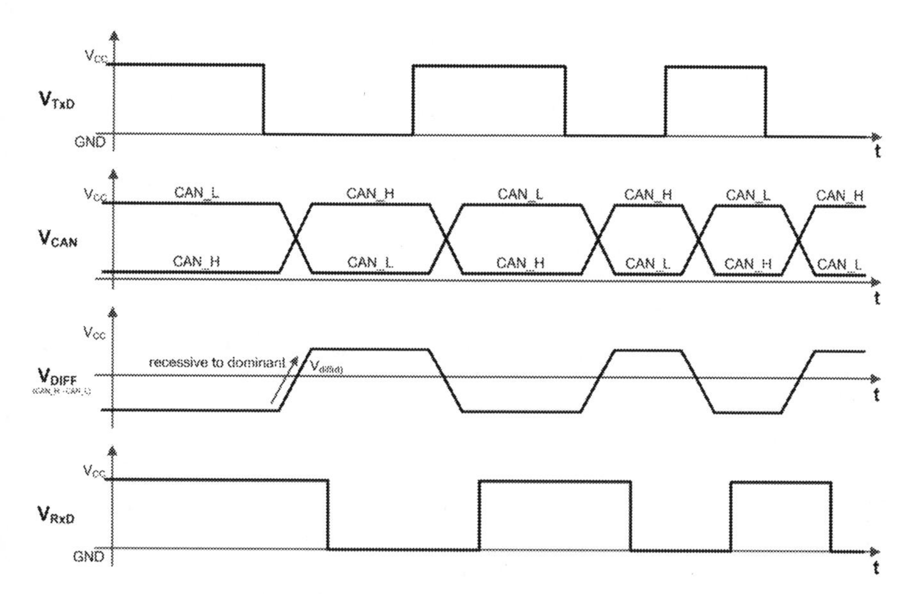

Fig. 2.3 Typical signals for a fault-tolerant low-speed CAN network

work? In principle, CAN_H and CAN_L work separately and in ant phase. CAN_H has a termination resistor to ground (via switch RTH) and CAN_L has a termination resistor via switch RTL to VCC (see Fig. 2.2). The signal will be transmitted via the CAN_H path and CAN_L low in ant phase. All receivers obtain the differential signal if there is no short or open wire on the bus. In this mode, the common-mode range and robustness is very high. If one short is detected, the transmitter deactivates this path (for example, if CAN_H is shorted to ground, CAN_H will be disabled and the communication will be transmitted over CAN_L only). For the termination, all nodes are complete. The value for the termination resistor can be between 500 Ω and 5 kΩ. In total, the value for all resistors (in a parallel connection) should be in a range of 100 Ω. Values less than 100 Ω are not allowed.

A high level on pin TxD corresponds with a recessive level on the bus. A recessive level on the bus means 0 V on CAN_H and 5 V on CAN_L. A low level on pin TxD corresponds with a dominant level on the bus. A dominant level on the bus means CAN_H and CAN_L are switched on and the levels are 4 V for CAN_H and 1 V for CAN_L. The absolute level depends on the busload. This ends in a differential level for recessive state of typical −5 V ($V_{CAN\_H} - V_{CAN\_L}$) and −3 V for the dominant state. In Fig. 2.3, the typical behaviour is demonstrated.

A fault-tolerant receiver consists of five different receivers which work in parallel. These receivers are:

- A differential receiver (for standard communication), with threshold voltage at −3 V
- A single-ended receiver for CAN_H (used in case of CAN_H short), with threshold voltage at 1.8 V
- Asingle-ended receiver for CAN_L (used in case of CAN_H short), with threshold voltage at 3.2 V

- Asingle-ended comparator for CAN_H (to detect CAN_H short to Vbatt), with threshold voltage at 7.2 V
- Asingle-ended comparator for CAN_L (to detect CAN_L short to Vbatt) with threshold voltage at 7.2 V

All receivers and comparators are also used to analyse failure cases on the bus. The failure management logic decides when there is a failure on the bus and changes from differential mode to single-ended mode. In the single-ended mode, the shorted transmitter is switched off (for example, in the case of CAN_H short to ground the CAN_H transmitter) and the single-ended CAN_L receiver is used. In the case of a CAN_L short to VCC or Vbatt, the CAN_L transmitter is switched off and the CAN_H single-ended receiver is used. The disadvantage of this single-ended mode is the lower noise robustness and the lower possible ground shift between the sender node and the receiver nodes.

In case of CAN_H short to Vbatt or VCC or a CAN_L short to ground, a high current flows through the termination resistors in sleep mode. This is the reason why termination resistor switches are implemented. In case of a short on the bus, the termination resistors are switched off. Pin RTL is the switch for the CAN_L termination resistor and RTH is the switch for the CAN_H termination resistor. In sleep mode, normally the 5 V supply VCC is switched off, floating or 0 V. To have a positive termination voltage, CAN_L will be high ohmic terminated to Vbatt. When CANS communication is started, the transmitter of the transmitting node pulls the CAN_L to ground and all other detects this as a remote wake-up event and the transceivers change into standby mode. In standby mode, INH is activated and switches on the voltage regulator and ramps up the microcontroller. After the successful ramp-up of the microcontroller, the transceiver should be set to normal mode for normal communication on the bus. In transceiver standby mode, the termination switches RTL and RTH on and terminates the CAN_H and CAN_L wires.

2.1.4.1 High-Speed Physical Layer

An ideal high-speed CAN physical layer has a termination resistor of 120 Ω on both ends of the wire. This reduces the echo on the wire to a minimum. All other nodes are connected in between. This concept allows a data rate of up to 1 Mbaud. The ringing, especially after the dominant to recessive edge, is minimized and the high data rate is possible. After switching on the CAN_H and CAN_L output stages, a current flows from CAN_H to CAN_L over the termination resistors. The result is a voltage drop over both termination resistors between 1.5 and 3 V. This is called dominant level. If both output stages are switched off, the voltage drop over the termination resistors is zero. This is called recessive level. The receiver thresholds are between 500 and 900 mV. A voltage drop higher than 900 mV will be detected as dominant level on the bus and a voltage drop smaller than 500 mV will be detected as recessive level. The common-mode range for the receiver is from −12 to+12V. If the bus common-mode voltage is higher or lower, the receiver can detect wrong signals. This concept is not proven against CAN_H shorts to ground and a CAN_L short to battery voltage. In this case, the communication can be corrupted.

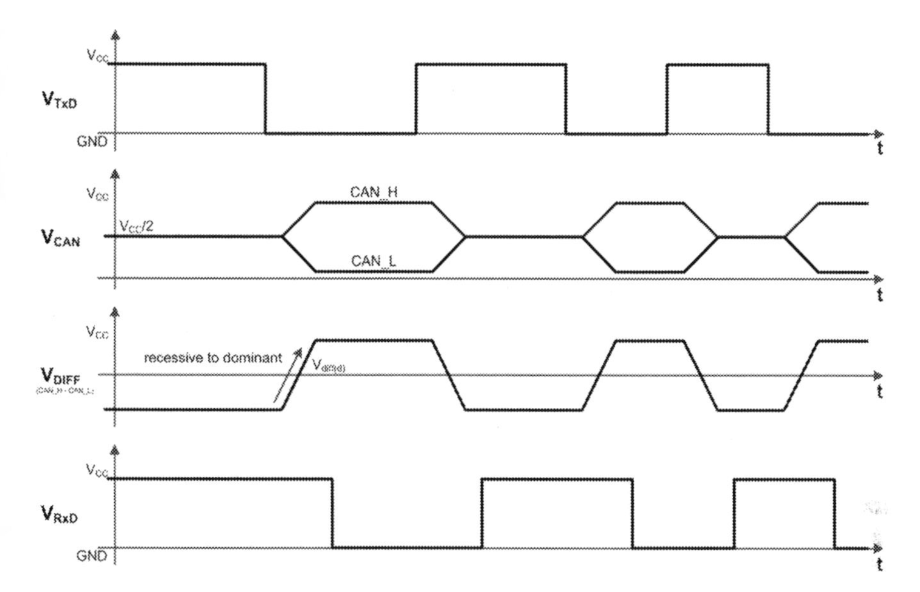

Fig. 2.4 Typical signal for a high-speed CAN physical layer

Figure 2.4 shows a typical high-speed CAN bus signal.

High-Speed Transceiver

Two different kinds of transceivers are in the market now:

- ISO 11898-2 transceiver
- ISO 11898-5 transceiver, with remote wake-up feature

The typical high-speed CAN transceiver consists of a transmitter and a receiver. Two 20-kΩ resistors, connected to an internal voltage source of 2.5 V , stabilize the bus voltage to 2.5 V in recessive state. The receiver is a differential comparator with electromagnetic compatibility (EMC) filter. This comparator monitors the bus levels and transforms the differential signal of the bus levels to a logic signal on pin RxD. High level on RxD is recessive level on the bus and low level on RxD is dominant level on the bus.

Figure 2.5 shows a block diagram of an ISO 11898-2 transceiver (left) and the ISO 11898-5 transceiver with remote wake-up (right).

Transceiver according to ISO 11898-2

A transceiver according to ISO 11898-2 is a transceiver with a transmitter and a receiver to transmit and receive data only. In the first generation of this kind of transceiver, no additional function was implemented. In newer generations, the ESD robustness is dramatically increased up to 15 kV and the emission is reduced to very

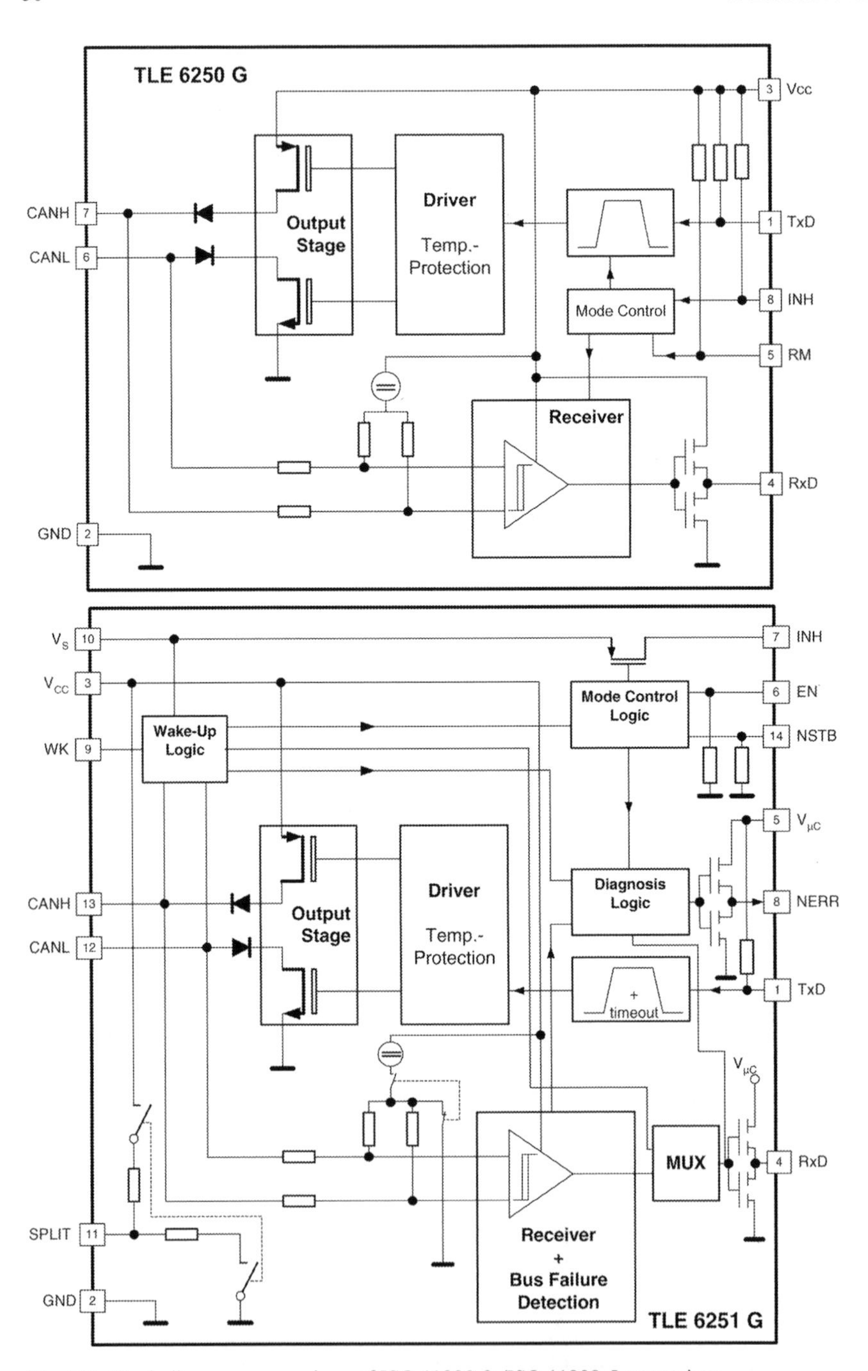

Fig. 2.5 Block diagram: comparison of ISO 11898-2-/ISO 11898-5 transceiver

low values. In addition, a TxD time-out function is added to lock the transmitter in case of failure on the TxD pin. This kind of transceiver is active if the 5 V VCC pin is supplied. Such transceivers are used in industrial applications and not permanently supplied applications. The Infineon TLE 6250G is a typical ISO 11898-2 transceiver and will be described in Sect. 2.5.

Transceiver According to ISO 11898-5

In some applications, especially in the car body, the transceiver is permanently supplied. If the bus communication is inactive, the current consumption should be reduced dramatically. For this mode, a new function was implemented and called sleep mode. This mode is described in ISO 11898-5. The new functions are:

- Sleep mode with a very low current consumption
- Remote bus wake-up function
- Reduced busload in case of unsupplied device

The ISO 11898-5 is based on ISO 11898-2. All bus parameters are the same; only the remote wake-up function was added.

2.1.5 Termination Concepts

The termination concept has an impact on the signal integrity and the emission in the network. A high signal integrity is necessary for high baud rates. A stabilized recessive level is necessary for low ones. In the International Organization for Standardization (ISO), two 120-Ω resistors are recommended for the termination, but some newer concepts are used to improve this concept.

2.1.5.1 Standard-Termination Concept

The termination resistor at the end of the wire reduces the echo on the bus and increases the signal integrity for a reliable communication. In the ISO, two 120-Ω resistors are recommended (see Fig. 2.5).

2.1.5.2 Termination with Centre Tab

To reduce the emission, the sum of the CAN_H and CAN_L level must be constant for a recessive and dominant signal. Deviation comes from different CAN_H and CAN_L propagation delays or different switching times. A ground shift can increase the emission as well. To reduce this impact, a termination with a centre tab is used in different applications. The centre tab, normally 4.7 nF or higher, increases the stability of the recessive level and reduces the emission (see Fig. 2.6).

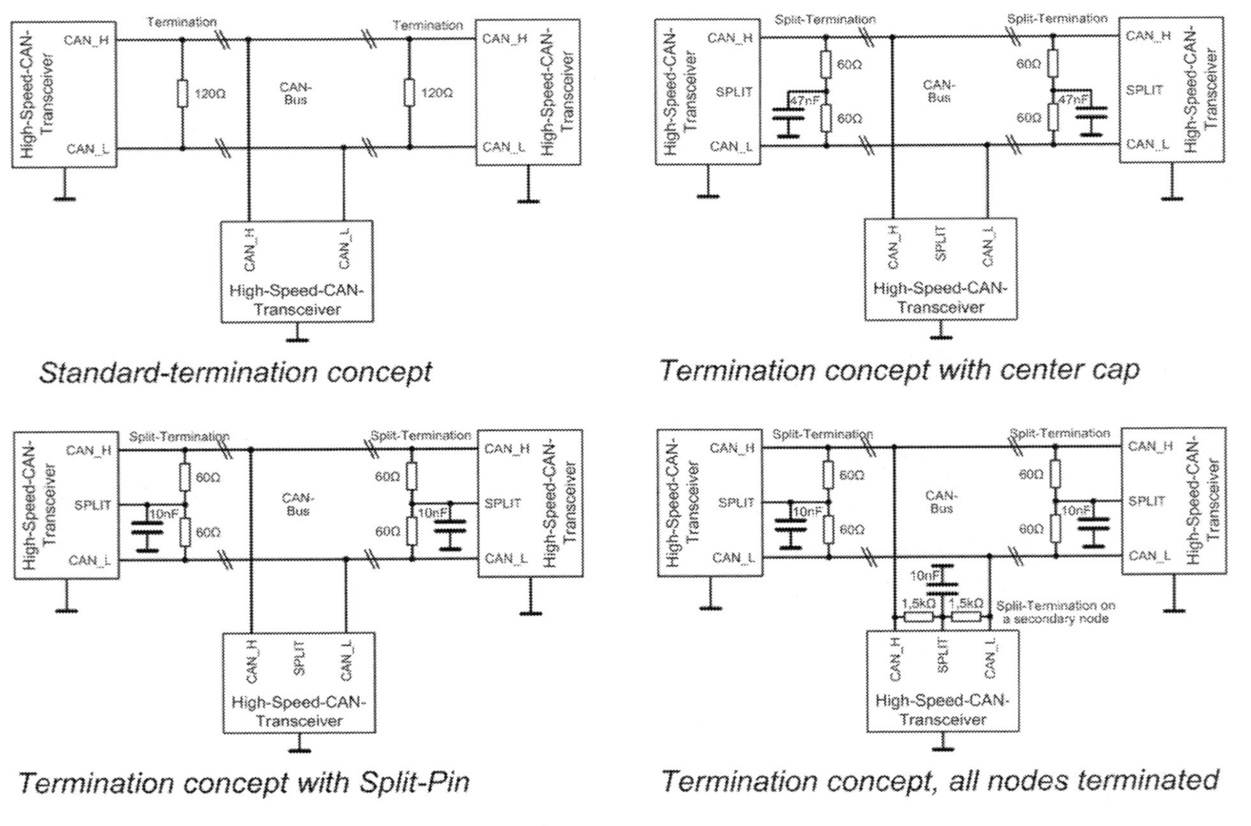

Fig. 2.6 Termination concepts in high-speed CAN networks

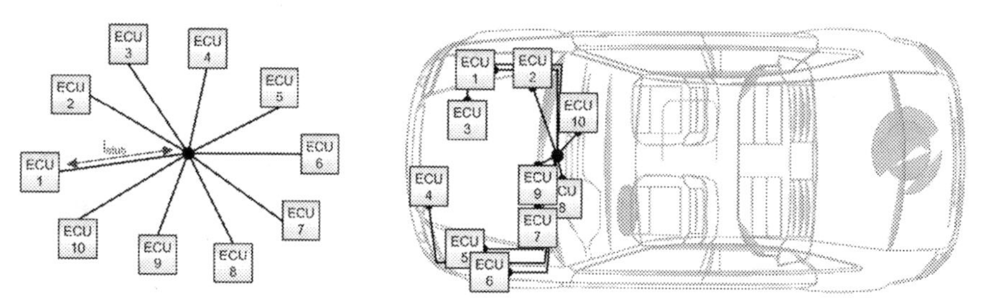

Fig. 2.7 Schematic and application of a single star topology

2.1.6 Network Topologies

In modern cars, different topology concepts are used. In industrial applications, the linear bus topology is preferred. The different topologies are:

- Single star
- Twin star
- Linear bus
- Hybrid topology

2.1.6.1 Single Star

In the single star topology (Fig. 2.7), all stabs are connected in the centre of the star. Only one termination resistor (60 Ω) is used in this single star architecture. The advantage of this topology is the flexibility and the possible high number of nodes in a network. The length of each wire can be up to 9 m. The maximum baud rate in this topology is 500 kbaud. Higher bus rates are not possible. The ringing at the end of a dominant bit is very long and dominates the maximum baud rate.

2.1.6.2 Twin Star

The twin star topology consists of two single stars, which are connected to each other (Fig. 2.8). With this kind of topology, an optimized adaption wiring harness is possible. The termination resistors are located in the centre of each star. The maximum baud rate is also 500 kbaud.

2.1.6.3 Linear Bus Topology

The linear bus topology is used in the industrial world and in cars (Fig. 2.9). For cars, it is not flexible enough, but it allows baud rates up to 1 Mbaud. In the ISO 11898-2, the recommended stub length is maximum 30 cm. The total length should be below 40 m.

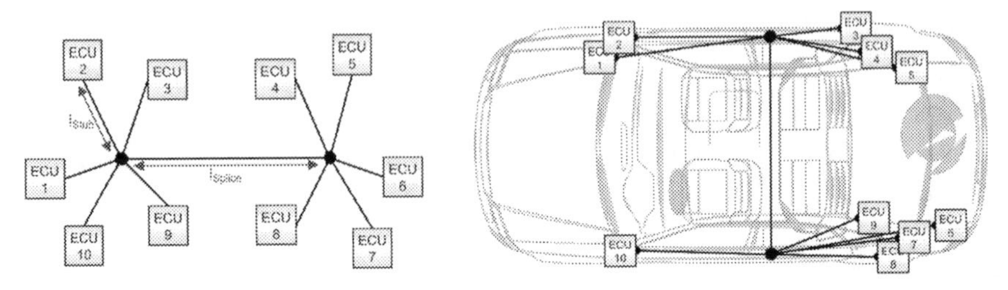

Fig. 2.8 Schematic and application of a twin star topology

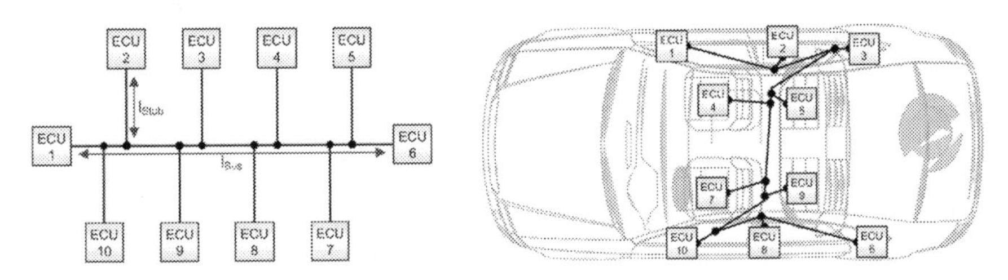

Fig. 2.9 Schematic and application of a linear bus topology

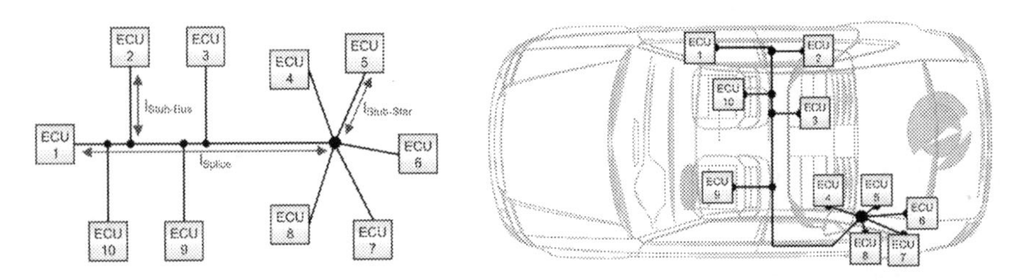

Fig. 2.10 Schematic and application of a hybrid technology

2.1.6.4 Hybrid Topology

The hybrid topology is a combination of linear bus and single star (Fig. 2.10).

This technology combines the advantage and disadvantage of both topologies. The baud rate can be up to 1 Mbaud and the maximum length of the wire can be smaller than that in the other solutions. However, the cost for this topology is higher than for the others.

2.1.6.5 Network Propagation Delay

The calculation of the network propagation delay (Fig. 2.11) in a network is necessary to guarantee a reliable arbitration. If more than one node starts to transmit a message, the higher ID will win the arbitration phase.

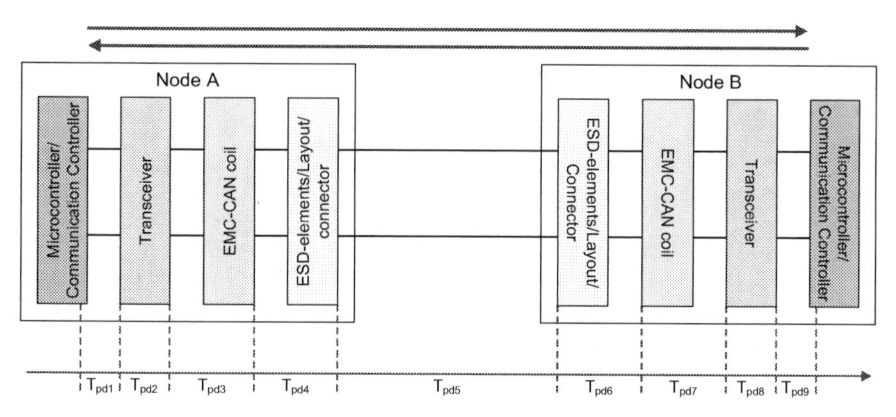

Fig. 2.11 Network propagation delay in a CAN network

In the worst-case scenario:

- The maximum propagation delay from node A to node B is calculated with microcontroller.
- Propagation delay node A = 30 ns.
- Transceiver propagation delay node A = 280 ns.
- Wire propagation delay for 40 m with 5 ns per metre = 200 ns.
- Transceiver propagation delay for transceiver node B = 280 ns.
- Propagation delay microcontroller node B = 30 ns.
- Wire propagation delay from node B to node A 40 m with 5 ns per m = 200 ns.
- The total propagation delay is 1040 ns.
- This is the reason why in long networks only 500 kbaud are possible.

2.2 Network Topologies—Design Constraints

The term "topology" can be deduced from the Greek words "topos" (place, location) and "logos" (theory, knowledge). Thus, topology means the science of the location. In a CAN system, several bus connections or nodes are connected to a system or network. As the way of connecting the nodes exerts influences (e.g. on the signal integrity or the signal propagation) or reacts on external influences, the architecture of the network is termed as the topology. Thus, the science of the placement and its corresponding impacts are considered in a CAN topology.

The impact of CAN topologies, discussed in this chapter, focuses on three different categories. First, the impacts and properties of CAN topologies are discussed, followed by the structure of the bus connection of a CAN node and the characteristics of its integrated bus connecting circuitry. Based on examples, different interactions between bus connection circuitry and the topology are discussed with the focus on the signal integrity. Analytic considerations and optimization measures shall give possibilities of a specific judgement and enhancement of the signal quality of the CAN network.

2.2.1 CAN Network Architecture

Although CAN bus is not one of the high-speed data networks, some phenomena show that the laws of physics apply here. Especially at speeds above 250 kbit/s and at cable lengths above 50 m, it must be analysed more closely and will require some calculation rules. The actual physical properties of the CAN bus lines as well as the circuitry of the connected control devices have to be considered. If the electrical specifications of these components are available, the quality of the received signals can be derived from the transmit signals. To reduce the complexity of calculations, some assumptions can be made, which are explained below.

Simulations may be used to avoid the need for such calculations. However, fundamental knowledge of all influencing parameters is important to predict the impact of changes or modifications on existing CAN networks.

2.2.1.1 Transmission Line Theory

Theoretical considerations of the electrical lines are based on the electromagnetic waves forming the signals, which spread along the line with a characteristic speed. Therefore, voltage and current pulses of CAN messages also move from transmitter to receivers with a characteristic propagation speed. For very slow operations, i.e. switching off light, this can be neglected. For the fast CAN signals, the propagation speed limits the allowable cable length. Signals are classified as fast, if the rise time is shorter than the *propagation time* of the corresponding line.

The specific propagation delay results from the material properties of the cable in use. The dielectric constant of the insulating material and geometric parameters plays an essential role. Hence, a wave impedance can be determined, which describes the relationship between current and voltage at any point on the line, and the propagation speed of pulses on the line. On closer inspection, these parameters are frequency dependent, which can have an impact on very fast signals (above about 5 Mbit/s). For CAN signals, this effect can usually be neglected. The impedance thus can be assumed to be a resistance.

The transmission of CAN signals is usually performed in a differential method. Two wires are used, where the data signal is derived from the voltage between the conductors. To simplify the calculation of the transmission signals, the signals of the two conductors can be decomposed in a *differential signal* and a *common-mode signal*.

The differential signal—also called odd mode signal—and the common-mode signal can be calculated if the voltage of each single CAN signal line is known, as shown in [2.1]:

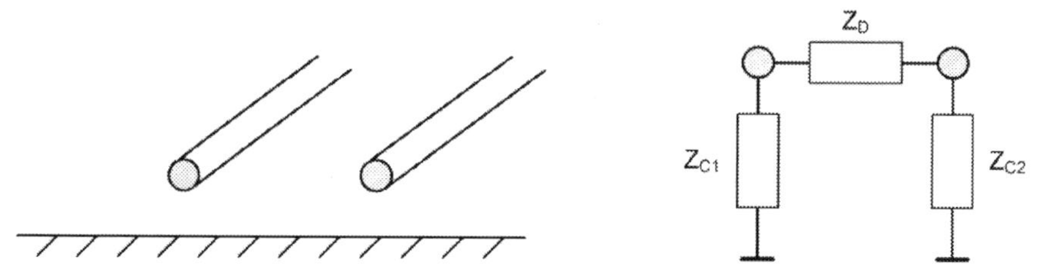

Fig. 2.12 Impedances of UTP cable as three-wire system

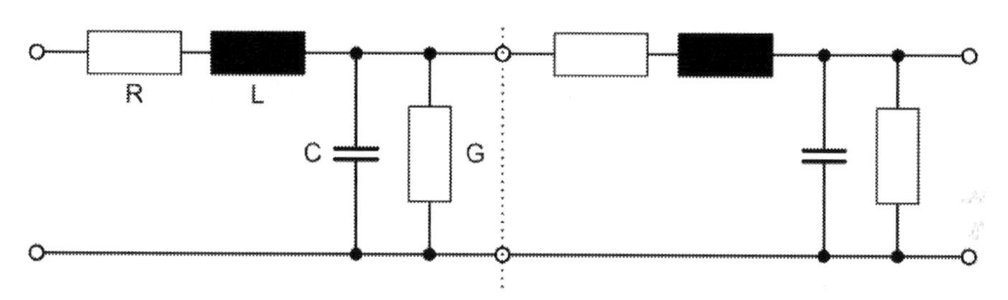

Fig. 2.13 Schematic of the elements, representing a transmission line segment

$$U_{diff} = U_{CAN\_H} - U_{CAN\_L}$$
$$U_{com} = \frac{U_{CAN\_H} + U_{CAN\_L}}{2} \tag{2.1}$$

All considerations can, therefore, be performed for the two CAN lines separately or alternatively for common-mode and differential signals. The partial results finally have to be superimposed. The common-mode signal can often be neglected if interference-free signals are considered. For signal propagation, however, both components are important. Since the circuitry and the line impedances for both components are very different, very different considerations arise.

In most cases, CAN systems use unshielded twisted pair (UTP) cables. The electrical equivalent is shown in Fig. 2.12. To calculate the electrical field, it can be assumed that both wires and a ground plane form a three-wire system.

The impedance can be calculated based on the material properties or by measurement. Driving the two wires by a pure differential signal allows calculating the impedance of the circuit arrangement as a parallel connection of $2 \times ZC$ and ZD. For pure common-mode stimulation, ZD is not effective and ZC can be measured.

Another method is to describe the line by short segments, each represented by concentrated electric components (R, L, C) as shown in Fig. 2.13. The length of each segment must be considered to be short versus wavelengths of the driving signals.

The concentrated circuitry describes the losses (R, G) and the energy-storing elements (L, C) per segment length—also known as primary line constants.

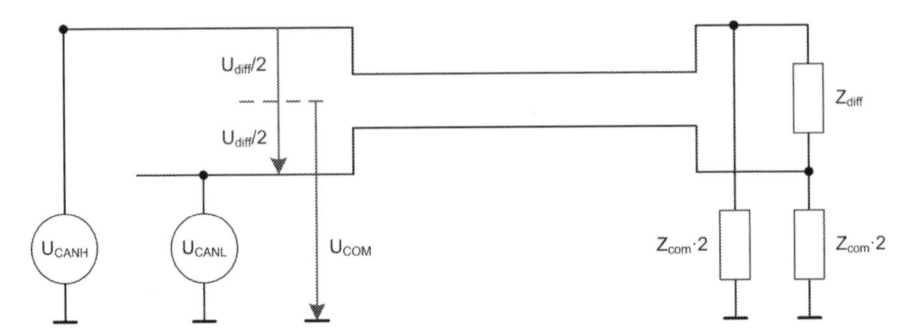

Fig. 2.14 Separation of CAN signals in differential-mode and common-mode components

The resistivity may lead to voltage drops along the line. In case of long lines, the loss along the line can be so large that the received signal by the voltage drop is no longer sufficient.

The capacitive characteristics—along with the input capacitances of the connected control units—can lead to significant switching times in extended systems. For the transition between recessive and dominant, these can be estimated by the internal resistance, the transmitting transceiver and the sum of the above-mentioned capacities. On the transition to recessive state, the system will be discharged via the termination resistor. The two periods are so different and must be separately determined. Additional capacitive portions can also play a role, if shielded wire is used (shielded twisted pair—STP).

In the following, only the differential signal is considered, which can later be supplemented by separate consideration of the common-mode behaviour (Fig. 2.14).

If we observe the propagation of a switching pulse along a line, then an unexpected behaviour occurs at the end of the line: If we consider an open end of the line, the pulse may not continue to move. To ensure the continuity of the energy, a wave is reflected back, which compensates for the incoming wave. This overlaps with the end of the wave and leads to signal distortion. Only if the end of the line is terminated by a resistor equivalent to the line impedance, the energy is absorbed and thus the overlay can be avoided. In all other cases, signal reflections occur, whose size is described by the reflection factor.

For differential signals, there is a typical characteristic impedance of about 120 Ω for a twisted-pair cable which is usual in CAN systems. This corresponds to a propagation speed of 4–5 ns/m. The CAN bus should be terminated at both ends with this resistance, as required, for example, in the standard ISO 11898-2. In real systems, the end of the line often cannot be determined unambiguously, like in a star-shaped or branched system topology. In these cases, a single termination or multiple distributed termination resistors may be used instead—as a result of an optimization process.

Reflections—as mentioned above—occur at each crossing point, because, at these points, the line impedance changes due to parallel connection of line impedances. The reflections of branches which are not accordingly terminated will overlay and distribute in the overall system. The result of these interferences depends

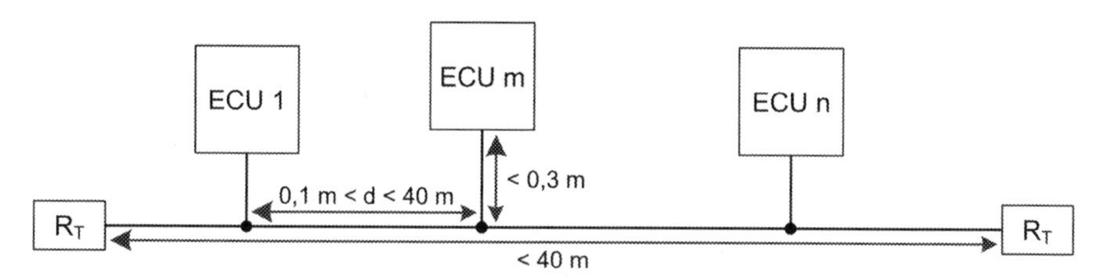

Fig. 2.15 Limits of system extent according to ISO 11898-2

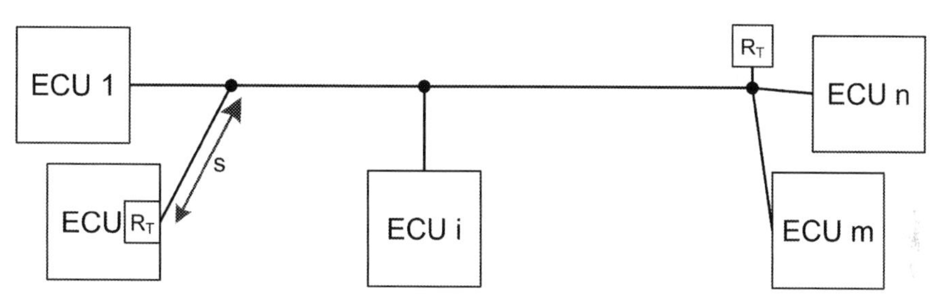

Fig. 2.16 Stub length in a network

on the location of the observation; it must be determined separately for each node of the system. Even simulation-based analyses result in a very large amount of data.

In many cases, it is unfortunately not possible to establish the system as a bus with short branch lines, as required by ISO 11898-2 (maximum bus cable length, 40 m; maximum stub length, 0.3 m) (see Fig. 2.15). Although this topology would cause least disturbances by reflections, the maximum elongation from 40 m at 1 Mbit can hardly be achieved.

Due to the arrangement of the electronic control units (ECUs), for example, in a motor vehicle, a multiple star configuration results quite often. Stub lengths of several metres may be required as shown in the example stub “s” in Fig. 2.16.

At the junction point of multiple parallel stubs, reflections increase, depending on the number of branches. For symmetric structures, the returning reflected signals will again come together and intensify. Symmetrical topologies are especially critical. For large networks with low symmetry, superposition of reflected signals may erase each other. In spite of the increasing complexity of the network topology, the resulting signals may be uncritical. The “golden rules” for the configuration cannot be given, only trends and experiences. These are particularly problematic for networks with a large number of optional ECUs. Problems can occur by adding or omitting devices in very critical combinations—even if the system works initially stable. Figure 2.17 shows such a case. To the system made up from the ECU 1 to ECU 5, the sixth is added. The propagation delay is increased in this particular case—by the ringing at the transition from dominant to recessive—from 280 to 422 ns.

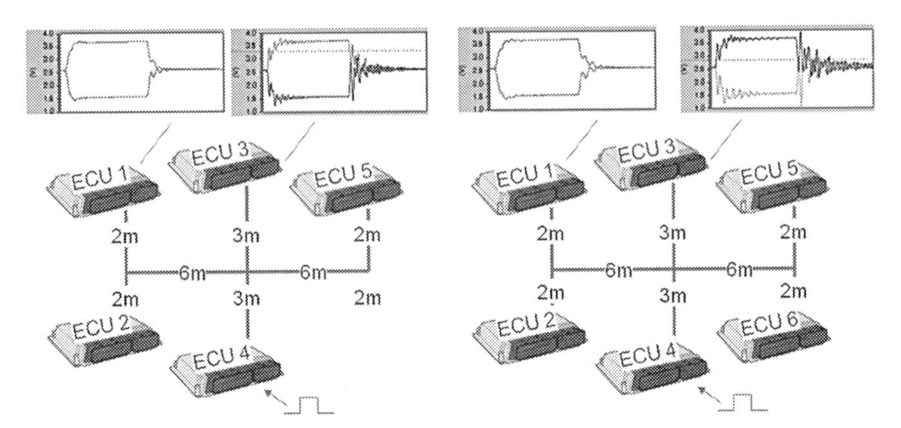

Fig. 2.17 Optional ECU 6 increases the propagation delay (220 ns → 480 ns) due to the symmetry of the topology

The waveforms also show very clearly that the signal quality is heavily influenced by the position in the system. It is not sufficient to analyse a single position in the network. Since the access of measuring points in the final system is very limited, the signal analysis by means of simulations is often the only solution.

2.2.2 Architecture of CAN Nodes

The previous chapter addressed the influences of the characteristics of network architectures. In this chapter, the architecture of a CAN node and its components shall be handled. In this context, all components taking part in the signal transmission from a digital to an analog CAN bus signal and vice versa are inquired. Functioning and implementation reasons as well as signal integrity crucial attributes are explained.

2.2.2.1 CAN-Bit-Timing and Oscillator Tolerances

The interface between the physical layer and the data link layer are the digital signals RxD and TxD. With the help of these signals, information is given over to the other respective communication layer. This is only possible if both sides are using the same comprehension of the information content. In case of the RxD and TxD signals, this is reduced to the binary logic with both its defined states logic "1" and logic "0". However, there is a further physical unit which is present at the interface between the physical and data link layer and shall be considered: the time. In this case, the timing aspect is interesting regarding the oscillator tolerances inside the protocol layer and different delay times in the physical layer. Through the configuration of the bit timing, the sample point, the synchronization jump width (SJW) and the data rate are defined. Tolerances of the clock sources influence the CAN clock directly. Today, used crystal oscillators have tolerances of about 50 ppm.

In case of two communicating CAN nodes, each has oscillators with a maximum deviation (one in positive and one in negative direction). This results in a maximum divergence of the data rate. Therefore, there is one fast-running and one slow-running CAN node. Taking delays of the physical layer (of the network architecture, of the signal path as well as delay times of state transitions from dominant to recessive and vice versa) into account, the correct information exchange shall be guaranteed even with the discrepancy of the two local bit rates of the CAN nodes.

Section 2.2.3 continues this aspect in combination with the analytic consideration of the maximum length of the signal path and how the clock tolerance influences the physical layer and the CAN communication.

2.2.2.2 Transceiver

Section 2.1 described transceivers in detail, however, this chapter focuses on aspects and properties given by the transceiver which influences the signal integrity evaluations. The task of the transceiver is to convert digital signals to analog bus signals and vice versa. Due to the fact that the transceiver is integrated into the signal flow directly, the influences given by the transceiver are intense and shall be evaluated precisely. Of particular importance are the dynamic processes during the change between the two logical states of the CAN bus, especially in the case of sending. Particularly static characteristics are important in the case of receiving. However, delay times are important characteristics in the case of sending or receiving. Delay times are next to signal propagation delays of the network crucial for the calculation and evaluation of CAN topologies. There are delay times due to transforming digital information to analogue bus signals (from a state change on TxD until the corresponding state change on the CAN bus) and there are delay times due to transforming analogue bus signals to digital information (from a state change on the CAN bus to the corresponding state change on RxD). These delay times occur at a state change from dominant to recessive (dom_rec) approximately from logic “0” to logic “1” and from recessive to dominant (rec_dom) approximately from logic “1” to logic “0”. Therefore, there are the four following listed important delay times of the transceiver:

- $t_{TXD\_dom\_rec}$: Transceiver acts as sender, state change from logic “0” to logic “1”.
- $t_{TXD\_rec\_dom}$: Transceiver acts as sender, state change from logic “1” to logic “0”.
- $t_{RXD\_dom\_rec}$: Transceiver acts as receiver, state change from dominant to recessive.
- $t_{RXD\_rec\_dom}$: Transceiver acts as receiver, state change from recessive to logic dominant.

Information for measuring these delay times is given in detail in ISO 11898-2 (for high-speed CAN).

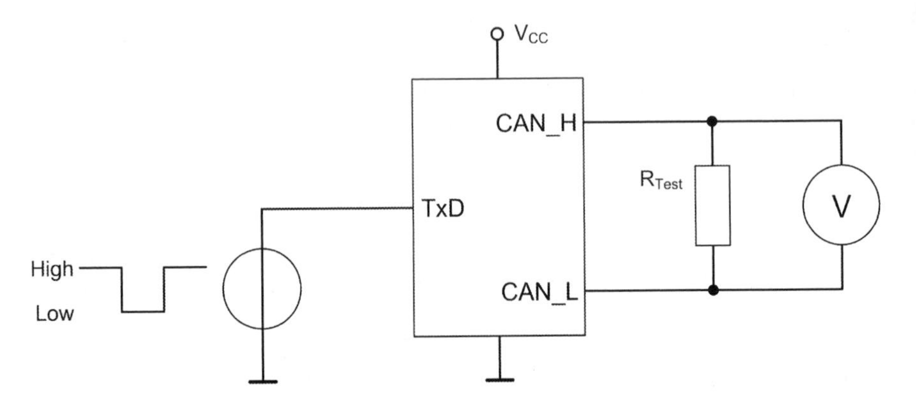

Fig. 2.18 Measurement set-up—dynamic behaviour of the transceiver output stage

As mentioned above, the dynamic characteristics of the transmitter part of the transceiver are important. There, the transceiver drives the bus signal which will be distributed over the network. The characteristics of the signal shape are not only relevant if it is possible to distribute the information over the whole network but also crucial for noise immunity and sensitivity against interference with (or of) other network components. Furthermore, the signal shapes of dynamic processes have a huge impact on EMC behaviour of the network. Important aspects for the consideration of the signal flanks are the symmetry between the flanks of the CAN_H signal and the CAN_L signal, the rounding of the flanks as well as the slew rate of the signal flanks. Through the rounding and the limitation of the slew rate of the signal flanks, high-frequency (HF) signal parts are limited to increase the EMC robustness and immunity. The symmetry of the CAN bus signal, which describes the line symmetry of the signals CAN_H and CAN_L, with the axis of symmetry lying on the mid-level (typically 2.5 V) is important to avoid asymmetries. Asymmetries are one cause for common-mode disturbances which enable the emissions of electromagnetic interference.

In case of validating the dynamic processes of a sending transceiver, the dominant to recessive state change is of particular importance because the transceiver actively drives the CAN bus with a differential voltage in the dominant state (providing energy into a system) and is turned off during the transmission to recessive level (high ohmic output). Discharging processes may result, regarding the disabling of an active energy source which may lead to disturbances. Such processes are analysed in detail further below.

Figure 2.18 shows an example of a measurement set-up to measure the flanks of a transceiver.

On the receiver side are on the one hand the thresholds and on the other the resistive and capacitive load against the bus from interest. The ISO 11898-2 standard defines the dominant state as a differential voltage greater than 0.9 V and the recessive state as a differential voltage smaller than 0.5 V. This means that the maximum possible threshold for the dominant state is equal to 0.9 V and the minimum possible threshold for the recessive state is equal to 0.5 V. Consequentially, the

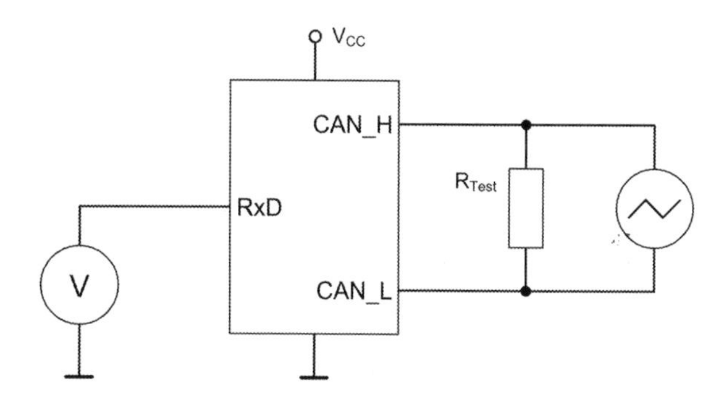

Fig. 2.19 Measurement set-up—thresholds of the transceiver receive stage

real thresholds lie between these boundaries. The real thresholds are interesting for signal integrity inspections, delay measurements or the calculation of the maximum possible network length. Figure 2.19 shows a measurement set-up to measure the thresholds of a CAN transceiver.

The characteristics of the receiver part of the transceiver depend on the differential resistance and the differential capacitance which acts between the bus signals CAN_H and CAN_L. The differential resistance helps in building abstract equivalent circuits for the calculation of the maximum possible network length. The voltage drop over this differential resistance shall be at least as big as the minimum defined differential voltage for the dominant level to make communication exchange possible. The capacitive load between the bus signals of a CAN node influences the signal shape during a state change. With an increase of CAN node, the capacitive load of the CAN bus increases as well and may result in a deformation of the slew rates in form of an approximation of a capacitive charging curve. Measurement techniques for the observation of the differential resistance and the differential capacitance of a transceiver are defined in ISO 11898-2.

2.2.2.3 EMC Circuits

To increase immunity against electromagnetic emissions and interferences, common-mode chokes (CMCs) are typically part of the CAN connecting circuits of a CAN node. In the following, the structure and operating principles are given to be able to judge and evaluate influences of the CMCs on signal quality. Due to the twisting of the CAN cables for the CAN_H and CAN_L signal electromagnetic field induced by the differential signal annuls each other. However, the problem of the emission or interference of common-mode signal parts is not solved by twisting the signal cables. This is the task of the CMCs. As the name indicates, the task of a CMC is to attenuate common-mode signal parts while differential signal parts (containing the information content) are passed without any attenuation. This task is realized by two chokes, one for CAN_H and one for CAN_L, which are coiled contrary around a common used core. Concerning the differential mode, the current directions in the CAN_H and CAN_L signals are in opposite directions. As a

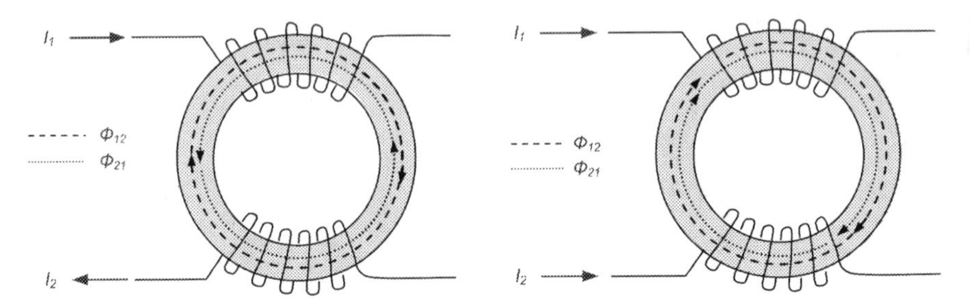

Fig. 2.20 Description of magnetic flows inside a common-mode choke

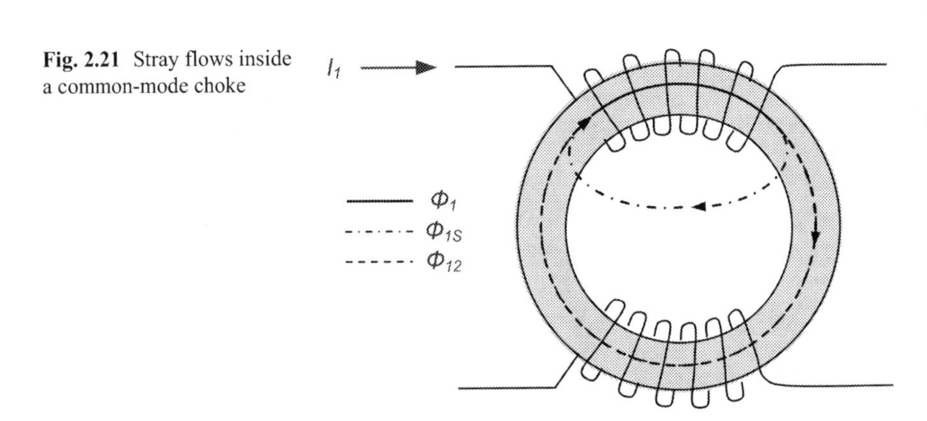

Fig. 2.21 Stray flows inside a common-mode choke

result of this, the magnetic flows induced by the chokes and coupled by the core are in opposite directions as well and cancel each other out. Without a magnetic flow, no magnetic field can be induced which would lead to an attenuation of the signal. In case of the common mode, the current directions in the CAN_H and CAN_L signals have the same directions. This results in an addition of the magnetic flows which induces a magnetic field which attenuates the common-mode signal parts. Figure 2.20 depicts the magnetic flows which are proportional to the signal currents. The magnetic flow Φ_{12} depicts the flow induced by the current I1 in choke 1 which influences choke 2. The magnetic flow Φ_{21} depicts the flow induced by the current I2 in choke 2 which influences choke 1.

The above shown function of the CMC does not contain the fact that, due to straying, the complete magnetic flow will not be transported through the core. The so-called stray flows are the parts of the complete magnetic flow which do not influence the magnetic flows induced by the other coil. Figure 2.21 explains the formation of stray flows. The magnetic flow Φ1 depicts the flow induced by the current I1. The magnetic flow Φ12 depicts the part of the flow Φ1 which acts in the other coil, where the magnetic flow Φ1S depicts the flow which does not influence the other coil.

The strength of the straying is depicted by the stray factor which is standardized on the rated inductance. The complementary of the stray factor is the coupling factor

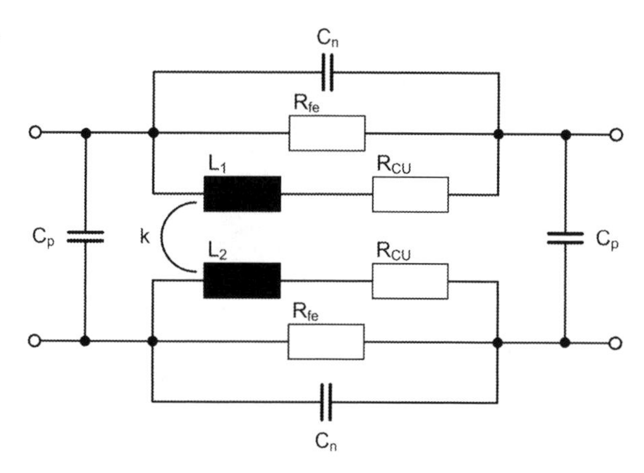

Fig. 2.22 Simplified equivalent circuit of a common-mode choke

or coupling coefficient, which depicts the strength of the impact (coupling) of the magnetic flow to the magnetic flow of the other signal. Through the normalization on the rated inductance, the sum of the stray factor and coupling factor is always 1.

This means that the stray factor or coupling factor gives indications for the effective working inductance in case of common-mode operation. Because the impact of the CMC on the differential signal shall be as small as possible, the coupling coefficient deals as a quality characteristic of the CMC and shall be as big as possible.

In practice, it must be distinguished between two different types of winding which shows different coupling coefficients. The chokes with sector winding are similar to the figures shown above (see Fig. 2.20). Each coil has its own part or sector of the core. The CAN_L and CAN_H signal lines are first twisted and then wind up together around the core in a CMC with bifilar winding. Through the sector-based layout with a distance between both coils, sector-based CMCs have a smaller coupling coefficient (typically $c \approx 0.97$) than CMCs with bifilar winding ($c \approx 0.99$).

Chokes with bifilar winding have, caused by the small distance of the CAN_H and CAN_L signal lines, bigger capacitances between the CAN signal lines. Different types of cores (in practice, there are ring cores and I-cores) do not result in significant differences regarding signal quality aspects in the time domain (working frequency range).

Figure 2.22 shows a simplified but sufficient equivalent circuit for signal integrity considerations in the time domain.

The coupling of the magnetic flows is mentioned as the coupling arch between the inductances L_1 and L_2 in the equivalent circuit. The practical values of the main inductance are in the range of 11 and 100 μH (automotive industry). Typically, CMCs with 51 μH are used in CAN systems (also the automotive industry) and 100 μH in FlexRay systems.

The resistance in serial to the main inductance describes the ohmic load of the coil. The capacitance Cn represents the sum of the capacitances between coil input and output (which should be as small as possible). The particular capacitances occur by distances between the signal lines to each other, distances between signal lines and the core, and distances of coil start and end. These parasitic capacitances influence the attenuation curve and the position of the resonance maximum of the CMC.

The capacitance Cp represents the capacitances between coil input of coil 1 and coil input of coil 2 as well as the capacitance between the outputs of both coils. The capacitance depends on the distance between the ends of coil 1 to the ends of coil 2 and the distance between the coil wires to each other.

Hysteresis losses ($\sim f$) and eddy current losses ($\sim f^2$) are represented by the high ohmic resistance Rfe in parallel to the inductance L inside the equivalent circuit. In case of a saturation of the core, this resistance will decrease and bridge the frequency-dependent resistances, and the attenuation would collapse.

The ohmic loads of the coil are frequency dependent caused by the skin effect. However, these influences are neglectable for signal integrity analysis in the time domain and, as a result of this the ohmic load, is implemented as the constant resistance R_{CU} in serial to the inductance L.

2.2.2.4 Cable Termination, Supporting Resistances

As already introduced in the former sections, there are several possibilities of terminating a CAN network with the help of termination resistances. The simple termination and the split termination were introduced. In this section, further aspects for network termination focused on the topology layout are considered.

In dependence of the topology, there are different aspects leading to an optimized termination concept. One important point is that the overall line termination should have a value of approximately 60 Ω. Another important point is that stubs with a long length in comparison to the wavelength generate reflections; termination resistances should be located at long stubs as well.

Therefore, the termination resistances should be placed in a star topology in the two stubs with the longest distance to each other. Alternatively, it is possible in an example with a star containing four short stubs and three long stubs to terminate each of the three stubs with 180 Ω (180 Ω || 180 Ω || 180 Ω = 60 Ω) or to terminate the star point with a single termination with 60 Ω.

It is possible to implement supporting resistances improving the EMC characteristics and interference immunity, even in the case of no usage of a split termination. These supporting resistances are implemented between CAN_H and 5 V and between CAN_L and ground (GND) to improve the symmetry characteristics and to stabilize the average potential to 2.5 V. The resistances have a typical value of 1.3 kΩ.

2.2.2.5 ESD Protection

To protect the bus connecting circuitry against current pulses by ESD, ESD protection components are implemented. Modern transceivers already have internal ESD protection measures; thus, the location of the ESD protection components should

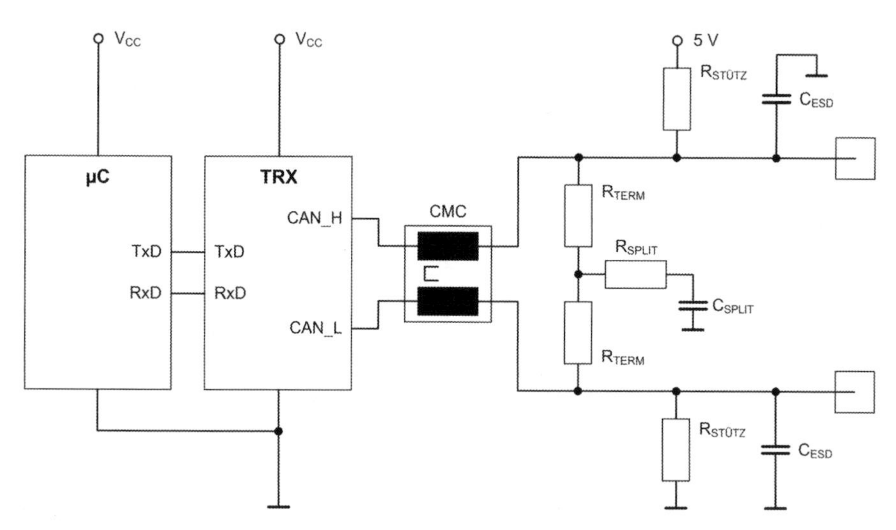

Fig. 2.23 CAN bus connecting circuit of a CAN node

be as close as the injection location which is normally at the bus connector pins of the node. Usually, ESD protection components are realized with varistors which are implemented between the bus signals and GND. Relevant for signal integrity inspections are the parasitic capacitances which are normally depicted in the data sheet of the ESD protection component.

To sum up all components of a CAN node bus connection, Fig. 2.23 implies, depicted for signal integrity valuations, important equivalent circuit.

2.2.2.6 Isolation and GND Shift

In case of applications with huge topologies and high current consumptions, GND shifts (or shifts on the supply line) may occur. In the following, an example case explains which measures shall be taken into account to avoid influences of the application's current load on the CAN communication. Further praxis-relevant solution approaches, related to industrial used CAN networks, could be found in the DeviceNet specifications [ODVA94].

In this example, the CAN nodes have a huge distance to each other and thus the cable length is relatively long. To illustrate, for example, a CAN control of industrial motors on a conveyor belt is used. GND line and supply line are installed together with the CAN lines. In case of a common use of supply and GND lines for application and CAN communication, the voltage drop over the GND cable will be dependent on the cable resistance per unit length, the length of the cable as well as the current load of the application (which could be huge in comparison to the CAN communication current load). This voltage drop over the GND cable affects as a GND shift which may influence the CAN communication and the related application.

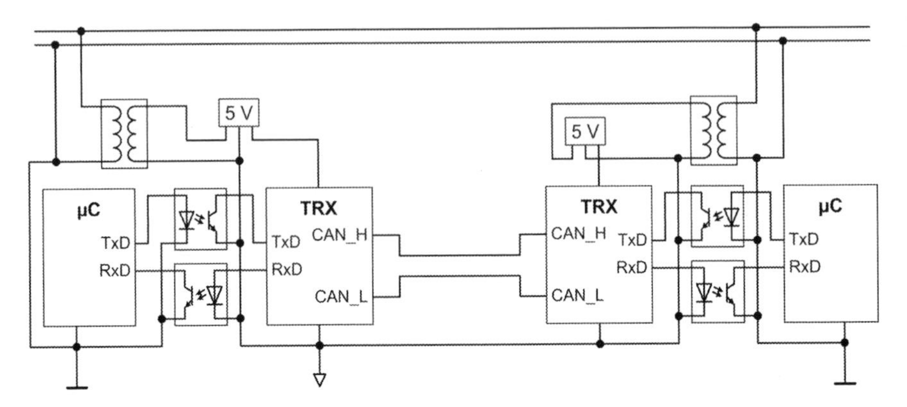

Fig. 2.24 Separated grounding of communication and application

GND shifts are critical because they induce common-mode currents which lead to worse EMC behaviour of the network and, in the case of huge GND shifts, the information exchange itself is endangered by wrong bit decoding.

One solution to avoid GND shifts, caused by the current consumption of the application, is to separate the GND line for the CAN communication and for the application as well as a galvanic isolation between the CAN communication and the application. Regarding the supply of the CAN communication parts, the galvanic isolation can be realized by direct current (DC)/DC converter; the galvanic isolation of the logic signals RxD and TxD can be realized by optocouplers. EMC behaviour and interference immunity are enhanced by the isolation of the CAN physical layer. The explained concept is depicted in Fig. 2.24.

2.2.3 Interactions of Components and Analytic Signal Integrity Inspections

An elementary problem at signal integrity valuations is the interactions of influences of different components on the overall behaviour of the system. That only the sum of all influences can be measured is problematic. A systematic approach is needed to identify and localize particular influences and to be able to evaluate these influences.

Sections 2.2.1 and 2.2.2 describe the influences of network components as well as those of the network architecture. The characteristics were analysed isolated from influences of other network component or influencing parts.

This section, however, describes, taking examples from praxis, how the different components interact with each other and how these interactions result in changes of the signal integrity of the overall system. Furthermore, analytic procedures are introduced to allow a specific consideration of the signal quality.

2.2.3.1 Ringing at the Transition from Dominant to Recessive

For signal quality evaluations of CAN network topologies, it is crucial to know where disturbances occur and which properties of the network architecture or properties of the CAN node architecture amplify these disturbances. In case of signal integrity evaluation of CAN networks, the state change from dominant to recessive is of particular importance. Abstracted, it could be said that during the state change an energy source driving the differential signal is taken from the bus (which results in the recessive state) and instead a high ohmic receiver circuit is added to the bus. The existing energy inside the system degrades over the passive components of the network. Implemented capacitive and inductive energy storages discharge and influence each other. Recharging between inductive and capacitive stored energy may occur which results in oscillations of the bus signals which is called ringing. These overshoots and undershoots occur at different strengths dependent on the characteristics of the different used components. These interactions can be clarified by formula [2.2] of the fundamental field of the electro techniques:

$$u = L \cdot \frac{\Delta i}{\Delta t} \tag{2.2}$$

First, to assume a constant stray inductance, which is operative in the differential mode, it is easy to see using formula [2.2] that the overshoots and undershoots will be more intense in case of a transceiver which produces faster slow rates; thus, the $\Delta i/\Delta t$ is higher. Furthermore, different CMCs have different stray inductances. CMCs with higher stray inductances (smaller coupling coefficients) invoke higher overshoots and undershoots. CMCs with sector-based winding have in general smaller coupling coefficients and approximately higher stray inductances. CMCs with bifilar winding have in general higher coupling coefficients.

Figure 2.25 compares both extreme cases of these interactions of different components. The solid curve refers to a CAN node with a transceiver with a fast slew rate, corresponding to a high $\Delta i/\Delta t$, and with a CMC with sector winding, thus a high stray inductance. The dotted line belongs to the same CAN node but in the case with a transceiver with a slower slew rate while applying a CMC with a higher coupling coefficient. It is easy to see that the ringing is less intense with the second equipped case (dotted line). The measured CAN node is located inside a symmetric CAN topology with two stars where all stubs have the same length to the star points. This topology is typical for a lot of ringing at the state change from dominant to recessive caused by the symmetric characteristics which leads to an adding of particular reflection parts.

As already mentioned, the energy distribution in the system is of importance. Therefore, two relevant scenarios shall be taken into account: On the one hand, the signal flow and behaviour at the sending node are important to consider because this is the point of the energy input. On the other hand, the consideration of the ACK bit is important. In that case, all receiving nodes start transmitting the dominant

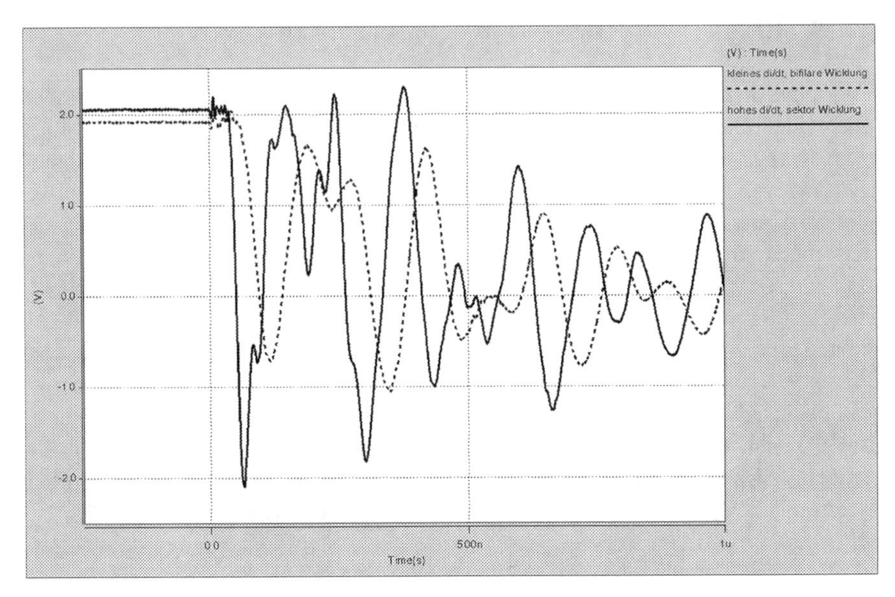

Fig. 2.25 Different intense ringing at the "dominant-to-recessive" transition

acknowledge except the frame sending node. As a result of this, there are many energy input points in the system and with it a lot of energy in the system.

Ringing occurs more heavily when the topologies consist of stars or of stubs with long stub length as well as when the topology has a symmetric shape. In this case, the more intense ringing is a result of reflections and adding of reflection parts in the case of symmetric shapes.

The effects of the ringing can be seen in the delay times. A view on the thresholds helps the valuation of the ringing: If the differential bus signal exceeds 0.9 V, caused by an overshoot, this will lead to a change of the logic RxD to the dominant state until the differential bus signal falls back below 0.5 V (worst-case threshold values). In this case, the ringing drives a change on the logic signals which results in a prolongation of the related system propagation delays.

2.2.3.2 GND Shift, CMC and Arbitration

With the help of a further example, it is possible to observe the effect of self-induction on the CAN bus. In the case of a lost arbitration, the state of the node changes from sending to receiving. Based on a scenario with two nodes sending a dominant bit followed by a dominant bit of the arbitration winning node and a recessive bit by the arbitration losing node, the direction of the energy flow changes at the losing node (from driving energy by sending a dominant bit to receiving energy by receiving a dominant bit of a foreign node). As a consequence, there is a reverse of the energy flow in the CMC as well.

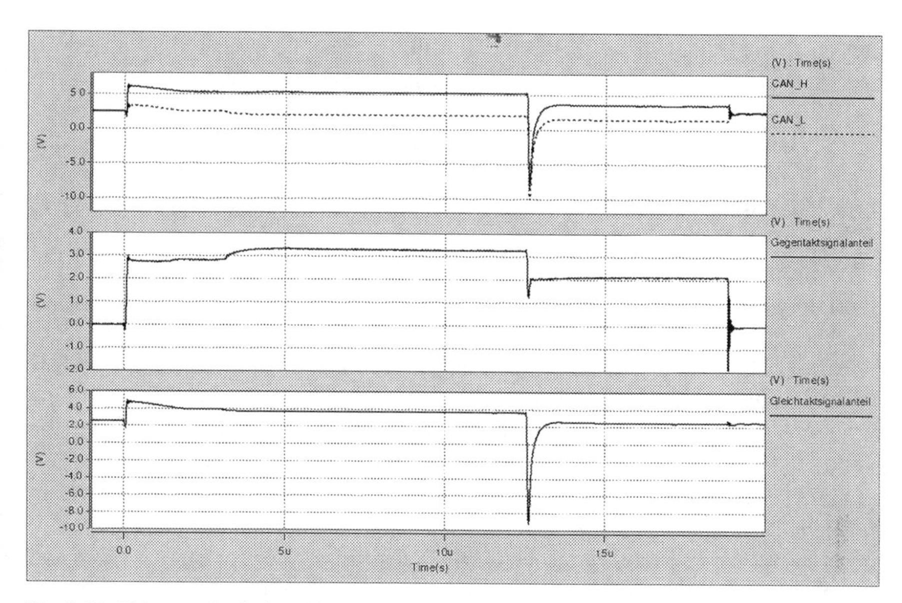

Fig. 2.26 Voltage pulse induced by common-mode chokes and GND shift

A self-induction describes the induction of a voltage by the inductance which tries to hold the current and the magnetic current constant in case of a change in the current flow. The proportionality factor L describes the intensity of the self-induction as well. A high inductance results also in a high inducted voltage in the case of a self-induction. Based on the above-described operating principle of the CMC, it clearly shows that in differential mode only the stray inductance is effective which is much smaller than the main inductance which works only in common mode. After a node lost arbitration it switches off the transmitter stage and the current in the CMC. An inducted voltage peak results, which is higher if the main inductance is active – in case of common mode signal. A GND shift is a source for common-mode currents which raises the effective working inductance and with it the intensity of the self-induction which will be evoked by the loss of arbitration.

As a consequence of this, it is possible to observe that CMCs induce voltage peaks on the CAN bus after the above-described scenario. Measurement resulting from these peaks is shown in Fig. 2.26.

2.2.3.3 Limitations of the Cable Length in Consideration of the Bit Timing

In principle, two different factors limit the maximum network distribution. On the one hand, the data rate, bit timing settings and oscillator tolerances influence the maximum possible cable length. On the other hand, resistances and capacitances of the cable and transceiver limit the possible cable length. The last condition refers to the signal amplitude where the limitation is that at least the threshold for the dominant state shall exceed overall the network independent of the sender's position. Data rate and bit timing are influenced by the cable length-dependent delay times.

The section "CAN bit timing and oscillator tolerances" already described the appearance of different fast-running nodes inside a CAN system. In this section, differences of oscillator tolerances and how they influence the CAN physical layer are discussed.

An example shall show how the CAN physical layer and bit timing configuration interfere with and limit each other. For this, the communication flow is analysed over the time space between two soft-synchronization events.

At the time of the first soft-synchronization (a slope from recessive to dominant), the communication nodes synchronize its clock to the bus signal. From now on until the next synchronization event, all timing-dependent influences shall be considered. These influences are mainly delay times of the communication controller and the transceiver at sender and receiver and eventually further components in the signal flow as well as differences in the node clocks. In addition to this, there are the propagation delays of the cable which depend on the propagation speed and the cable length. These delay times occur at the first considered slope as well as the second considered slope. The basic condition for a successful communication is that each bit is sampled correctly at the position of the sample point.

The sample point can be shifted to a later position to enable long cable distances. Thus, a longer time period from the beginning of the bit to the sample point is provided to allow enough time for all the delay times. However, the margin for compensating clock differences is limited by shifting the sample point to a later position because with it the SJW size is limited as well.

It is possible to construct worst-case scenarios where two nodes are considered which have a huge distance to each other and are operating on opposite boundaries of possible clock ranges. In this case, the sending node has a fast clock and the receiving node has a slow clock. During a long period without synchronization, the receiver sampling will fall behind the transmitter. The deviation increases with the difference of the clock rates and with the duration since last resynchronization. The margin for propagation delay times narrows in dependence of the oscillator times.

Based on an error-free communication scenario, time spaces without resynchronization are limited by the bit stuffing rule. A simple example can be derived with five dominant bits followed by a recessive stuff bit followed by four recessive bits—a time space of 10 bits without resynchronization. For error-free communication, it is required that the slower receiver (propagation delays of the network have to be considered as well) has to sample the tenth bit as recessive. Therefore, the following inequation can be derived [2.3]:

$$10 \cdot NBT \cdot \left(1-\Delta f\right) > t_{RD} + 10 \cdot NBT \cdot \left(1+\Delta f\right) - \text{Phase\_Seg2} \cdot \left(1+\Delta f\right) \quad (2.3)$$

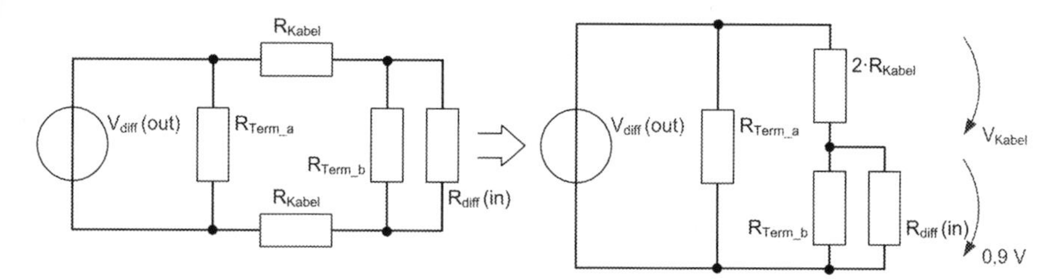

Fig. 2.27 Simplified equivalent circuit for calculating the maximum allowed cable length at a point-to-point connection

with NBT=nominal bit-time, t_{RD}=propagation delay recessive-to-dominant, $NBT \cdot (1 - \Delta f)$ =“fast” clock and $NBT \cdot (1 + \Delta f)$ =“slow” clock.

Furthermore, it shall be considered that the evoked time gap, caused by the propagation delays, and the different fast-running clocks shall be smaller than the SJW to avoid a drifting of the clocks over longer time spaces.

2.2.3.4 Limitations of the Cable Length in Consideration of the Line Losses

In the following, static influences are analysed which impact the cable length and signal quality. The maximum possible cable length is limited mainly by the resistance per unit length and capacitance per unit length and the number of implemented nodes. Through an implementation of too many nodes containing very high differential capacitances, bit deformation may occur. The dominant bit shape becomes close to a capacitive charging shape (slope to dominant) respectively discharging shape (slope to recessive).

Latest, at the sample point position of a dominant bit, the differential voltage shall have (even in worst case) a voltage of 0.9 V or more to guarantee the communication. However, the recessive bit shall have at the same position a differential voltage of less or equal to 0.5 V. The worst-case consideration with the maximum/minimum thresholds of 0.9 respectively 0.5 helps here again in analysing the influences of resistances of cable, transceiver and bus connecting circuitry. Abstractions to simple equivalent circuits using resistances help the right dimensioning of the network.

At a simple point-to-point connection, the differential voltage of the sender acts as a simple voltage source. Parallel to this voltage source is located the termination resistance. The receiver is to substitute by the second termination resistance and the differential resistance of the receiver stage. The cable length-dependent cable resistances connect the sender and receiver equivalent circuit. As mentioned above, the basic requirement making a communication possible is that at the receiver side the differential voltage shall be equal to or higher than 0.9 V. Therefore, the equivalent circuit shown in Fig. 2.27 can be used for calculations.

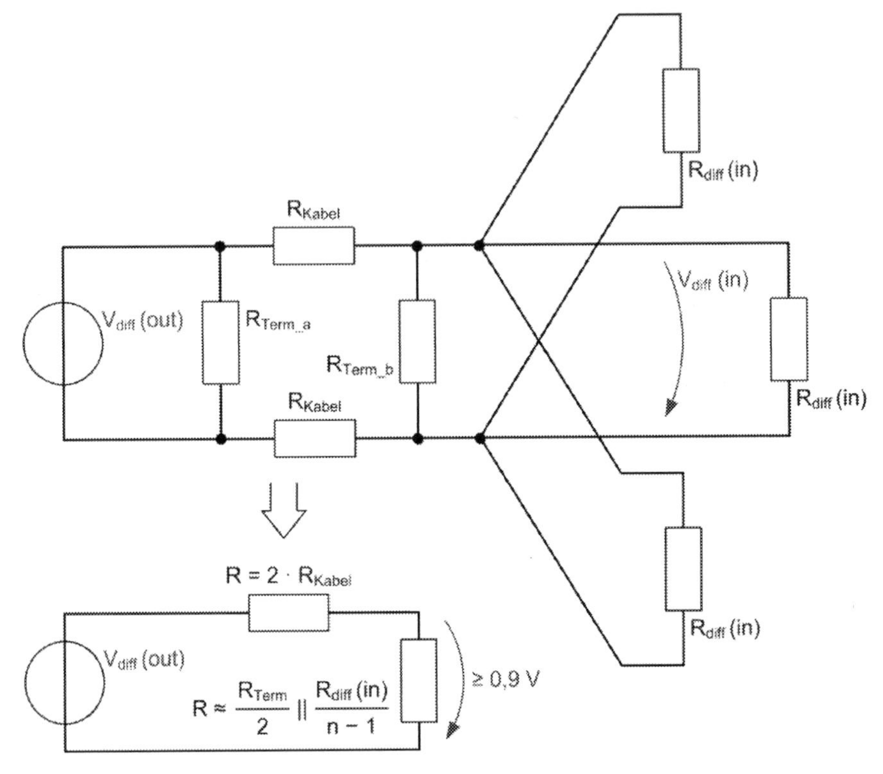

Fig. 2.28 Simplified equivalent to calculate the maximum allowed cable length of a star topology

It is possible using the voltage divider formula to calculate the maximum allowed cable resistance (and with it the maximum allowed cable length) under the consumption that the voltage drop over the receiver resistances is 0.9 V or higher [2.4]:

$$R_{Kabel_{\max}} = \frac{\frac{V_{Kabel}}{V_{diff}(out)} \cdot R_{Term\_b}}{\left(1 - \frac{V_{Kabel}}{V_{diff}(out)}\right) \cdot 2} \tag{2.4}$$

Considering a star topology, the resistance arrangement and with it the simplified equivalent circuit change to that shown in Fig. 2.28.

At a daisy chain topology, thus a linear topology with stub length of 0 m, an iterative calculation helps calculating the maximum allowed cable length as shown in Fig. 2.29.

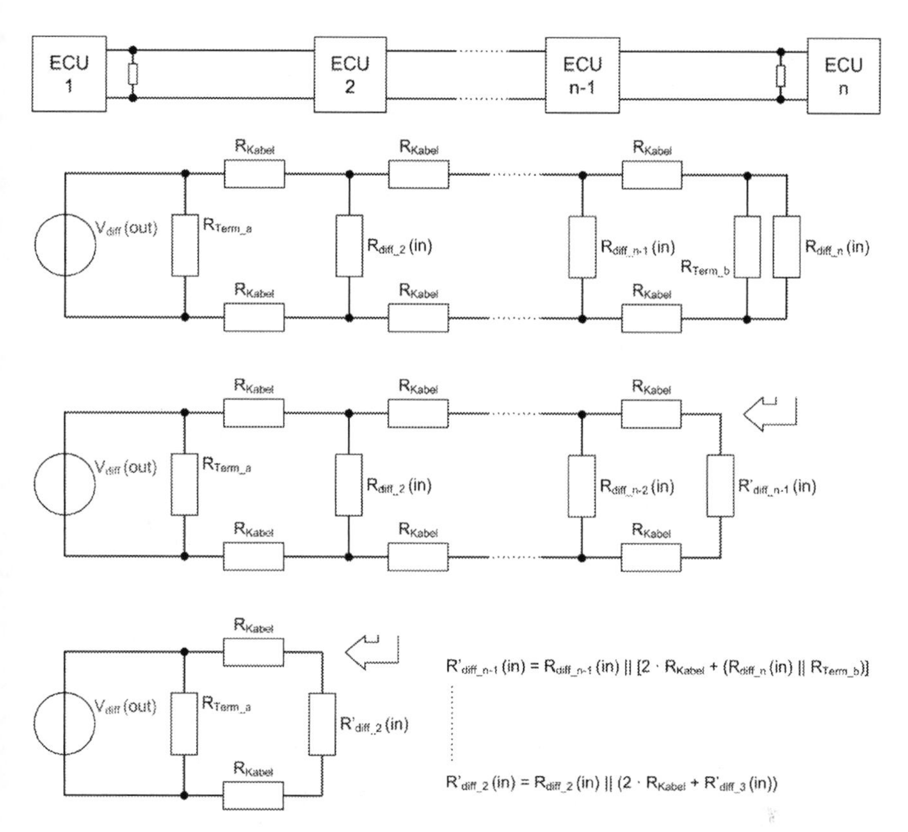

Fig. 2.29 Simplified equivalent circuit to calculate the maximum allowed cable length in a daisy chain topology

2.3 Network Topologies—Design by Simulation

2.3.1 Development of Automotive Networking Topologies

Due to the ever more stringent cost, quality and time-to-market requirements, the development of automotive vehicle networking systems has become a very challenging task. When specifying new networking systems or revising existing implementations, a major challenge is to validate that an implementation will work properly and robustly over the entire operating range. In-vehicle networking systems can be divided into two domains:

- *Logical Network Architecture*: This domain deals with the actual data communication between ECUs.
- *Physical Network Architecture*: This domain deals with all aspects of the physical implementation of the in-vehicle network.

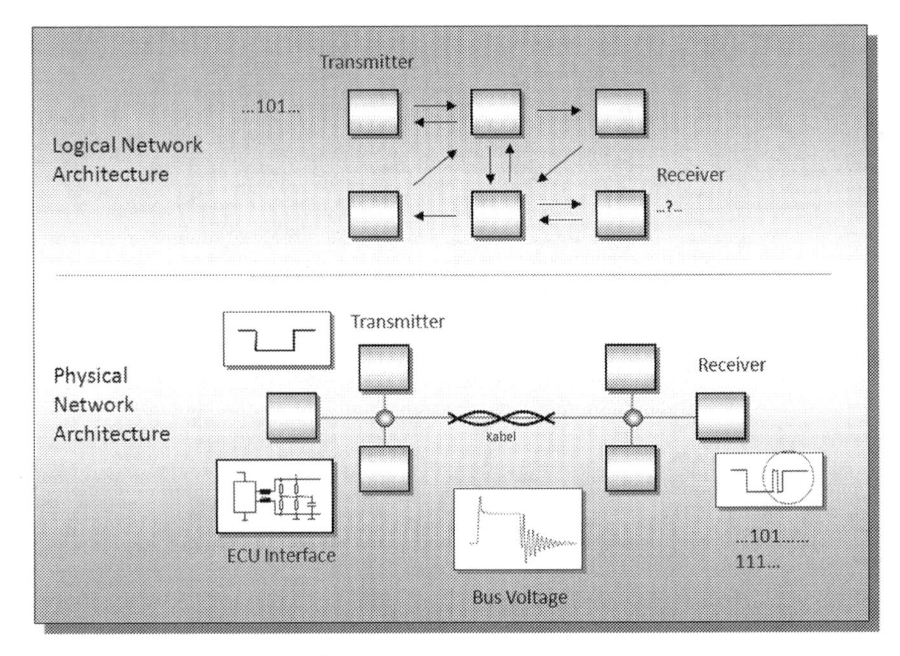

Fig. 2.30 Logical versus physical network architecture

This chapter focuses on the physical network architecture or "physical layer". Due to the increasing complexity and size of automotive networks over the past years, more rigorous exploration and analysis of the physical layer have become necessary. In order to address this very challenging task, simulation-based methodologies have proven to be the methods of choice. The technical necessity for simulation-based methodologies is driven by the following issues typically encountered by the network developer during the development process:

- Complexity (e.g. number of ECUs and wiring harness)
- Variety of vehicle platform variants
- Reproducibility of test conditions (e.g. worst case)

The complexity of modern in-vehicle networking systems makes it extremely difficult for developers to derive any a priori conclusions about the correct functionality and robustness of the system before prototyping. This problem will be explained by means of an example. Figure 2.30 depicts a networking system composed of six ECUs that are connected through a high-speed CAN bus. If one of the ECUs is transmitting a data stream to the other ECUs in the network, it is important for the network developer to know what data have been sampled by each of the receivers. The reasons for problems during communication are particularly related to the analog electrical signal behaviour between the different ECUs. Although CAN is a digital communication protocol, the electrical behaviour along the bus and between the ECUs is analog. This behaviour is significantly impacted by the following elements:

- Type of topology
- Physical interface of ECU and number of ECUs
- Cable type and length

The cumulative impact of these elements is so complex that it is impossible to manually predict the electrical behaviour of a vehicle topology and analyse its robustness through a pure paper specification. This is particularly important during the specification phase since any problems discovered during the later stages of development can be very costly to correct. This problem is even more severe if problems related to network implementation are uncovered close to the start of production (SOP) and the timely delivery of the vehicle to market is at risk.

Aside from the complexity, the huge diversity of vehicle platforms related to different sets of vehicle equipment adds another dimension to the challenge. It is necessary to ensure that the network implementation of all variants associated with each vehicle platform function properly. However, the effort to test all variants would be tremendously high and time consuming and in some cases not even feasible since certain dedicated variant types are not available for physical testing before SOP.

Since vehicles are exposed to harsh environmental conditions in everyday operation, it must be ensured that the communication network functions robustly over all of these possible operating conditions. Testing this would mean running the tests under authentic and repeatable environmental conditions. These may relate to temperature, vehicle component tolerances, cable impedance variations, etc. Traditional tests using a real vehicle prototype typically only allow for quite limited coverage of the previously mentioned criteria and thus make it difficult to reach a final conclusion about the robustness of the vehicle topology.

2.3.2 System Simulation as a Tool for Network Developers

To meet the growing demands and challenges that network developers face, simulation-based approaches have become the preferred and dominant methodologies. Initially, semiconductor companies adopted simulation methodologies but nowadays automotive original equipment manufacturer (OEMs) and their Tier 1 suppliers have understood the necessity and benefits of a "virtual prototyping"-based development approach. As opposed to the classic approach of designing the network topology on paper and then later realizing it as a physical prototype (e.g. breadboard) for testing purposes, the virtual prototyping approach allows one to create an executable specification by means of a simulation model. The benefits of using this methodology include the following:

- Proof of-concept of the topology specification without a physical prototype.
- Flexible adjustment of topology specification to explore different implementation options.
- Analysis of dedicated operating conditions and reproducible test results.
- Reduction of cost through optimization and accelerated test procedures.

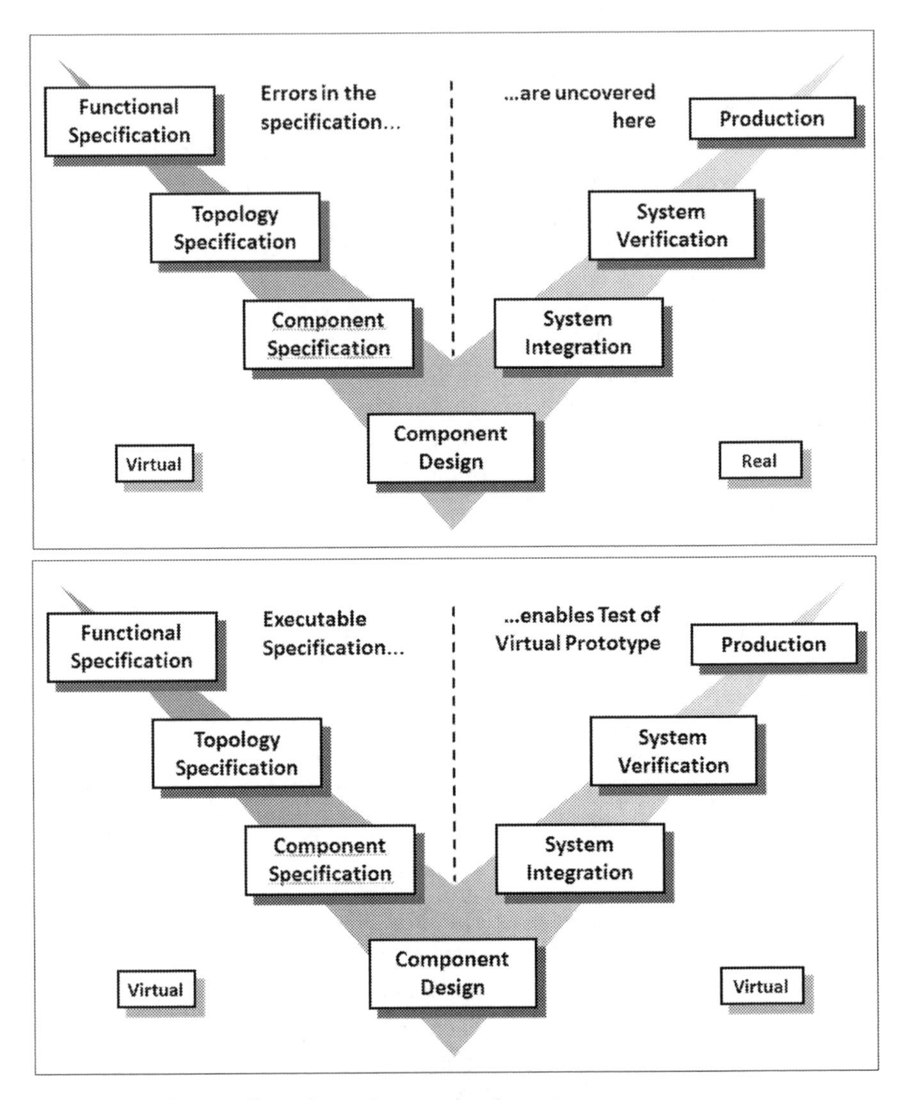

Fig. 2.31 Development flow using real versus virtual prototype

Figure 2.31 compares the classic development approach and the simulation model-based methodology using the V-diagram. The difference between the two approaches is that the model-based methodology uses a virtual prototype along the entire path of the V-diagram. The developer can validate the topology at a very early stage of the development process and does not need to wait until he has obtained all ECUs from suppliers to assemble the real prototype through breadboarding. If the simulation-based approach uncovers failures in the topology specification, alternative options can be quickly analysed by making adjustments to the virtual prototype and re-simulating. This approach increases the efficiency throughout the development process and reduces risk early in the process by allowing designers to assure that development is heading towards a reliable and robust network topology implementation. This helps to eliminate problems being discovered late in the development process. However, simulation's benefits extend beyond the conceptual phase. Even for network implementations that are already used in production vehicles, simulation helps to ensure the required level of quality if changes to the topology need to be applied. For example, suppose the on-board vehicle network department applies changes to the wiring harness system due to routing considerations. Naturally, this will also impact the electrical behaviour of the networking system (e.g. CAN bus). Using a simulation-based approach, the network developer can quickly validate how significantly the changes impact the networking behaviour and then work with the on-board vehicle network department to reach a viable solution.

Before an executable specification (or virtual prototype) of a CAN topology is built up, the following questions should be answered:

- What questions will be answered through simulation?
- What data need to be created to answer these questions (e.g. analog and digital)?
- What type of simulation is required to create the data (e.g. transient and AC)?
- In which format should the data (results) be prepared (e.g. signal curves and tables)?

The importance of exploring and answering these questions becomes immediately obvious when one tries to develop the topology model and characterize it with data. The approach described in this chapter focuses on the signal integrity of CAN topologies which partially answers the first question. To complete the answer to the first question, it is necessary to use typical signal integrity evaluation criteria. This encompasses quantities such as propagation delay which quantifies the timing delay between the transmitting CAN node and receivers. Another important criterion is the settle time after the transition from dominant to recessive state and vice versa. In particular, the second criterion is closely associated with the robustness of the system since a long settling time may lead to sampling errors by the CAN controller. To validate the robustness of the network system in detail, it is desirable to determine the available timing margin. This means figuring out how close the sample point of the CAN controller is to the critical ringing area after the transition of the differential bus signal.

This automatically leads to the answer of the second question since the described criteria above required a certain set of specific signals to be available after the simulation. Evaluation of the settle time requires analogue signals. The propagation delay is measured between the falling edges of the TxD signal at the transmitting node and the RxD signal at the receiving nodes. Since both of these signals are digital, it is understood that in order to model the entire topology both analogue and digital models are required. Accurate simulation of the propagation delay also requires inclusion of the propagation delay through the transceiver chip to complete an accurate system behavioural model.

Having the first two questions answered, it becomes apparent that simulating automotive networking topologies requires a holistic system level approach since

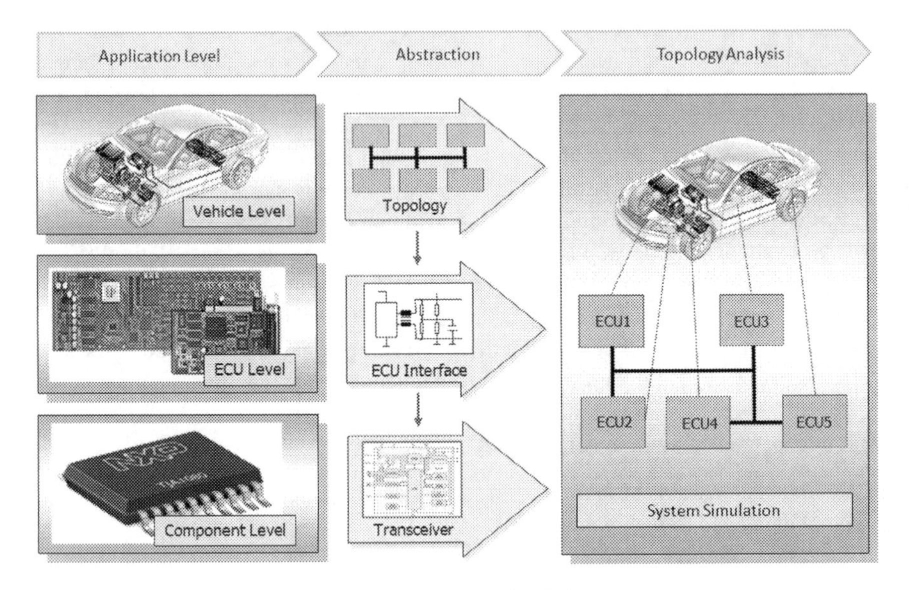

Fig. 2.32 Topology analysis in the context of system simulation

the validation of the network robustness requires information from all layers of the system.

As shown in Fig. 2.32, in order to model a CAN topology there are three different layers to be taken into account. From a system/vehicle-level perspective, the type of topology is important. Is the topology a linear topology with stub nodes or a passive star? The electrical signal behaviour of these topology types is very different and the related implementation cost for each topology type is a major differentiator.

The cable model of the bus between the ECUs plays an important role for the analogue system behaviour of the topology and has a significant impact on the accuracy of simulation results. The same applies to the physical ECU bus interface. This in particular applies to high-speed CAN topologies as the integrated CMCs can cause problems for the analogue system behaviour due to stray inductance. Finally, the simulation model of the topology needs to contain a behavioural model of the transceiver chip.

One major objective in using simulation is to make design decisions in the very early development phase. This objective is very closely tied to the abstraction level of the overall system. At first glance, it seems desirable to model the vehicle including all effects impacting the electrical signal behaviour in order to accurately predict signal integrity. This approach would not be very reasonable from a simulation perspective as the effort to develop a model with such a high level of sophistication will not be feasible and this level of accuracy is not really necessary. It would be very difficult to gather all required data to characterize the model. Especially during the early concept phase, this set of data is usually not completely available. Furthermore, it may also be problematic to leverage required compute resources to simulate such a highly sophisticated vehicle networking system and the required amount of time to simulate the model would most likely go beyond the available timing budget during the concept phase. The benefit of time savings during the development phase using a virtual prototype would disappear. Therefore, it is important to target an appropriate abstraction level and to focus modelling work on the actual required set of data and constraints. This means that, for instance, any routing and vehicle packaging information should be disregarded to model the network cabling. Only the cable length and the electrical properties of the cable should be taken into account. As far as the model of the ECU is concerned, a proper level of abstraction is required. Since the CAN control unit may not yet exist during the conceptual stage, and, therefore, the electrical effects of the printed circuit boards (PCBs) on the wiring interface are unknown, a pure component-based model of the ECU interface has been found to be sufficiently accurate. In general, even in the later stages of modelling the parasitic effects are disregarded since acquiring such data requires running complex and time-consuming finite element analyses. The semiconductor chips should also be modelled using a behavioural modelling approach. Applying this to the transceiver chip would mean modelling the analogue behaviour of the transceiver bus pins. Internal state machines can be modelled as simplified idealized digital state machines. Using full transistor-level models from semiconductor vendors would not be appropriate as this would not provide any additional accuracy benefit while simultaneously causing a significant increase in simulation time. Automotive manufacturers and semiconductor vendors have joined forces to standardize requirements for transceiver models. The GIFT-consortium has already leveraged comprehensive work together with automotive companies and made significant progress in this regard.

The answer relating to the question of what simulation analyses should be applied can be quickly ascertained based upon the answer to the first question. Since the nonlinear behaviour of the entire topology is modelled and the criteria of interest are measured within the time domain, a transient simulation is the analysis type of choice. A simplified frequency domain simulation would only be possible if the nonlinearities were ignored, for instance, in the behaviour of the transceiver chip. In order to evaluate the robustness and worst-case behaviour, simulation models that include the component data sheet-specified tolerances over the entire operating range are needed. For this type of analysis, a continuous thermal model is not necessary but rather specific corner case conditions are sufficient. These are much easier to model and represent the actual critical cases for automotive networking applications. Usually, network developers start validating the quality of the topology implementation using the nominal behaviour of the system to develop a basic understanding about the system robustness. Starting with the worst-case behaviour from the beginning does not make sense since the network performance may already show some problems under nominal conditions, and an analysis of the worst-case behaviour for this topology configuration would be redundant.

The previously described item builds the transition to the question around the format of the generated simulation results. A major benefit using a simulation-based approach is to fully automate the entire testing process. This applies to both simulation as well as post-processing. The standard testing procedure for CAN networks is called Round-Robin communication. During this procedure, each network node acts once as transmitter and all other ECUs act as receiver. Thus, a network including ten nodes requires ten transmission sequences. The discrete electrical measurement results are consolidated into a signal matrix which shows the transmitter–receiver relationship. This usually is a three-dimensional matrix representation since multiple validation criteria are applied and need to be evaluated for each transmission sequence. It is highly desired to fully automate the creation of this matrix including a validation and reference scheme. This automated approach allows the network developer to quickly gain an insight into the quality of the topology implementation with respect to the signal integrity. A validation using this discrete data is very helpful; however, it is still necessary to have the analogue signal curves available in order to qualify particular signals in detail or to generate eye diagrams.

The previously described methodology contains the general aspects and objectives and builds a solid foundation for the simulation of automotive networking topologies. These methodological requirements are next implemented using a simulation tool.

2.3.3 Saber—A Development Tool for Simulation and Analysis of the Electrical Physical Layer of Networking Topologies

Driven by cost pressure and compatibility, it is often required and beneficial to apply open standards. This also applies to the simulation of automotive electrical physical layer topologies. It has become common practice that required simulation models (like transceivers) are delivered along with their hardware component by semiconductor vendors. However, it would be difficult and cost intensive for semiconductor companies to support creation, testing and delivery of simulation models for each of their customers with various dedicated model types. In this particular case, it is prudent to use a tool-independent open standard. Very high-speed integrated circuits (VHSIC) hardware description language-analog mixed-signal (VHDL-AMS), defined through the Institute of Electrical and Electronics Engineers (IEEE) standard 1076.1 (now IEC 61691, Part 6), is such a tool-independent modelling language that is very much applicable to model supply chains. VHDL-AMS is an extension of the VHDL-IEEE 1076 standard and was created for the purpose of hardware design. This means that the language supports modelling constructs to describe the behaviour of physical systems. One of the tools supporting this modelling language is Synopsys' Saber. In addition to VHDL-AMS, Saber also supports other diverse modelling languages and software development tools providing developers with great modelling flexibility. Figure 2.33 shows a consolidated overview of the features supported by the Saber product.

In order to model a CAN topology, the developer can either use Saber's component library transceiver models or use models provided directly by component vendors. The Saber tool allows the integration of externally developed models into the tool. The simulation model of the topology can be easily built up by selecting

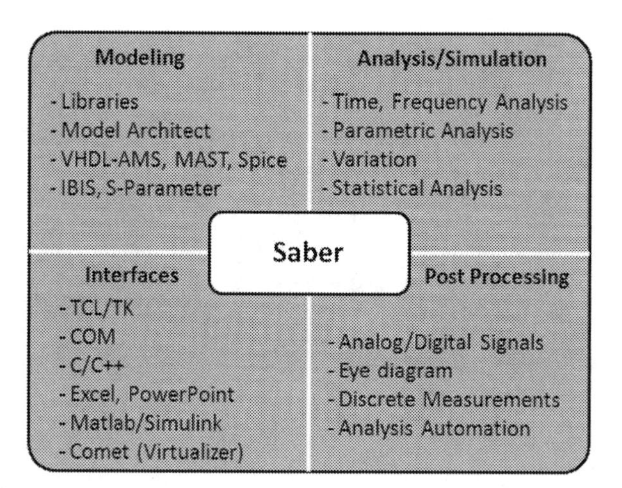

Fig. 2.33 Overview of the Saber product features

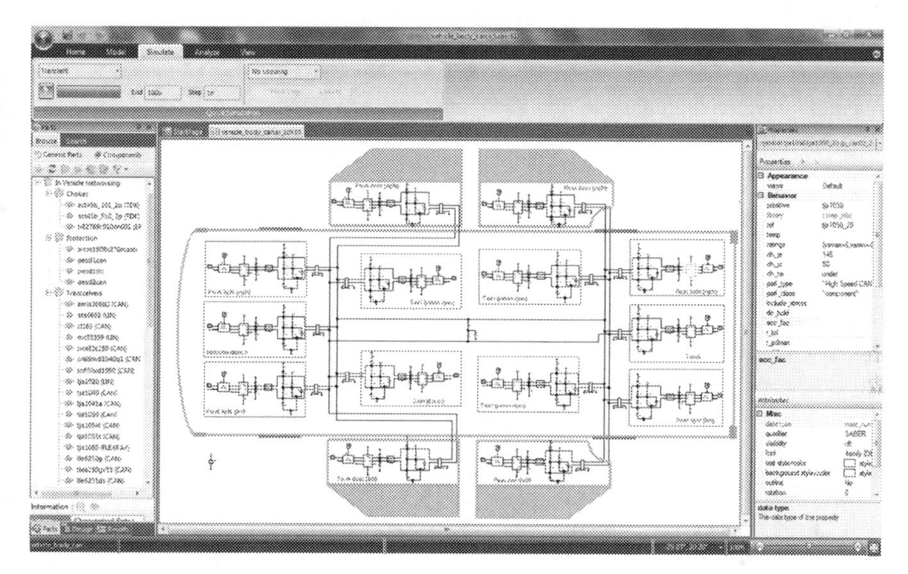

Fig. 2.34 Simulation model of high-speed CAN topology in Saber

component models from Saber's parts gallery, placing them on the schematic editor and connecting corresponding topology instances with each other.

Figure 2.34 shows an example for a high-speed CAN topology modelled in Saber. This simulation model allows a convenient way for the network developer to analyse the signal integrity behaviour of the topology. How detailed of a level at which the system is evaluated and what the specification requirements are depend very much on the automotive OEM. This means that the level of detail for the topology model can also vary. In order to qualify the signal integrity, the network developer now has, amongst other things, access to the differential bus voltages as

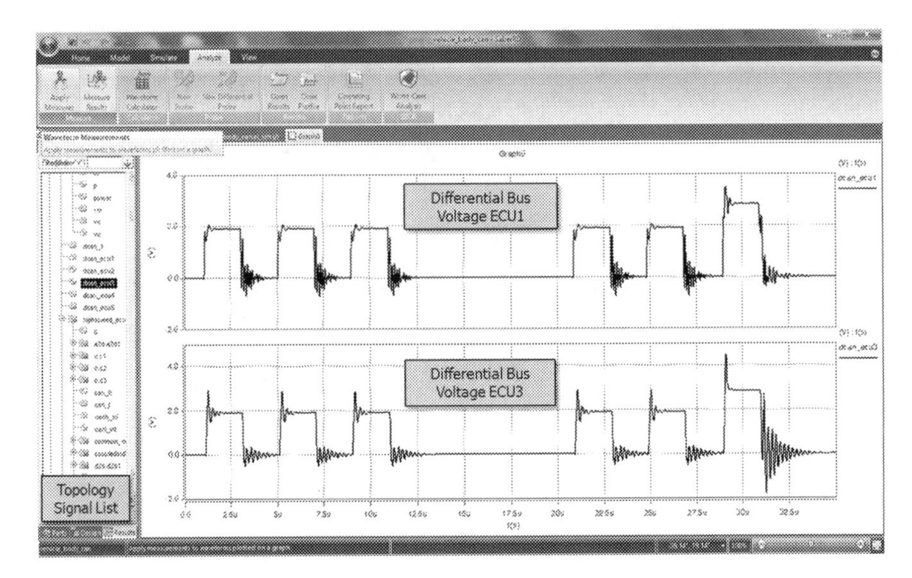

Fig. 2.35 Simulated differential bus voltages

well as the digital RxD signals at the transceiver. Figure 2.35 depicts the differential voltage for two of the ECUs. Using Saber's post-processing capabilities, measurements such as slew rate can be easily extracted with a single mouse click.

Moreover, the simulation performed in Saber affords the developer a deeper insight into the CAN topology in order to evaluate the robustness of the physical layer implementation. Aside from the analogue bus signals, it is possible to integrate a digital CAN controller model into the simulation to take into account the bit timing behaviour. Saber's model library includes a simplified CAN controller model which is optimized for performing signal integrity analysis of the physical layer implementation. The origins of this model started with an initial specification from Volkswagen in Germany. The model contains the following features:

- Transmit and receive functionality
- Arbitration
- Hard and soft synchronization
- Acknowledgement
- Bit timing register settings

In addition to the optimization of the signal integrity mentioned previously, the model allows for exploration of the impact of the controller configuration on the overall system behaviour. Figure 2.36 gives an overview of the available post-processing data using the CAN controller model in Saber.

By modelling the bit timing, the developer can determine exactly the time between the sample point and the critical transition area after the change of the digital signal. If the sample point is too close to the critical ringing area, the developer can apply changes to the controller software and adjust the location of the sample point or fix the problem in the hardware topology. The first solution is often easier to

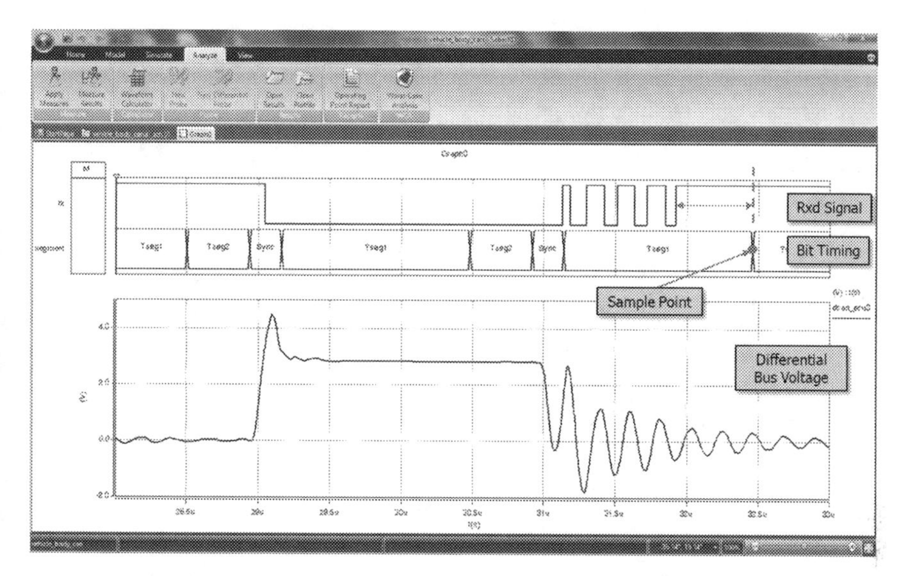

Fig. 2.36 Analysis of the bit timing in Saber

implement as it is only a software parameter that needs to be changed. Simulating a virtual prototype quickly provides an overview of the quality and the robustness of the CAN topology implementation. Problems related to the implementation can be discovered early and the developer has a powerful solution available in order to compare and validate different implementation options.

The previously discussed methodology demonstrated a simple case with just a handful of data to evaluate. If the number of signals or amount of data is much greater, it would be very time consuming to manually analyse all of the simulated signals. Let us assume we had a topology consisting of 14 ECUs and just three validation criteria (propagation delay, slew rate and settle time). This would mean the network developer would have to analyse $14 \times 14 \times 3 = 588$ discrete measurements. It may also be necessary to visually inspect other signals in the system. And if variations of stub lengths as well as worst-case configurations for the transceivers or cable impedance were to be simulated, then the amount of data would increase significantly. Obviously, a manual evaluation of the simulated data would be very time consuming, error prone and would reduce the time-saving benefits of using a virtual prototype. In Saber, the network developer can fully automate the entire simulation and post-processing steps. Figure 2.37 shows a possible scenario.

The starting point of the entire process is an automated Round-Robin communication for which the developer specifies the desired communication pattern. The entire simulation process is then put into a script and can be run via a single mouse click. In the second step, the generated simulation data will be further processed/modified according to the specific requirements of the automotive OEM. Due to the fact that each OEM has different topology requirements and prefers a different representation of the final data, it is infeasible to provide an automated post-processing solution off the shelf that addresses everyone's needs. Therefore, Saber

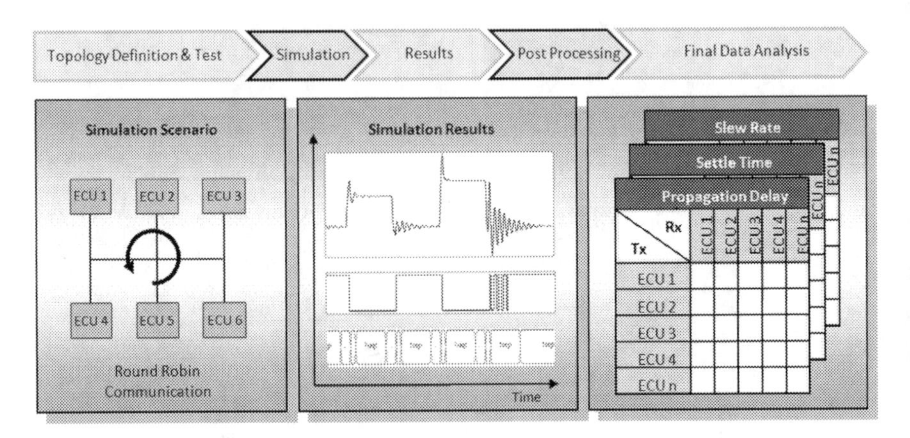

Fig. 2.37 Example of an automated analysis

provides a flexible platform that allows users to customize the output data to the desired format. Saber provides an interface to Excel that can be used to export data and transform it into any sort of desired representation. As mentioned previously, often the signal parameter matrix is applied which shows the relationship between transmitter and receiver. Using the Excel interface, the data can be compiled into a workbook where each worksheet includes a two-dimensional matrix related to a dedicated discrete signal parameter. Standard functionalities of MS-Excel can be used to extract the extreme values from these matrices. Such an automated analysis and post-processing procedure is particularly valuable for frequently used simulation scenarios, since the bulk of the time spent is related to preparation and validation of the resulting data.

2.4 Electromagnetic Compatibility

Technical system design actually is conceived on a distributed architecture concept. When implementing such a distributed system, a powerful data exchange technology for communication between the system components is required. Among others, CAN communication protocol is a powerful means for a highly effective data exchange. As such, CAN is widely used for distributed control communication in automotive and industrial applications.

Data bus systems have high significance regarding system EMC consideration because data bus systems represent one of the physical interfaces of the electronic modules to the wiring harness, thus forming a vastly distributed network structure. As this network is stretched all over the system, the network must be considered as the major EMC coupling media. This implies that through the wiring harness most of the electromagnetic noise is either injected into the system or emitted from the system. This theoretical consideration is proven by practical experience.

These considerations apply to a high extent to automotive electronics as in automotive electronics mostly unshielded bus cabling is applied other than in other application areas such as industrial control or avionics where typically shielded cables are used. These characteristics result in much higher immission and emission of electromagnetic stray fields and transient interfering pulses. These very harsh EMC conditions must be compensated by correspondingly increased constraints for bus interface design. In order to ensure this, automotive manufacturers are required to coordinate any of the EMC activities and design constraints as well as to specify the related qualification criteria. The layout of bus topology, the choice of protection circuitry for the bus interface and the applied transceiver have great impact on the EMC characteristics of the whole system. Therefore, characterization of these semiconductor devices requires EMC measurement already in an early development stage of cars. This task is performed by vehicle manufacturers in various ways. German car manufacturers have agreed upon a common procedure in 2009. This comprises the specification of the basic structure of the bus interface as well as its passive and active components.

2.4.1 EMC Requirements, Specifications and Guidelines

For CAN systems, there is no explicit EMC specification in the form of a national or international standard or guideline. Nevertheless, this is subject to the corresponding legal or manufacturers internal requirements of the related devices or assembly groups, for which these specifications would apply to.

But as a speciality for vehicle electronics applications, car manufacturers specify specific requirements on the basis of which semiconductor devices are to be tested. For this purpose, the technical specification IEC 62228 TS was developed and published. This standard is the basis for EMC release of CAN transceivers for German vehicle manufacturers.

2.4.2 Factors Effecting EMC of CAN Buses

The major factors influencing EMC of CAN buses are bus topology, termination concept, bus lines, bus filter and transceiver. All these areas require an optimal layout with regard to their EMC behaviour in order to meet the higher level requirements for safety-relevant applications as they are, e.g. for automotive electronics.

2.4.2.1 Bus Topologies and Termination Concepts

Passive bus topology is the mostly preferred topology in industrial automation with both ends of the bus lines being terminated by line impedance resistors. Any further bus node is connected to the bus by short stub lines (high-impedance terminated connection) or by the so-called daisy chaining. The advantage of this topology is the avoidance of reflections on the bus lines, which may make the EMC get worse.

A passive bus implies a reduced flexibility in a certain way. That is why the passive star is another preferred bus topology, especially for cars. This topology is characterized by all nodes being connected in a star-shaped form, or even a combination of multiple stars is applied. However, the termination of all segments with low reflection is not feasible. In order to achieve an optimum between flexibility and reflection characteristics, two distributed terminations are applied, which are located at the ends of the two segments with the largest geometric distance. Another termination method is implemented in the star node itself. In this method, the termination of the longer lines as well as the HF decoupling of the lines are applied. All these result in optimized EMC characteristics of the CAN system as a whole.

In general, there is no simple relation between bus topology and related EMC characteristics. However, there is a unique dependency between signal integrity—that is, the characteristics of CAN_H and CAN_L bus signals—and the EMC characteristics of the system. In comparison with an optimal bus layout, a decline of signal quality coming, e.g. from reflections from the line ends or attenuation by passive components results in a reduction of system safety margin, which, in conjunction with external noise, may cause system failure. Furthermore, it can be stated that shortening the length of the bus system normally improves EMC due to a reduced effective antenna surface, and thus less noise is emitted or immitted.

2.4.2.2 Bus Lines

For CAN bus cables, twin lines with or without shielding are used. Nominal differential line impedance shall be 120 Ω. Typically, the value is in the range of 90–130 Ω. Shielded twin lines typically are applied in industrial control applications and avionics. In current cars only UTP cables are used with various twist pitches between 20 and 50 mm.

When using shielded bus lines, immission and emission of noise are widely reduced or even eliminated. The efficiency of shielding depends on the cable characteristics itself as well as on the applied concept of shielding (cable, connector and module box). High shielding effect in the relevant frequency range (f=150 kHz up to a couple of GHz) can only be achieved if the shielding concept is strictly applied to all components.

These twisted pair bus lines do not have any shielding. The advantage of twisted pair cables is that external noise (e.g. HF electromagnetic fields) affects the system as common-mode interference. Therefore, the differential-mode CAN signals are not disturbed. In that respect, the same is true for interference fields emitted by the bus lines. At a high-level symmetry of the bus lines, the HF interference parts compensate each other and thus they are not radiated as common-mode signals.

2.4.2.3 Bus Filter

Besides the applied transceiver, the implementation of the bus filter has great influence on the EMC characteristics of the systems as a whole. According to the EMC requirements of the application, there is a different emphasis on the implementations:

- Industrial control: High priority on protection against burst pulse, surge pulse and ESD.
- Automotive: Focus mostly on protection against HF interference and ESD.

2.4.2.4 Transceiver

CAN transceivers represent the physical interface between the CAN bus system and the bus lines. Therefore, the transceiver itself has great impact on the EMC characteristics of the system as a whole. Especially for in-vehicle applications, the transceivers must meet increased requirements, as in this application only UTP bus lines are applied. Therefore, higher level noise amplitudes may get to the transceiver.

The most important requirements for CAN transceivers are as follows:

- Immunity against common-mode interference with high amplitudes due to the induction of interference fields and transients into the bus lines.
- Immunity against differential-mode interference with relatively low amplitudes, coming from the induction of noise into the non-symmetrical bus lines (or bus segments) or induced by the common-mode–differential-mode conversion caused by non-symmetries in the filter electronics or the transceiver by itself.
- Immunity against common-mode and differential-mode noise in conjunction with ground offset between different bus nodes.
- Sufficient protection against damage caused by ESD.
- High degree of symmetry of the bus signals in order to avoid emission of common-mode noise through the bus lines.

These most important requirements led to the development of an EMC test specification for CAN transceivers. These requirements are mostly relevant for car electronics and they partially represent the constraints for EMC qualification. The test specification is described in more detail subsequently.

2.4.3 EMC Evaluation of CAN Transceivers

2.4.3.1 Ways for EMC Evaluation for Cars Applications

For the evaluation of the EMC characteristics of semiconductors in car applications, there are analysis methods available covering all phases of the development process. Table 2.1 depicts the environmental conditions and the range of dynamics for all three test levels.

Table 2.1 Environmental conditions, dynamics range and potential ICs test verdicts

	Car measurement methods	Component measurement methods	IC measurement methods
Environment	Finalized vehicle	Components, special test board	Special test board
Test verdict	Pretest, release	Pretest, (prerelease), comparison of semiconductors	Pretest, comparison of semiconductors
Dynamics range for comparison	—	Corresponding to transmission function of measurement method	Very high, broadband measurement
Validity range	Corresponding to the particular car	Corresponding to the particular constraints of the component measurement	General
Point of time of measurement in the development process	Only applicable on availability of the particular car	Only applicable on availability of the particular component or always applicable on the application of a dedicated test board	Always

Measurements of the integrated circuit (IC) level can be performed already on early development samples of the semiconductor. They result in general test verdicts and they are characterized by a high dynamics range. As such, they are very well suited for comparison of semiconductors from different semiconductor suppliers or for comparison of different sample versions of one specific kind.

Analysis on module level is advantageous, because it can be performed in an early phase, independent of the vehicle itself. When knowing the correlation between the applied component measurement method and the car measurement method, the car manufacturer can evaluate the semiconductor device correspondingly to the component's requirements. There are some restrictions though, coming from the partially lower bandwidth noise transmission function of the component measurement method and the dependency of the layout of the components, the applied passive components and the ground layout.

When performing measurements in cars, typically there is sufficient EMC environment available and thus a decision can be taken on the release of the transceiver. Unfortunately, this kind of measurement can only be performed when the car is available, which is relatively late in the development process. Furthermore, the information on EMC capabilities can only be derived for the tested type of vehicle. Any statement on a noise immunity range margin, which may be required for the application on another type of car, cannot simply be derived.

A reasonable step-by-step combination of all three levels of measurements results in an effective process for the development of CAN transceivers with sufficient and even excellent EMC characteristics. The comparison of test results of the individual levels for transceivers, which have already undergone this test process, allows the specification of test requirements on IC level, which correlates with the later stage car constraints. In conjunction with the test specification for CAN transceivers, an EMC measurement specification for the EMC evaluation on IC level had been developed and published as the international standard IEC 62228 TS. This test specification specifies measurements on the basis of international standards for semiconductor components and car components, adapted to the corresponding EMC requirements for later application in cars.

2.4.3.2 EMC Evaluation on Semiconductor Level

Test Philosophy

Basic Concept

The basic focus of the EMC test specification of CAN transceivers is exclusively on the evaluation of the hardware functionality of the transceiver in conjunction with the electromagnetic environment. For this purpose, the signals are directly analysed at the CAN transceiver pins without the CAN protocol chip connected. This is applied to stand-alone CAN transceivers and, if applicable, to CAN cells which are integrated in application-specific integrated circuits (ASICs), multi-chip modules or embedded systems.

EMC Test Requirements

Based on the EMC requirements derived from measurements on components and cars, the following EMC characteristics of transceivers are tested:

- Noise immunity against transient noise, coupled through lines
- Noise immunity against radiated noise
- Noise immunity against ESD
- Noise emission of radiated noise

When performing measurements on noise immunity, a separation of tests into destroying tests and tests on malfunction is required. The measurements on noise immunity checking a malfunction are performed on any possible operational mode (normal, standby and sleep), while a—within limits—defined undisturbed functionality of the transceiver is to be specified as evaluation criteria.

The evaluation of the noise emission is performed in the frequency domain. For the optimization of the analysis, there is a complementing measurement in the time domain.

EMC Pin Classes for Semiconductors

The analysis of the relevance of individual parts of the circuitry of semiconductors leads to a classification into the so-called pin classes (Table. 2.2).

Table 2.2 Pin classes for connections of CAN transceivers

Pin classes	Characteristic features	Example for pins at CAN transceivers
Global	Direct connection to peripherals of the component or only simple filtering, respectively	CAN_H, CAN_L, VBAT, Wake
Local	No direct connection to peripherals of the component or very strong decoupling by filtering, respectively	RxD, TxD, Mode, VCC

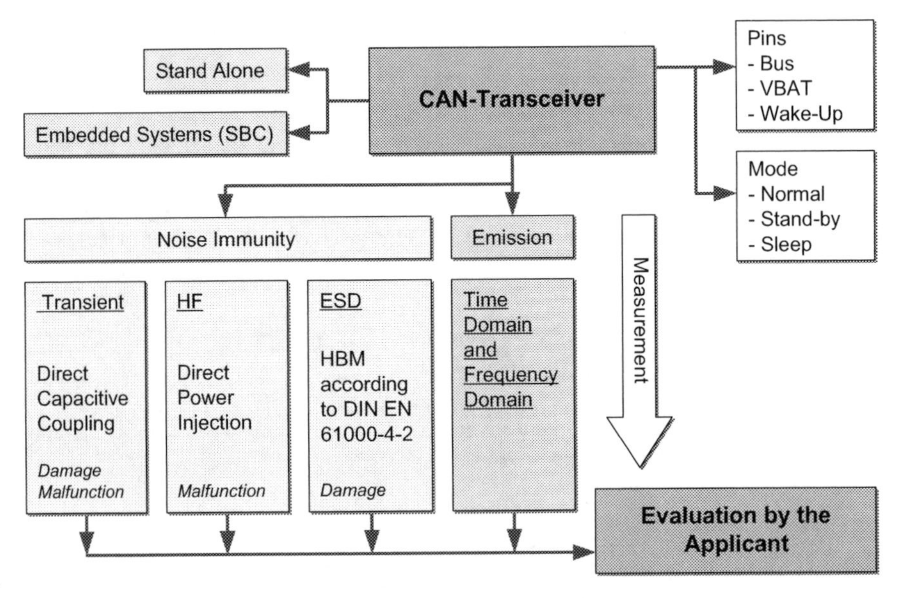

Fig. 2.38 Test philosophy for EMC evaluation of CAN transceivers

Corresponding to this classification into EMC pin classes, only those pins of a CAN transceiver are tested with respect to noise immunity and noise emission, which, as global pins, feature either a direct or only a weakly decoupled connection to peripherals of the component and thus to the connected wiring harness.

EMC Test Specification for CAN Transceivers

Figure 2.38 depicts the relation of the EMC test philosophy for CAN transceivers according to IEC 62228 TS.

This philosophy defines the EMC measurement method for ICs as well as the therefore-required constraints for the measurements according to the above-described scheme. The result of the measurements according to this test specification does not allow deriving of any generalized EMC conformance though. Only when making use of application-specific boundaries, any decision on the applicability of the transceiver can be taken. In 2008, for the first time German car manufacturers jointly had published such a specification of requirements, which, from that point of time on, represents the basis for EMC release.

Measurement of Emissions in the Frequency Domain—1 Ω/150 Ω Method

The measurement process of the emission spectrum (electromagnetic emission (EME) test) of semiconductor devices is described in the standard IEC 61967: "Integrated circuits—Measurement of electromagnetic emissions". This standard is subdivided into a general part I, which contains general definitions for all measurement methods, as well as into five further parts for each of the individual measurement methods, respectively. The 1 Ω/150 Ω method (IEC 61967, Part 4) is based on the direct measurement of lines-based noise of semiconductor devices. This method assumes that any internal noise, which is generated by high-speed current or voltage swings, will proceed over the pins to the electronic circuit board or to the wiring harness, from where the noise is radiated over cable loops or any other antenna-like structure. For the most commonly used body shapes of ICs (with the exception of large microprocessors), the internal parasitic antenna structures, which are able to radiate as a function of frequency, are significantly smaller than those of PCBs. As such, the external radiation is predominant in the application. In this case, the measurement method is applicable for the comprehensive characterization of the noise emission of the semiconductor device. For the analysis of noise emission of the global pins of CAN transceivers, the HF measurement technique of noise voltage is applied, using a voltage probe of 150 Ω. This technique represents one of the two potential measurement methods of the standard.

Measurement of Noise Immunity in the Frequency Domain—Direct Power Injection

The measurement of noise immunity of semiconductor devices against narrowband radiated noise is performed analogously to noise emission described in the standard. IEC 62132 "Integrated circuits—Measurement of Electromagnetic Immunity" is divided into a general part I and four further parts describing the individual measurement methods. In this standard, the test methods also are subdivided into those ones for direct field immission into the IC body and those ones for methods of noise injection over the pins. In the frequency range below f=1 GHz, the direct immission can be neglected for a multitude of semiconductor devices. Applied to CAN transceivers, the direct power injection (DPI) method (IEC 62132-4) emulates those parts of noise, which are generated by field immission into the wiring harness or into antenna-like structures of the PCB of a component, from where the noise is line-borne fed into the semiconductor device. For this purpose, a noise power corresponding to this field immission is directly injected into the global pins of the CAN transceiver.

Noise Immunity against Transients—Direct Capacitive Impulse Coupling

The direct capacitive impulse coupling method is similar to the noise immission method into signal and data lines using a capacitive coupling calliper corresponding to ISO 7637 Part 3 or IEC 61000-4-4, respectively, which is applied for immunity measurement of electronic components. In this case, though the 100-pF distributed capacity of the standardized coupling calliper is substituted by a discrete capacity in the form of a capacitor component in order to perform the noise immunity measurements while using test boards.

Based on this approach, the measurement process is very similar to the DPI method and to the 150-Ω method. Therefore, this approach can be applied in combination with the latter methods for the evaluation of CAN transceivers with respect to radiated and impulse-based noise.

ESD Measurement

For ESD tests, the test directive specifies an analysis of the destruction immunity of CAN transceivers. For this purpose, an ESD is required, using the "human body" model (HBM) corresponding to IEC 61000-4-2 (R=330 Ω, C=150 pF and tan ca. 700 ps) and applied onto the global pins of the CAN transceiver with no power supply. This test is oriented towards the requirements of the so-called packaging and handling tests as known from automotive industries, which requires the ESD with this network to the connector pins of the component.

2.4.3.3 Basic Test Procedure for EMC Evaluation of CAN

Basic Thoughts

The basic approach to an EMC evaluation of CAN transceivers shall result in a test close to application while using IC measurement methods which are oriented on the test constraints and requirements for the higher level measurements of components and vehicles. For this purpose, the transceiver basically is tested by connecting to a minimum circuitry as specified by the semiconductor supplier (e.g. stabilizing capacitors at the supply pins or filter circuit at wake pin).

Another approach though is applied concerning the CAN bus lines circuitry: In this case, the specification of the applicant (car manufacturer) is applied, who only specifies a specific CMC as bus filter. In order to comply with this influence, the following commitment was made:

- Test without bus filter which results in high dynamics measurement for the comparison of different types of transceivers.
- Test with bus filter (CMC for noise suppression) for the adaptation to the real application constraints, if they have been implemented EMC optimal.
- Corresponding to the single-test requirements at first a grouping into two main test parts is done:

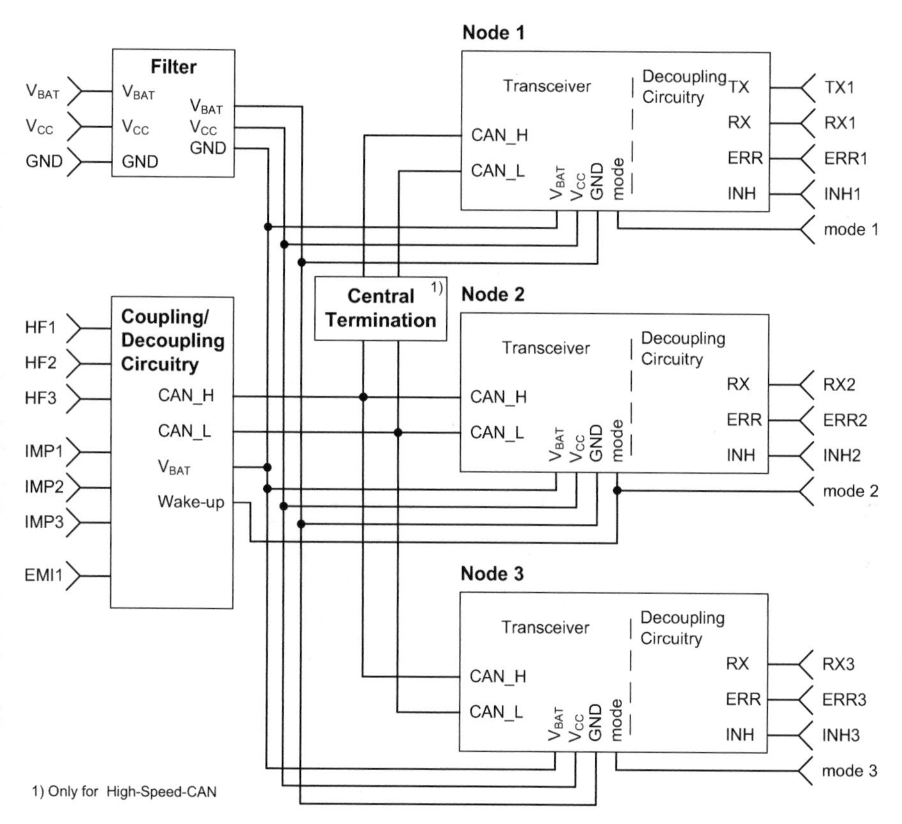

Fig. 2.39 Minimum network for measurement of noise immunity and noise emission of active transceivers

- DPI tests, transients tests and EMI tests on a minimum network consisting of three active transceivers
- ESD tests for the destruction of a transceiver with no power supply

Minimum Network for DPI Tests, Transients Tests and Emission Tests

Figure 2.39 depicts an overview of the minimum network for tests on an active transceiver. The network consists of three CAN nodes. Each of the node circuitries are identical, consisting of the transceiver which is to be evaluated, its minimum circuitry according to the data sheet, the bus filter and a decoupling circuitry which connects the signal for stimulation and control of the functions. The required voltage supplies (VBAT and VCC) are connected to the test network through well-specified filters. Noise injection and noise observation are performed through specific coupling networks, which are compliant to the requirements of the individual IC measurement method (see Table 2.3).

Table 2.3 Coupling circuitry-compliant IC measurement methods

Coupling circuitry	Purpose	Implementation
HF1	Symmetric DPI coupling into CAN	Parallel circuit of two times R=120 Ω+C=4,7 nF
HF2	DPI coupling into VBAT	C=4,7 nF
HF3	DPI coupling into wake	C=4,7 nF
IMP1	Symmetric transients coupling into CAN	Parallel circuit of two times C=1 nF
IMP2	Transients coupling into VBAT	D (standard diode)
IMP3	Transients coupling into wake	C=1 nF
EMI1	Symmetric 150 Ω voltage measurement CAN	Parallel circuit of two times R=120 Ω+C=4,7 nF with matching resistor of R=51 Ω

Table 2.4 Transceiver operational modes, pins for noise injection and evaluation

Operational mode	Noise injection pins	Evaluation pins	Purpose
Normal	CAN, VBAT, Wake	RxD	Evaluation: Communication
		ERR	Evaluation: Error
		INH	Evaluation: Inhibit output
Standby	CAN, VBAT, Wake	RxD	Unintentional-wake-up observation
		INH	Evaluation: Inhibit output
Sleep	CAN, VBAT, Wake	RxD, INH, respectively	Unintentional-wake-up observation

Communication Specifications

Corresponding to the individual test parts, communication will be performed by various sequences of signals which are transmitted by node 1. The transmitted information will be received by all other nodes and monitored through their RxD outputs:

- Test signal 1: Symmetric square wave with a 50 % duty cycle (CAN signal with continuous 0–1 data swing) and a frequency of 250 kHz or a bit rate of 500 kbit/s, respectively.
- Test signal 2: Non-symmetrical square wave with 90 % duty cycle and a frequency of 50 kHz.

For noise immunity tests with communication, only test signal 1 must be applied. Noise emission measurements are to be performed using test signals 1 and 2.

Definitions for Test Modes and Error Criteria for Tests on Malfunction.

Table 2.4 shows the relation between operational modes of the transceivers, pins for noise injection and evaluation as well as the intended test purpose for measurements to be executed for noise immunity checking a resulting malfunction.

Table 2.5 Error criteria for the evaluation of RxD, ERR and INH

Operational mode	Test	Transmit signal	Maximum amplitude deviation			Maximum timing deviation		
			RxD	ERR	INH	RxD	ERR	INH
Normal	DPI, pulse	With	±0.9 V	±0.9 V	−5 V	±200 ns	–	–
Standby	DPI, pulse	Without	±0.9 V	–	−5 V	–	–	–
Sleep	DPI, pulse	Without	±0.9 V	–	+2 V	–	–	–

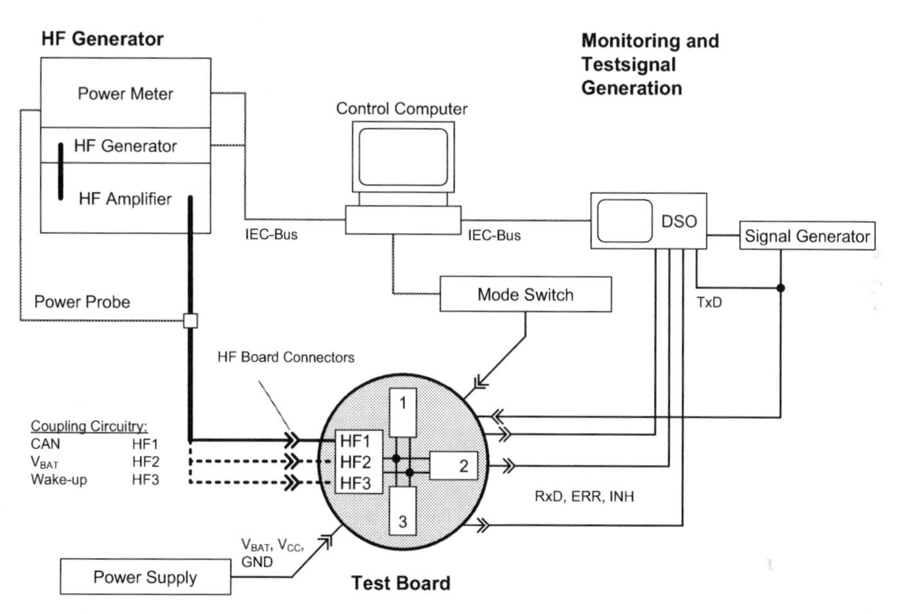

Fig. 2.40 Measurement set-up for DPI tests

The error criteria for the evaluation of pins RxD, ERR and INH is listed in Table 2.5. These error criteria are applied to all three nodes of the minimum network simultaneously, which means that any signal of any one of the transceivers exceeding the maximum specified deviations for amplitude or timing is interpreted as malfunction. No timing criteria are applied for evaluation in standby or sleep mode and basically for error pin and INH pin, because in these cases static signals are to be evaluated. For evaluation of non-disturbed communication, an RxD mask test is applied. The applied mask is derived from a non-disturbed RxD signal of the minimum network.

Test Mock-Up and Execution of DPI Tests, Transients Tests and Emissions Tests

The above-described parts of the minimum network are implemented on a test board, which is applied to all three test groups (DPI, transients and EME).

Figure 2.40 shows the measurement set-up used for DPI tests. The HF noise is induced into the minimum network through one of the DPI coupling circuitries in

Fig. 2.41 Example for an EMC test board (here an SBC with integrated CAN transceiver is shown)

each case. The function of the network under the influence of noise is controlled by devices for monitoring and generation of the test signals. The digital storage oscilloscope (DSO) is used for the mask tests.

As the result of an automatic measurement process, the noise immunity-limiting curve of the individual test function part is determined.

The transient tests are basically performed in the same way. A noise impulse generator, which generates the standard pulses to be tested according to ISO 7637 part 1, is applied to the test mock-up. Additionally, a complementing test for destructibility stability is required.

For emission spectrum analysis of the bus lines, output EMI1 is used and connected to a measurement device according to CISPR 16.

As an example for a physical implementation of the test board for EMC tests, refer to Fig. 2.41.

Test Mock-Up and Execution of ESD tests

The ESD test on destruction stability of the non-powered CAN transceiver requires another test board, because for this test the constraints are different. This test board has specific layout characteristics and it realizes the required ESD minimum circuitry for the CAN transceiver, as well as the ESD points. The test board is connected to the ESD ground reference plane by an adapter (ground block) at low impedance. Figure 2.42 depicts a basic measurement as well as a physical implementation.

For test execution the tip of the ESD simulator with the contacting discharge module according to DIN EN 61000-4-2 is directly put onto the discharge points while increasing the ESD voltage until the transceiver will be destroyed. The destruction

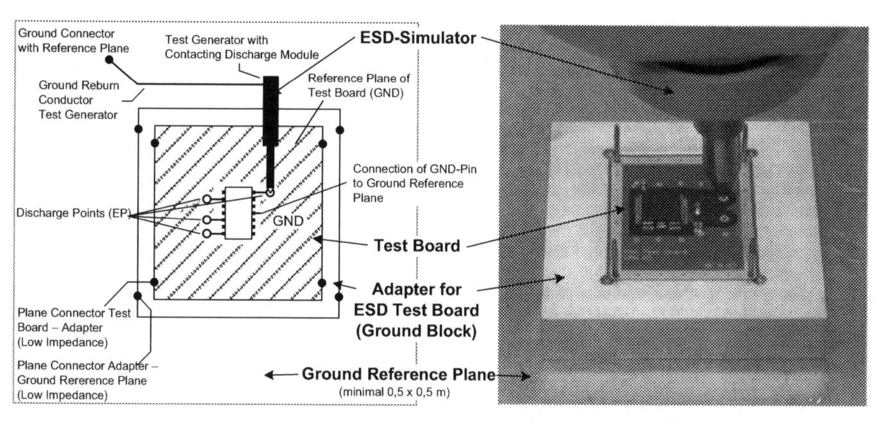

Fig. 2.42 Measurement set-up (on the left) and test implementation (on the right) for ESD tests with ESD test board (here an SBC with integrated CAN transceiver is shown)

or damage, respectively, is detected by the analysis of the input-voltage–current characteristic of the pin under test as well as by checking the functionality of the CAN transceiver.

2.4.3.4 Comparison of EMC Evaluation Results for Measurements of IC, Components and Cars

A direct comparison of EMC measurement results derived from measurements of cars, components and ICs typically is difficult to do. The reasons come from the big differences in measurements of cars. Depending on the location where a control unit is installed as well as depending on the length and the position of the bus lines and ground lines, the EMC load for transceivers varies significantly which, e.g. results in different immission immunities even for one unique type of transceiver. The same is true for radiated noise emission through the bus lines. The interference effect depends on the transceiver characteristics and also on the coupling strength of the bus lines to the car antennas for radio, TV and radio communication.

Concerning both of these important aspects, subsequently some results of systematic analysis are presented, comparing measurement results. They are supposed to help to discover limits for IC measurements or to relate them to a broader database, respectively.

Noise Immunity—Radiated

The listed results of measurements in cars represent the worst case of the measured noise immission immunities of the analysed transceivers (CAN high speed, types A, B and C) implemented in various control units and thus for various installation locations and constraints in an upper middle-class passenger car. The immission

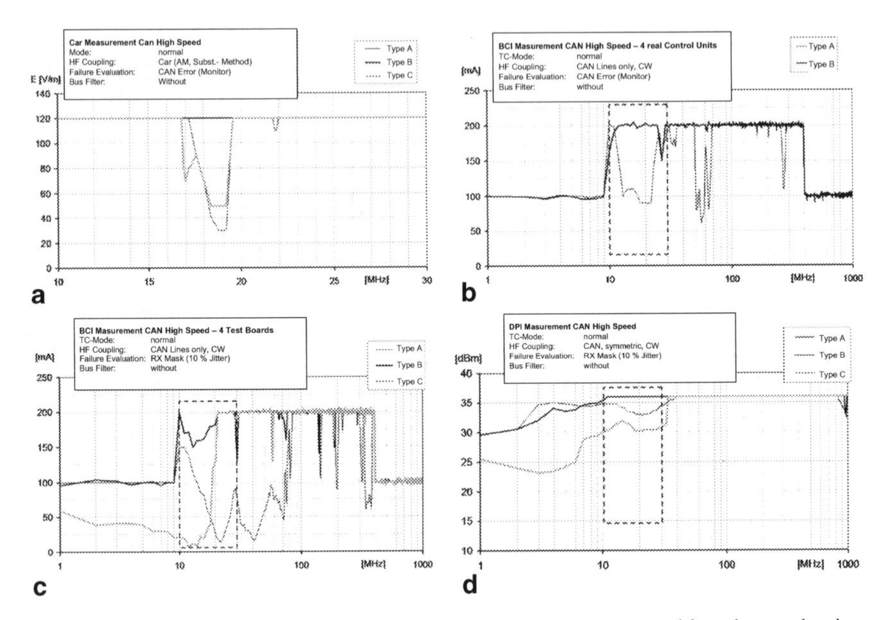

Fig. 2.43 Emission immunity: **a** car measurement, **b** BCI measurement with real control unity, **c** BCI measurement with specific tests boards and **d** DPI measurements

immunities are presented as noise immunity-limiting curves with the test variables electrical field strength (V/m), interference current (mA) for the executed bulk current injection (BCI) measurement as well as interference power (dBm) for DPI measurement. All of the measurement results represent systems without any external protection circuitries (with the exception of car measurements where, in all cases, a split termination including a capacitor was implemented).

Figure 2.43a shows a narrowband interference of types A and C, measured in cars at shortwave frequency range. Narrowband failure is typical; it is caused by wiring harness resonance effects. Type B, however, possesses a sufficient noise immunity at $|E| = 120$ V/m (in the analysed case, this corresponds to a peak-to-peak value of approximately 200 V/m electrical field strength). The failure criteria were the occurrence of CAN error frames.

Figure 2.43b depicts the results of the BCI component measurement of installed control units in cars applying types A and B. In these cases, the same constraints apply as within cars (real CAN communication, failure criteria are the occurrence of CAN error frames). Common-mode interference exclusively is coupled through the CAN lines. The different behaviours of types A and B from measurement in cars can be observed in the same way.

The transition from real control units to specific test boards as well as the application of a communication mask for failure evaluation at RxD pin instead of doing a CAN error frame evaluation is done in the BCI measurements depicted in Fig. 2.43c. The effect of influence of type B is rising, which indicates an increasing jitter, which, however, is not recognizable in the same degree at the sample point in real CAN communication.

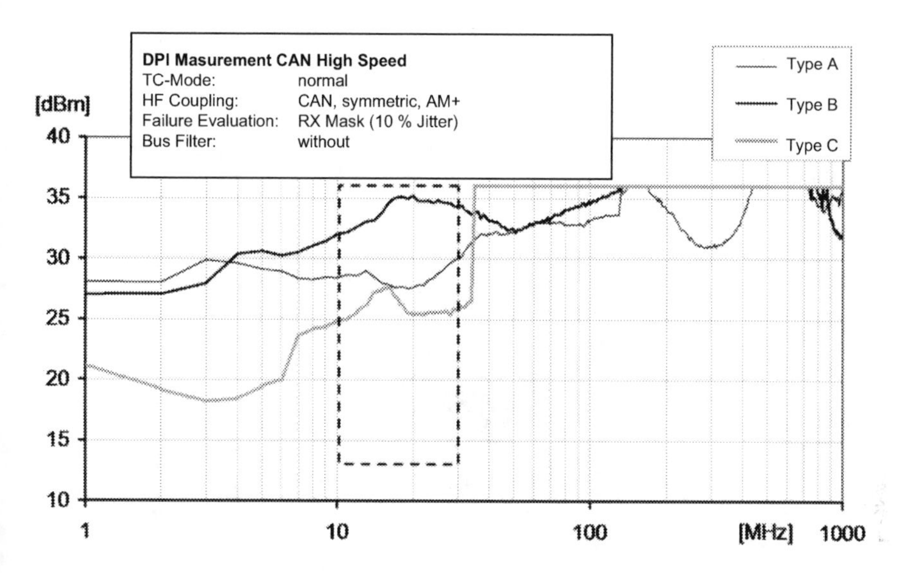

Fig. 2.44 Emission immunity, DPI measurement with increased requirements

Figure 2.43d shows the results of the DPI measurements for all three transceiver types. Type B providing sufficient noise immunity in measurements in cars achieves an immunity power of 36 dBm (4 W) in the critical frequency range (10 MHz<f<30 MHz). Both other types, however, only provide up to 3 dB lower values, which correspond to a halving of the immunity power.

Typical requirements for noise immunity of components are in the range of I=100 mA (chain line in Fig. 2.43 a and b) for BCI measurements at current coupling into the overall wiring harness (including power supply lines). According to these requirements, both of the BCI measurements show failures in the frequency range above f=30 MHz. These failures could not be observed when performing measurements in cars. When comparing all three types at DPI measurement with increased requirements (Fig. 2.44—test with upward modulation), immunity weaknesses of type A in the upper frequency range can be stated clearly. Furthermore, with respect to the considered frequency range, the difference in the immunity degree between type B, providing sufficient noise immunity, and both of the other types A and B gets bigger up to values greater than 6 dB.

Noise Emission—radiated

The approach to transceiver qualification is based on the application of basic test functions emulating real communication.

Figure 2.45 compares the emission measurements (150-Ω method) for CAN transceivers with a given TX test function (square wave signal with 50 % duty cycle and 500 kbit/s) and a real CAN message (500 kbit/s). Obviously, with respect to the

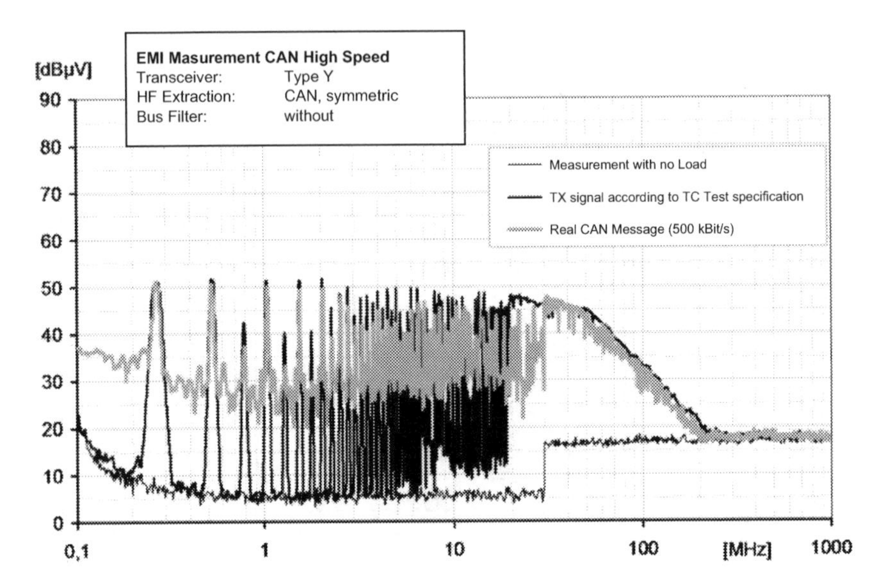

Fig. 2.45 Emission measurement (150-Ω method) at CAN transceivers, comparison of Tx test function and real CAN message

maximum values, the easy-to-generate TX test function is well suited to emulate the emission of the respective transceiver under real communication conditions.

The derivation of limits for the qualification of transceivers with respect to radiated emissions is comparably more difficult to do than for the noise immunity problem. This is demonstrated by the measurements shown in Fig. 2.46. These measurements compare the noise emission of a CAN network into a car-integrated amplitude modulation (AM) receiver in a passenger car for two different installations of a CAN network within the car. The set-up 1 represents an unfavourable layout with a short distance between a small CAN network and the integrated antenna. In set-up 2, a big CAN network was implemented, which, however, has a larger special distance to the antenna.

The comparison of both of the set-ups demonstrates the known problematic concerning the special separation of noise source and noise drain with a dynamic range of partially more than 30 dB. The process of specification of the limits for the emission from transceivers should be similar to the one applied for components. In this case, different classes for interference suppression are defined, which recognize the coupling to internal antenna systems.

Figure 2.47 compares cars and IC measurements for two CAN transceivers.

For measurements in cars, the unfavourable set-up 1 was chosen. Comparing the differences between the two transceivers for the particular partial measurement shows a very good resemblance between measurement in cars and on ICs.

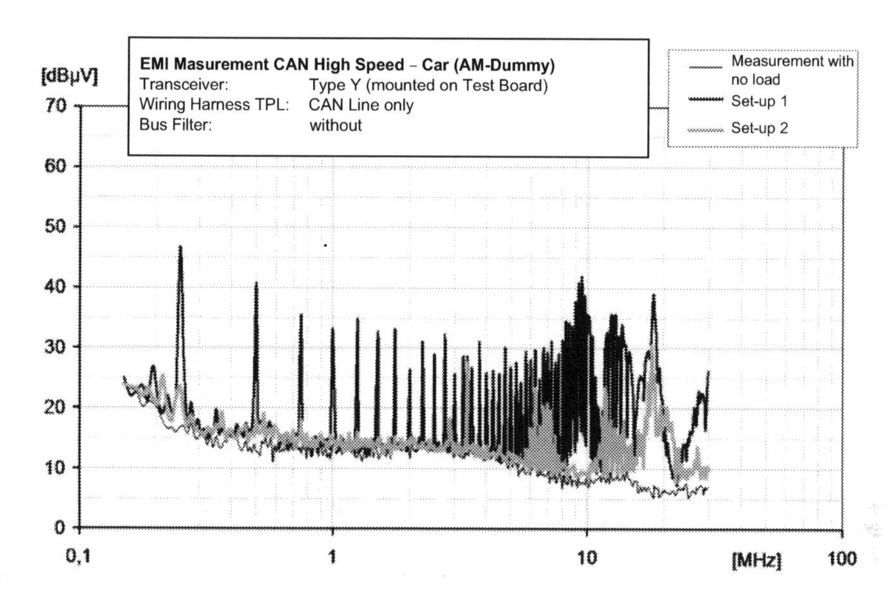

Fig. 2.46 Emission measurements in cars (integrated AM-Antenna with a measurement probe) with different layout positions of a CAN network

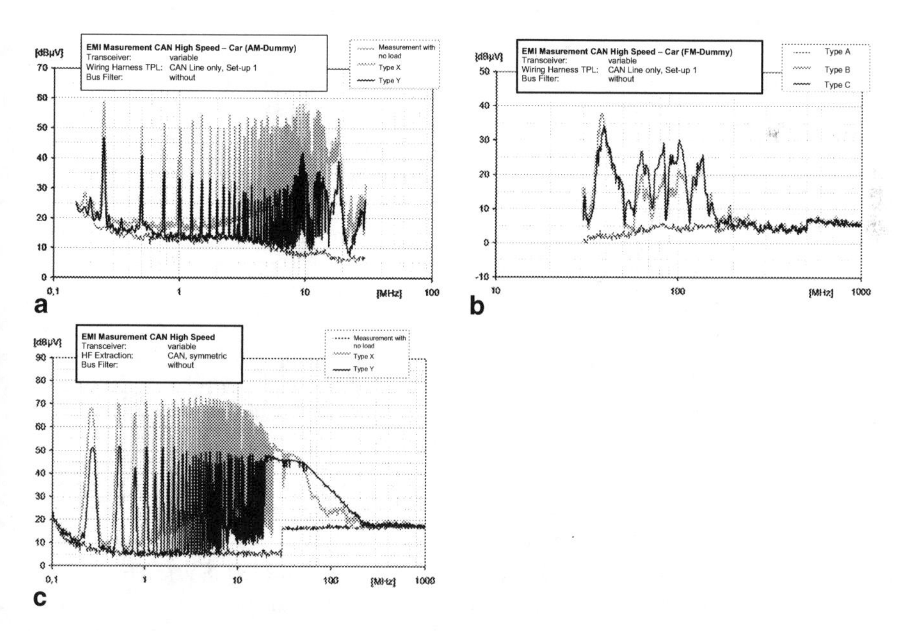

Fig. 2.47 Emission of CAN transceivers: **a** measurements in cars, AM range, **b** measurements in cars, FM range up to 1 GHz and **c** 150-Ω method according to transceiver test specification

2.5 Partial Networking

During the last years, the number of comfort function in a car is dramatically increased. New functions like seat position and heating, window lift and so on were added. These additional functions increase not only the comfort but also the current consumption. Every function consumes typically 50 mA (without actuators) independent of the fact that this function is used or not. During a start-up of a car, all ECUs will be ramped up in parallel. This ends in a high current peak (up to 25 A without actuators) but this is not necessary. Not all ECUs are required for the start of the engine. A distributed ramp-up of the ECUs will be much more. A 100-ms delay has no impact on the comfort but helps to reduce the load on the battery. To reduce the current consumption and the CO2 emission, it must be possible to deactivate the functions that are not needed during active CAN communication. An impact on the comfort is not allowed. Two different concepts are available and established. One concept based on a microcontroller modification is called pretended networking and the other concept is based on a transceiver solution.

The microcontroller uses the internal CAN communication controller to monitor the bus communication like established. However, all other functions in the microcontroller are deactivated. Through the detection of the CAN communication controller, an identifier, the rest of the micro will be awaked and the message can be analysed. For this solution, a CAN transceiver with optimized current consumption set into receive-only mode is recommended.

This technology requires a new generation of transceivers that can monitor the CAN bus traffic to detect the well defined wake messages. To reduce the current consumption of an ECU, voltage regulator and microcontroller are switched off. The current consumption in this solution is much lower than in the microcontroller solution. However, with this solution only one ID for wake-up is possible.

2.5.1 Motivation

In the last years, the requirements for eco-friendliness, drive engineering and CO_2 emissions have been drastically changed. In future, the vehicle manufacturer has to pay penalties for its vehicles which exceed the maximum allowed carbon dioxide emissions. Driven by ecological sensibility of the consumer and the limited reserves of conventional used fuel resources, the development of alternative drive engineering and the optimization of well-established drive concepts increased significantly. However, a second trend says that the vehicle manufacturers have to face an increase in applications and comfort which leads to an increase in electrification inside the car. An increment of electronic functions inside a car, and with it an increment of ECUs per car, leads to a higher power consumption which results also in a higher fuel consumption. An increase in functionality usually implies an increase ofelectric power supply which usually must be compensated with power from fuel. Therefore, optimizing the electrical power consumption is a necessary step to meet the expectation of both trends, increasing efficiency and increasing functionality.

With the help of a calculation example using conversion formulae, the statement and the potential for efficiency enhancement are confirmed. Using a battery supply of 14 V and an ECU with an idle-current of approximately 100 mA implies 1.4 W per ECU in idle state. The conversion formula says that 100 W electrical load are current to 0.1 l of fuel per 100 km, where 1 l fuel raises the emission of 2.5 kg of carbon dioxide. Thus, an additional ECU would raise 3.5 g CO_2 emissions per 100 km even when it is in idle state and is waiting for a functioning request and its actors are not active. The statistical department of the German government gives an average of 11,500 driven km per car for the year 2008. As a consequence of this, an additional ECU would result in an additional emission of about 400 g of CO_2 per year.

Regarding electric vehicles, the motivation lies in a more efficient power consumption, which is essential to bring up today's kilometre range of electric vehicles.

How is it possible to face these two none-harmonizing trends?

The fact is that many ECUs are responsible for functions that are not used very frequently. A back-view camera, for example, is not in use when the car is driving on the highway for hours. Other examples are seat ECUs, door ECUs or sun-roof ECUs. If these ECUs are connected to the CAN bus, the bus communication holds these ECUs in permanent awake state even though the communication is not relevant for these ECUs. A solution must be found to set unused ECUs to a sleep mode and to reactivate them only in case of use. This working principle is named as partial networking.

Figure 2.48 shows the potential of a CAN bus with partial networking. None-used functions result in sleeping of the corresponding ECU. In case of a functional request, the corresponding ECU is woken up. This can happen without any notice of the driver. The real-time requirements are fulfilled in such a way that the driver may not notice that a function is completely disabled due to its sleeping ECU.

Back to the above example calculating the impact of electrical power consumption on CO_2 emissions and considering an average of 12 ECUs which can be used for partial networking this would bring a potential of 17 W saving electrical power. This would result in approximately 0.42 g CO_2 per km and in 5 g CO_2 per year. Even high-saving potentials of more than a gram for partial networking are presented in the past. Considering penalties of up to 95 euro for exceeding emission boundaries forced by the government, a partial networking ECU can result in up to 3.33 euro savings. Even 17 W brings some effect on the range of electric vehicles.

In the following, several variants of setting particular ECUs to sleep and reawake them are described.

2.5.2 Realization Methods

2.5.2.1 Variant 1: Disconnecting the Supply Voltage

By disconnecting the supply voltage, ECUs can be turned off completely. This kind of separated working availability of ECUs already exists in the car. ECUs, which

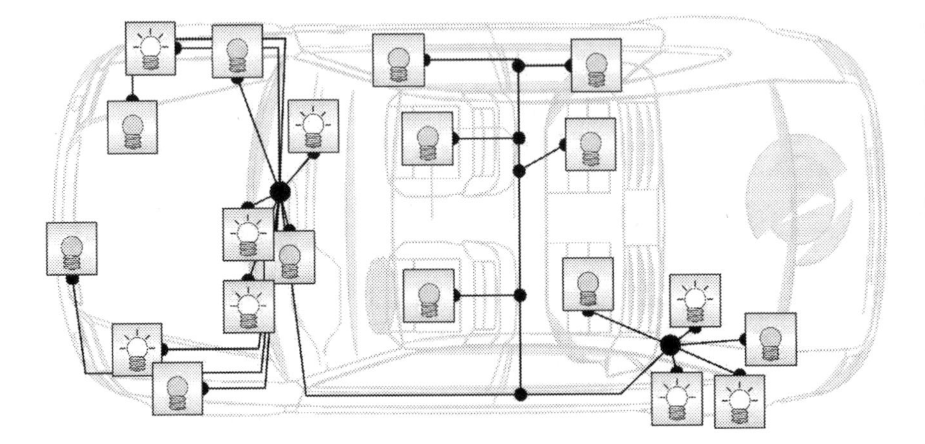

Fig. 2.48 Partial networking—ECUs can be aimed deactivated and aimed woken up

are able to operate every time, are connected directly to the battery (Clamp 30). However, ECUs, which are only able to operate after the start of the engine, are connected to the battery voltage via the ignition key signal (Clamp 15). With this method, the network is split into only two parts. Further parts can be realized by, e.g. relays circuits. The disconnection of the supply lines leads to an additional effort of cabling or to a rigorous splitting in Clamp 15 and Clamp 30 ECUs which does not meet the requirements of a dynamic wake-up behaviour.

2.5.2.2 Variant 2: Separate Wake Line

Through the integration of a wake line, ECUs can be woken up as well. It must be distinguished between two different ways of signalling on the wake-up line. In case of a signalling with voltage levels indicating between sleeping and awake, the same disadvantages (global wake-up, not selective) of the above-described variant 1 are given. The case of a pulse sequence indicating which ECU connected to the wake line shall be woken up brings the advantage of a selective wake-up method but implies high implementation costs. A simple protocol must be implemented to control the wake conditions of a complex protocol.

2.5.2.3 Variant 3: Bus Levels

Additional cabling effort, lack of flexibility and high implementation costs of the above-described variants lead to the conception that a solution is needed inside the existing system. Considering the open system interconnection (OSI) layer 1, the physical layer, special interpreted bus levels which are not affecting the already used bus level definitions may be a solution. One concept of using special bus levels is described in the SAE J2411 standard (Single-Wire CAN). High voltage levels can lead to a wake-up. The disadvantage of this concept is again the low flexibility because such pulses are seen globally as a wake-up event. A selective wake-up of particular ECUs is not possible. Another disadvantage is that due to the higher voltage levels, more intense EMC emissions are possible which result in worse EMC characteristics.

2.5.2.4 Variant 4: Bus Messages

Considering the OSI layer 2, the data link layer, special bit patterns can be used to decode wake-up events. Again, it must be distinguished between two different scenarios: On the one hand, the microcontroller can overtake the decoding and the decision if a wake-up sequence occurred. On the other hand, the transceiver can be added with the functionalities to be able to decode wake-up sequences.

If bit sequences are used like the content of two data bytes of a frame, the implementation effort inside the transceiver would be relatively low, but the disadvantage would be that due to application-dependent unknown data occurrence unwanted wake-ups can occur. To avoid this disadvantage the CAN ID should be considered as well. Thus, whole CAN messages should be considered to decode particular wake-up events for particular ECUs.

Through the consideration of whole CAN messages, the required flexibility is given and the risk of unwanted wake-ups is avoided. For a solution inside a transceiver, this implies that the transceiver functionality must be enhanced with decoding functions regarding the ISO 11898 as well as configuration functions (with ID and data content) to configure the wake-up frame (WUF) of the transceiver. Difficulties exist in the internal clock generation for sampling the CAN bus because external clock generation components would need additional transceiver pins (backward compatibility) and are expensive.

Actually, the solution of enhancing the transceiver functions is favoured. Because the previously mentioned necessary requirements for partial networking with selective wake-up can be fulfilled with the transceiver haven CAN frame decoding functionalities, this variant is described in more detail in the following section.

2.5.3 Partial Networking (Infineon)

In a normal CAN network, all CAN nodes are permanently active when the CAN communication is running, independent of the fact that the ECU is used or not. However, many applications are not used all the time, and these ECUs can be switched off to reduce the power consumption. A solution for such a realization must fulfil the following criteria:

- No negative impact on the physical bus (no disturbance)
- Can be awake with a dedicated CAN frame
- Low current consumption

In addition, this solution must comply with the following three operational modes of a vehicle and its networked system: the normal CAN communication mode, the start up phase of the system and the vehicle, and finally the parking mode of the vehicle with the silent system.

2.5.3.1 Normal Communication Mode

In normal CAN communication, ECUs that are not needed can be set into a special sleep mode. All other CAN nodes can communicate and will not be disturbed by the deactivated ECUs. With a dedicated WUF, one or more ECUs can be woken up with a small time delay.

2.5.3.2 Start-Up Phase

During the start of CAN communication, in a CAN network, all ECUs ramp up and, together, they will consume a lot of current. This is not necessary and with the new approach, only the needed ECUs should be ramped up. All other nodes change from sleep mode to a bus observation mode. After the successful ramp-up, the other nodes can be added one by one into the communication, if necessary.

2.5.3.3 Parking Cars

If you park a car, a very low current consumption is required to unload the battery. However, if, for example, the radio is on, all CAN ECUs located on this CAN bus will stay active and consume a lot of current. With this new solution, only the necessary ECUs are active (for example, wheel to control the radio and the radio itself) and all other nodes are sleeping or shut off. This reduces the current consumption dramatically.

2.5.3.4 Partial Networking

In the partial networking approach, the WUF detection unit is implemented in the high-speed CAN transceiver. This new unit contains:

- A high precision oscillator
- A CAN message decoding unit
- An error-handling management
- A WUF configuration

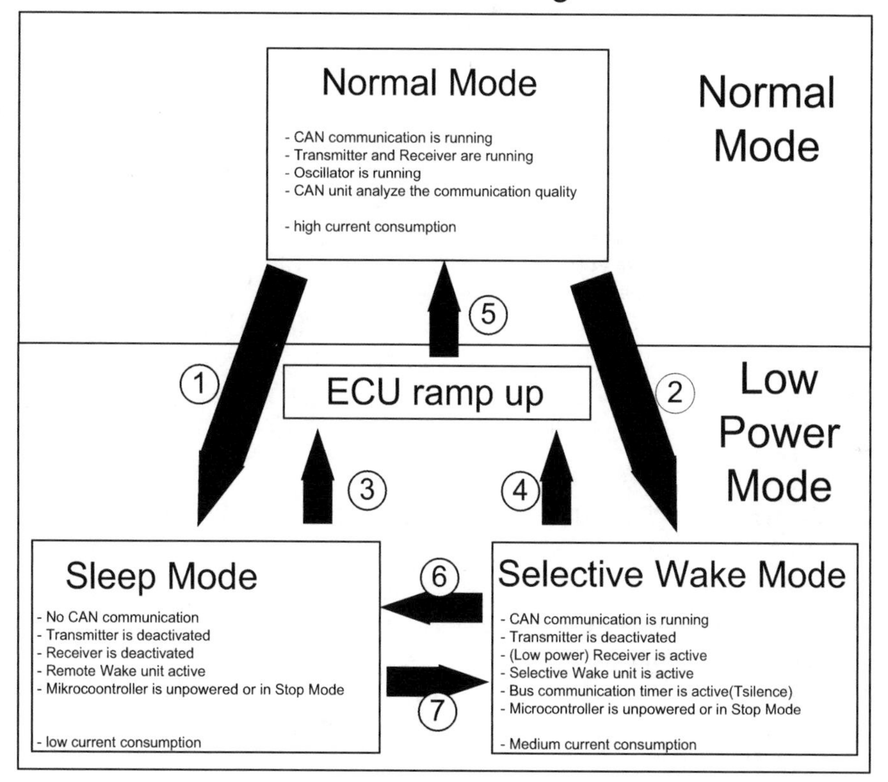

Fig. 2.49 ECU modes for partial network applications

- A compare unit

During selective wake mode, this unit is active and monitors the CAN communication like a watchdog. If a dedicated CAN frame is observed, the transceiver wakes up the ECU. These kinds of transceivers have now two modes in the so-called low-power mode:

- Sleep mode
- Selective wake mode

In sleep mode, the current consumption is reduced to a minimum and all functions in the ECU are disabled. Every message on the bus wakes up the transceiver and the ECU.

In selective wake mode, the current consumption is also low, but the WUF detection unit is active and monitors the bus. All other functions are disabled. With the dedicated WUF, the transceiver and the ECU will be woken up.

In Fig. 2.49, the ECU modes with this new implementation are demonstrated.

The mode transitions 1, 3 and 5 are well known from the ISO 11898-5 transceivers. New are the transitions 2, 4, 6 and 7.

For transition 2 (mode change into low-power mode with selective wake function) the transceiver must be checked, if the oscillator is synch, and the CAN frames were detected from the transceiver correctly. Before the transceiver can be set into low-power mode, the dedicated WUF must be sent to the transceiver. After this operation proceeds successfully, the transceiver can be set into this low-power mode. If the transceiver has detected a WUF, the transceiver changes into standby mode and ramps up the ECU (transition 4). If there is no communication on the bus in low-power mode with selective wake, the transceiver will change into sleep mode (transition 6s) and returns after the communication on the bus starts again (transition 7). The advantage from this approach is a very low current consumption in low-power mode. The disadvantage is the fact that only one dedicated WUF wakes up the ECU. In addition, the long ramp-up time is a disadvantage of this system. A first implementation from Infineon can be found on the TLE9267QX.

2.5.3.5 Trx_Standby Mode with Activated Selective Wake-Up Function

In this mode:

- RxD is decoupled from the receiver and RxD is set to logical 1.
- Transmitter is deactivated.
- VBIAS (voltage source for VCC/2) is active to improve the signal integrity.
- INH is logical "1" (to activate the voltage regulator).
- The WUF-detection unit is active.
- Tsilence timer is active.

The transceiver enters this mode after a valid WUF is detected.

2.5.3.6 Trx_Sleep Mode with Activated Selective Wake-Up Function

In this mode:

- RxD is decoupled from the receiver (RxD is logical "1").
- Transmitter is deactivated.
- VBIAS (voltage source for VCC/2) is active to stabilize the signal integrity.
- INH is logical "0" (to switch off the voltage regulator and the rest of the ECU).
- WUF-detection unit is active.
- Tsilence timer is active.

In this mode, the microcontroller is unpowered or in stop mode. After the detection of a WUF, the transceiver changes into the Trx_Standby mode with selective wake-up function and sets the INH pin to logical "1". With "1" on INH, the voltage regulator is activated and powers the microcontroller. After ramp-up of the microcontroller, the software is able to set the transceiver in normal mode. Through serial peripheral interface (SPI) of the transceiver, the wake-up source can be read out.

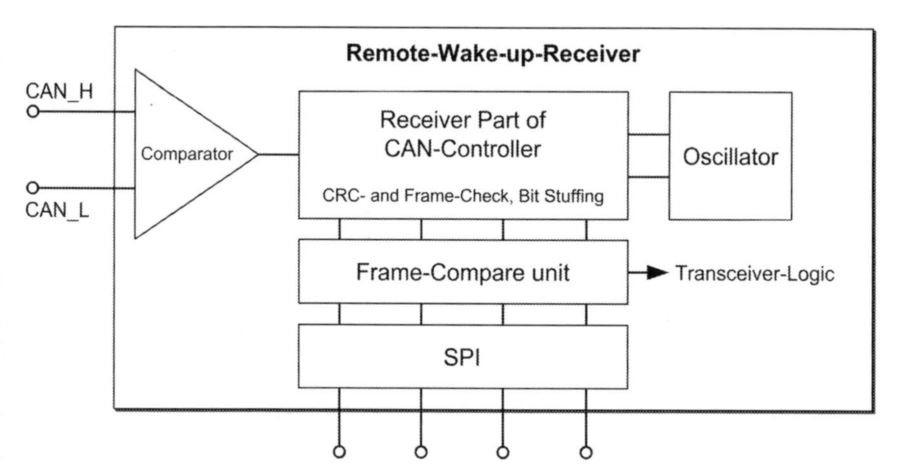

Fig. 2.50 Block diagram of the new selective wake function block

2.5.3.7 CAN Communication Timer (Tsilence Timer)

In Trx_Sleep mode and Trx_Standby mode, the transceiver monitors the bus permanently on bus traffic. If there is no communication for 1 s on the bus, the transceiver changes automatically into the normal sleep mode to reduce the power consumption. After restart of the communication, the transceiver awakes and changes into the Trx_Sleep mode to monitor the bus communication again. To be synchronized, five transceivers need up to five frames. After these five frames, a WUF will be detected reliable.

2.5.3.8 Wake-Up-Frame Detection Unit

The wake-up-detection unit monitors the bus communication like a watchdog. This block (see Fig. 2.50) contains the following:

- Low current consumption receiver
- Receiver part of the CAN communication controller according to ISO 11898-1
- High -precision oscillator (1 % tolerance)
- Comparison unit

Before a microcontroller sets a transceiver in the new PN_Trx mode, he/she hands over the WUF via SPI. The current consumption of the detection unit is reduced to a minimum. However, the requirements for oscillator tolerance, EMC performance and ESD robustness are as high as for the standard CAN communication. Before the transceiver changes into PN_Trx_Sleep mode, the following checks will be done on the WUF:

- Cyclic redundancy check (CRC)
- Frame check
- Bit stuffing

If the WUF-detection unit is not able to detect the CAN frames correctly, i.e. if more than 32 frames were not correctly detected, the new error counter-sets the transceiver into PN_Trx_Standby mode.

2.5.4 Transceivers for CAN Partial Networking (ELMOS)

The partial network operation presented here is an integral part of a large number of possible solutions the automotive industry pursues in order to reach the overall target for an average CO_2 emission of 120 g/km in 2015 as defined by the European Union (EU).

The most substantial contributions to CO_2 reduction are expected from measures focusing on engine and powertrain as well as from modifications in design and construction. Examples for engine-related measures are, among others, the improvement of conventional combustion engines, the needs-tailored control of ancillary units and the increased use of hybrid and electric drives. With respect to powertrain, the use of modern transmissions with six or seven gears is among the measures that can help realize the reduction of emissions. Moreover, CO_2 reduction is possible by changes in construction and chassis through the use of lightweight structures and aerodynamic improvements. All these approaches have a direct effect on CO_2 emission as recorded in the New European Driving Cycle (NEUDC). In contrast, the so-called complementary measures such as a gear-changing timing indicator, an efficient air conditioning, tyre pressure monitoring systems and smooth-running tyres are not directly taken into account by test cycle. Although they are credited with up to 10 g CO_2/km extra, they do not fully cover the reduction potential of efficient technologies that are beyond the test cycle. Further measures with an effect on the vehicle, the so-called eco-innovations, are therefore additionally added towards the target fulfilment of the manufacturer, on the amount of their minimum contribution yet with no more than 7 g CO_2/km.

In order to raise the potential of the above-mentioned measures, innovations in automotive electronics are necessary to a large extent. A case in point, it has been presented in [HUDI09] that operating currents can be reduced by up to 5.7 A with an energy-optimized E/E-architecture, equivalent to the reduction of CO_2 emission by 1.7 g/km in the real customer cycle. With an emission reduction potential of approximately 0.04 g CO_2/km per control unit and with more than ten control units per vehicle suitable for partial networking, this network operation mode as introduced in the previous chapter represents an eco-innovation with a high potential in this context.

2.5.4.1 Requirements for High-Speed CAN Partial Networking

In view of the intended levy of 95 € per gram of CO_2 beyond the limit [EC09], it is understandable that eco-innovations such as CAN partial networking are currently in the focus of vehicle manufacturers. Within the framework of a working

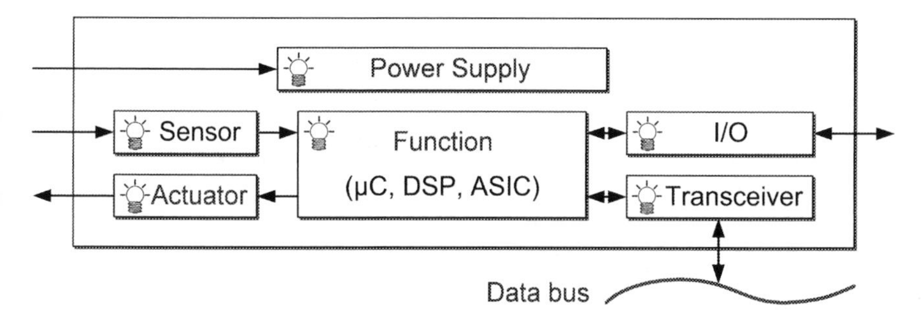

Fig. 2.51 Generic electronic control unit in active mode

group initiated by the Verband der Automobilindustrie, VDA), an ideas competition was launched by the German automakers, calling on semiconductor vendors to present their solutions for the support of this kind of partial network operation by a transceiver. The results were pooled in a requirement specification document [OEM11]. From the very start, however, this VDA working group had the stated goal to establish an international standard rather than a proprietary solution. For this reason, experts from automakers and semiconductor vendors worked together under the roof of the so-called selective wake-up interoperable transceiver in CAN high-speed (SWITCH) group in order to prepare the standard proposal ISO/NP 11898-6 "Road vehicles—Controller area network (CAN)—Part 6: High-speed medium access unit with selective wake-up functionality". This proposal has been forwarded to the ISO for decision and is currently in the stage "new project approved".

The essential outcome of the standardization efforts of the VDA working group and the SWITCH group so far is that partial network operation will be realized by selective wake-up of ECUs in partial network mode with the help of individual WUFs. The WUF is a valid CAN data frame in accordance with ISO 11898-1.

Figure 2.51 depicts a generic ECU in active mode. A typical current consumption of such a control unit can be several 100 mA.

The challenge is to minimize the current consumption of such a control unit while providing for selective wake-up capability via the data bus at the same time. One reasonable approach involves the implementation of the necessary additional functions for partial network operation in the transceiver as this device is the link for the connection between the control unit and the vehicle's data bus. If the transceiver is supplied directly by the battery in partial network mode, as is common practice today in sleep mode, all other components of the control unit can be deactivated. Therefore, this approach is distinguished by higher energy saving potential when compared to the implementation of partial network operation, for example, in the microcontroller. The control unit in partial network mode is depicted in Fig. 2.52, indicating inactive functional blocks in grey.

The transceiver now "listens" to the data bus and analyses the bus traffic. Once the WUF, the individual message for the respective control unit, is detected on the

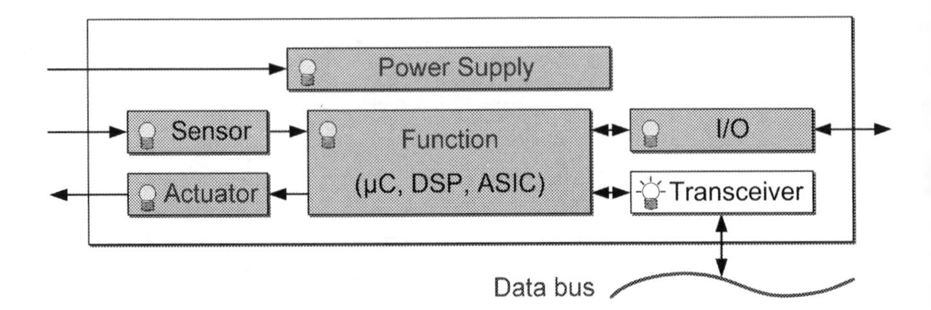

Fig. 2.52 Generic electronic control unit in partial network mode

data bus, the transceiver activates the power supply via the INH pin and the entire control unit is transferred to active mode again. In order to implement the behaviour thus described, a transceiver for CAN partial networking must meet (at least) the following requirements:

- Option for configuration to allow storing of the individual WUF in the transceiver
- Bit-accurate analysis of the bus traffic to make the detection of the WUF possible
- Signalling the status change from partial network to active mode for the entire control unit if the WUF detected in the data stream corresponds to the one that is stored in the transceiver

2.5.4.2 Implementation of High-Speed CAN Partial Networking

The requirements for transceivers supporting CAN partial networking as described above call for new functional blocks in the monolithically integrated circuitry of the transceiver device.

2.5.4.3 Clocking

The transceiver must be capable of receiving CAN messages sent on the data bus in partial network mode. For a bit-accurate analysis of the CAN message, it must be assured that the received physical analogue signal is sampled in such a way that the bit-timing requirements defined by the CAN protocol are met. The transceiver for CAN partial networking must feature its own on-chip clocking with a sufficiently low oscillator tolerance (see Sect. 2.2.2). Clocking, provided by an external quartz component, is ruled out for cost reasons.

2.5.4.4 Analysis of the CAN Message

For an analysis of the received CAN message, the transceiver digital logic must be enhanced such that complete CAN data frames can be decoded, consisting of an 11-bit (standard) or 29-bit (extended) identifier and up to 64 bits of reference data.

2.5.4.5 SPI-Compatible Interface

An SPI-compatible interface allows data exchange between microcontroller and transceiver for configuration and mode control. The interface is used for:

- The storing and readout of the individual WUF
- The configuration of parameters, e.g. for bit-timing (Sect. 2.2.2)
- The control of the transceiver's mode or mode changes
- The readout of diagnostic registers

Moreover, it has to be assured that high-speed CAN transceivers for partial network operation are downward compatible with conventional CAN transceivers according to ISO 11898-5 so that both transceiver types can be operated together in one network (interoperability).

2.5.4.6 Implementation in the Device E520.13 by ELMOS Semiconductor AG

The essential challenge to the realization of a high-speed CAN transceiver for partial networking is to meet the bit-timing requirements of the CAN protocol in partial network mode, with only an on-chip oscillator available for clocking in the transceiver. In order to solve this problem, potential clocking concepts were theoretically analysed at first; then concept verification was carried out together with a key OEM customer by means of a hardware demonstrator. After positive results had been achieved, the next step taken was the design of a monolithically integrated solution for a partial network high-speed CAN transceiver meeting the requirements and the standard proposal ISO/NP 11898-6.

The device E520.13 is a high-speed CAN transceiver according to ISO 11898-5 that additionally realizes selective wake-up for partial networking according to ISO/NP 11898-6. It features remote wake-up through an individually configurable WUF for supporting partial networking, in addition to the common wake-up sources "local wake-up via WAKE pin" and "remote wake-up according to ISO 11898-5 (pattern)". In partial network mode, the current consumption of the device is below 500 μA. A typical application circuit is depicted in Fig. 2.53.

Finally, results achieved with component E520.13 in a real network will be presented. The bus topology used for this purpose is a passive star with seven bus participants and a total wire length of 40 m. The data rate is 500 kbaud. The

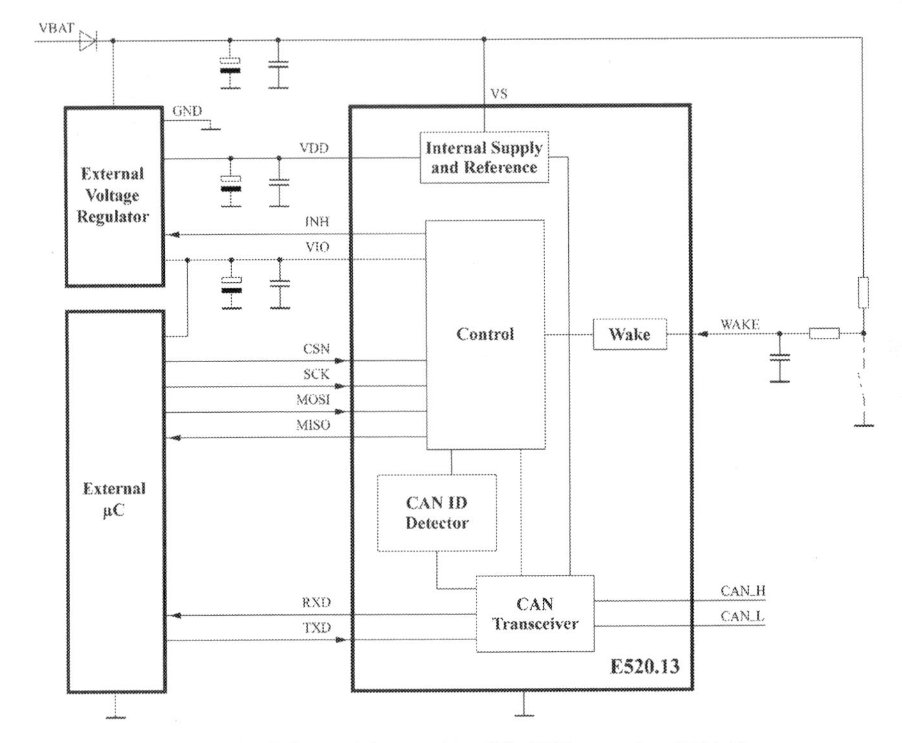

Fig. 2.53 Application circuit for partial networking HS-CAN transceiver E520.13

length of the CAN identifier is 11 bits and the length of the CAN data field comes to 64 bits.

Figure 2.54 depicts the selective wake-up of a control unit equipped with the device E520.13 by means of a WUF. At the start of the measurement cycle, the control unit is in partial network mode. Shown are the respective signal sequences at the two bus pins CAN high (CAN_H) and CAN low (CAN_L) and at the pin INH through which the wake-up call is signalled to the controller's internal power supply. The upper subfigure a) shows a detail of the bus traffic, where several hundred data frames with pseudo-arbitrary content are transmitted. The control unit in partial network mode "listens" to the bus communication and is capable of receiving, decoding and analysing the data frames in a reliable and robust way. The control unit remains in partial network mode. Only when the controller-specific WUF, consisting here, e.g. of the CAN identifier 0xFF and the CAN data field 0×00 00 00 00 00 00 00 FF, is transmitted on the data bus, the wake-up call is initiated by changing from low to high at the INH pin to get the controller out of partial network mode. The data frame setting off the wake-up event is shown enlarged in the lower subfigure (Fig. 2.54b). A very short delay in the range of a few tens of microseconds between WUF and state-change at the INH pin can be observed.

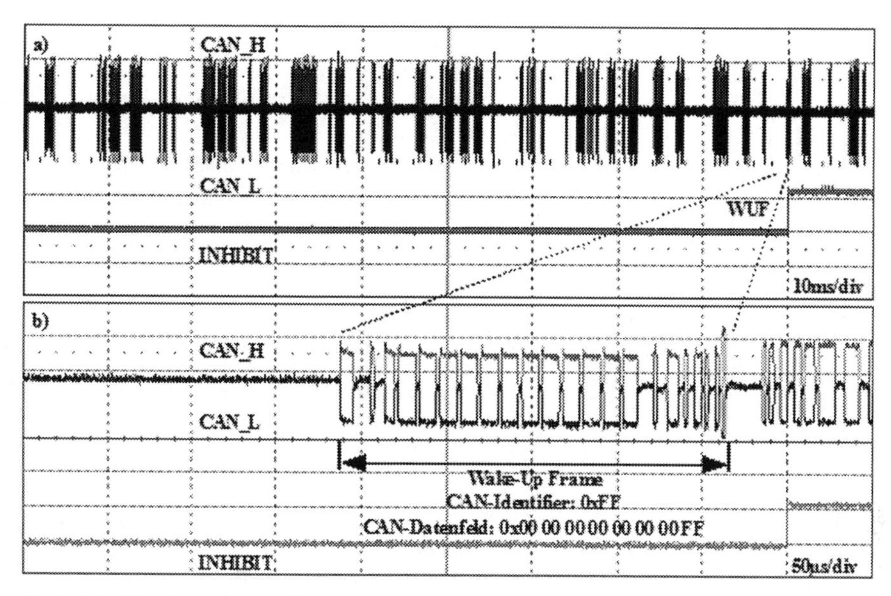

Fig. 2.54 Selective wake-up of a control unit from partial network mode

Engineering samples of the ELMOS device E520.13 have been available since middle of 2010 and have been tested by various vehicle manufacturers and automotive suppliers since then. Automotive qualification of the device will be finalized in 2012, while the first in-vehicle series application is assumed for 2014.

2.5.5 Pretended Networking—the Microcontroller Approach (Infineon)

2.5.5.1 Is Power Saving Possible with a Microcontroller?

Is power saving possible with a microcontroller? Since several years, microcontrollers do contain power-saving features. The *power modes* have been introduced, due to the fact that several ECUs continue to run, in a parked vehicle. For example, the XC2000/XE166, including the MultiCAN module, is having the so-called *Idle*-instruction, as its ancestors have, including its well-known grandfather the C167. The idle instruction allows switching off the clock supply of the central processing unit (CPU) and recovering the clock by an interrupt within one cycle. Since several years, all modules can be switched on–off and a cyclic wake-up can be performed by the real-time clock (RTC) or the standard timer module (STM).

2.5.5.2 ECUs Sleep During Driving?

To save power during driving, the main question is whether the existing power-saving features can also be used during driving.

Yes, it is possible. Most applications, which are currently discussed for partial networking, are activated by the driver. ECUs, for example, the trailer module, have to signal regularly to the bus if a trailer is attached or not set and can be switched off in case no trailer is attached by the transceiver. Applications, like the window lift, have to react so fast that the user will detect no difference to today's solutions. However, not only comfort but also safety play their role, e.g. in case of a crash, all doors have to open automatically. As during a crash, also the power supply cannot be guaranteed any longer, and the reaction time to the crash message has to be as fast as possible. Therefore, using partial networking is not possible, as otherwise rebooting the microcontroller by the transceiver might take too long to guarantee the power supply. However, these ECUs are able to save power.

The easiest way to save power is to include the idle instruction (HALT mode) into the operating system, so that all CPUs do not loop but save power in case of no operation. All interrupts will wake up the CPU without any delay. For example, a cyclic timer interrupt can cause such a wake-up.

If now additional peripherals of the microcontroller are switched off, not even touching the memories, the power consumption can be significantly reduced. If now the CPU is set into idle mode and only the MultiCAN module is kept running, the power consumption can be already reduced to round about 50 %.

A more radical approach is to reduce the operating frequency (*slow-down mode*) and, using different message catalogues, to reduce the bus traffic and the amount of wake-up events.

The XC2000 has a *standby random access memory (RAM) in addition*, so that a minimal program plus parameters can be saved within this RAM, in case memories shall be switched off. There are different levels of power saving, and the easiest are described above. Further power-saving modes are *"active stop over"*, *"passive stop over"* as well as the *"standby"* modes. In standby, including the using of the standby RAM, the power consumption is below 1 mA. A more detailed description can be found within the application note AP16170 "Power Management with SCU Driver". Under the same title, a driver can be found with the implementation of the power modes.

2.5.5.3 Outlook

In the future, more features will be integrated into the microcontroller, which shall support the application to save power. Nonetheless, with today's microcontrollers it is already possible to save a significant amount of power, without changing the networking management of a car platform.

As partial networking has only first samples in the market, the power saving within microcontroller can already be started today and even on existing ECUs.

This can be done without changing the network and network management of a vehicle or of machinery. In can be so easy to save power.

2.5.6 Comparison Between Partial and Pretended Networking (Infineon)

Both approaches do have their advantages and also their disadvantages. The microcontroller solution reacts fast and it is the right approach for time-critical applications. Therefore, it consumes more power and well-designed software to able to use the power mechanisms at the right time and to have the resources available if needed. The transceiver approach looks fairly easy in the first place, exchange the transceiver and you are done, but it needs a significant redesign of an existing networking architecture. This solution needs a long booting time and cannot be used for timely critical applications. The power consumption is lower than with the microcontroller solution. In addition, here the quality of the software and of the physical outline of the networks defines the quality and stability of the solution. Which solution will get the majority within the vehicle is currently not defined. The co-existence is possible and the further advantages and disadvantages will be shown in the future. During printing of this book, both solutions are still in evaluation phase and only in very few test vehicles.

2.5.7 NXP's Concept and Implementation

In order to achieve maximum robustness in a partial networking system and to avoid the limitations of wire harness design, NXP has incorporated a CAN protocol decoder into its new transceivers. The decoder is clocked by a very precise internal free-running oscillator. The ISO 11898-1 compliant CAN protocol decoder was originally developed for NXP's LPC229x microcontroller family. Very low temperature dependence and high immunity to changing supply voltage conditions in the oscillator make calibration of the frequency relative to the received bus traffic unnecessary. Consequently, the full capability of the CAN decoder to analyse received messages is available at any time after a start-up phase lasting several microseconds. As a result, reliable gap-free detection of wake-up messages can be guaranteed without the need to generate cyclically repeating wake-up or calibration messages. In addition, there is no need for a waiting period after start-up to allow the oscillator frequency to be adjusted relative to that of the received messages. So the transceiver is as robust as a CAN decoder integrated in a µC. The alternative to using such a stable oscillator is to calibrate the oscillator relative to the received messages as described in Sect. 2.5.3. NXP decided not to use oscillator designs that require calibration for a number of reasons:

- Jitter on the received edges caused by radio frequency (RF) fields lowers the precision of the calibration result; increasing the immunity of the transceiver would increase current consumption and, in any case, could only be achieved to a limited extent.
- Oversampling is needed to achieve a sufficient level of precision; the higher the oversampling rate, the higher the current consumption.
- Ringing on the recessive edges means that they cannot be used for calibration.
- Messages that occur after silence on the bus are not defined and thus can be transmitted in a number of different formats; in the worst case, there are only four dominant edges in a message (ID=0E0h, RTR=1, DLC=1).
- The ID field can be disturbed by arbitration and, therefore, cannot be used for calibration; moreover, the non-calibrated receiver does not know where the ID field ends (note that the ID can be 11 bit or 29 bit).
- Car makers require calibration to be completed by the time five messages have been received; otherwise, start-up will take too long and wake-up events might be missed.
- Errors such as "stuff error", "form error", "CRC error" or error frames can disturb the calibration process.
- Transmitter clock tolerances (up to ± 0.5 % according to SAE bit-timing requirements) lead to errors in the calibration results. This is critical when the receiver clock has been tuned to a slow sender and the wake-up message is sent by a fast sender (or vice versa).

NXP transceivers with partial networking have the same state diagram as standard transceivers like the TJA1041(A) or TJA1043. So the familiar "normal", "standby" and "sleep" modes are available in the new transceiver generation. The only difference is in the wake-up mechanism; the new transceivers will only respond to the configured WUFs.

To achieve good emission performance throughout the system, pins CAN_H and CAN_L are biased towards 2.5 V when the transceiver is in a low-power mode (standby or sleep) and bus traffic is being monitored for wake-up messages. Biasing is turned off automatically if the bus is silent for more than a second. Standard transceivers always terminate the CAN_H and CAN_L pins to GND when in a low-power mode, enforcing a common-mode step with every message transmitted on the bus.

2.6 Transceiver Implementations

2.6.1 Implementation Example TLE 6254 3G (ISO 11898-3)

TLE 6254-3G is a standard low-speed CAN transceiver of Infineon technologies with an excellent EMC performance and a high ESD robustness. All transceiver

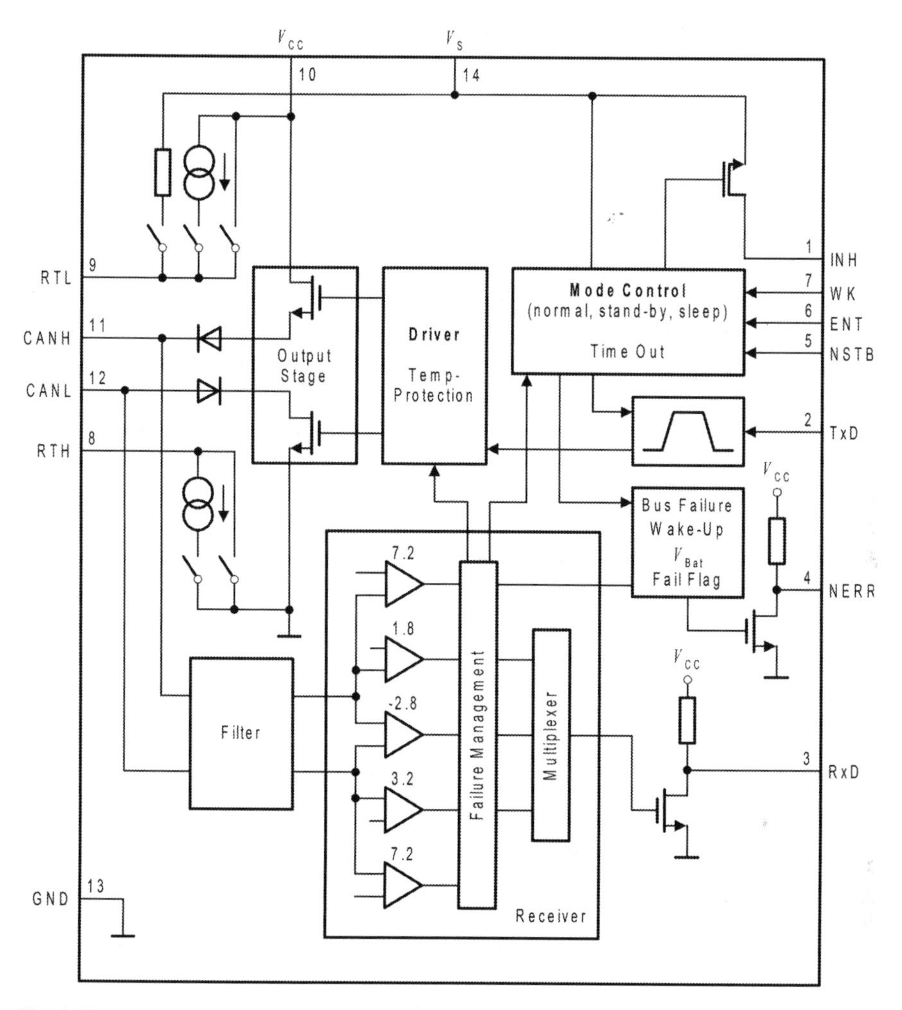

Fig. 2.55 TLE6254-3G block diagram

modes, described in the former chapter, are implemented in this transceiver. The block diagram in Fig. 2.55 shows the important blocks of this transceiver like:

- Transmitter block to transmit data
- Receiver block to receive the data from the bus
- Mode control unit to control the transceiver mode
- Bus failure management block
- Wave -shaping block to optimize the CAN_H and CAN_L waveform to reduce the emission

2.6.1.1 Pin Description

- WK: Local wake pin to wake up the transceiver from sleep mode via an external switch on the ECU.
- INH: High-side switch to control the connected voltage regulator. On in non-sleep mode and off in sleep mode.
- NSTB and ENT: Mode input pins to control the transceiver.
- NERR: Logic output pin to flag failure conditions.
- TxD: Transmit data input pin. The data on pin TxD will be transmitted on the bus via CAN_H and CAN_L.
- RxD: Receive data output pin. The received data are formed into logic-level signals for the microcontroller.
- CAN_H, CAN_L: Bus pins. Very robust high-voltage pins to transmit and receive data.
- RTL, RTH: Termination resistor switches. Off in case of bus failure and on in failure-free mode.
- VS: Supply voltage pin (reverse polarity-protected battery supply).
- VCC: 5 V supply pin.
- GND: Ground connection.

2.6.1.2 Transmitter

The requirements on the transmitter according to voltage range, ESD robustness and EMC performance are very high. To reduce emission during data transmission, the symmetry of the CAN_H and CAN_L signals must be very high and the corners must be rounded. This task will be done in driver of the output stage. The driver stages CAN_H and CAN_L have to be very robust against ESD pulses. An ESD pulse is very short, but during this time a current peak up to 20 A has to be handled without any reduction on the driver performance. In case of a permanent dominant signal on TxD (for example, caused by a short on pin TxD), the driver CAN_H and CAN_L will be switched off to release the bus. The transceiver can now receive data but cannot transmit data until the level on pin TxD is high. A high level on pin TxD release the look and the transmitter can transmit again. The transmitter will be also disabled in case of an over-temperature. After returning to the normal temperature range, the transceiver logic releases the transmitter automatically.

2.6.1.3 Receiver Block

The receiver block consists of five different comparators:

- Differential receiver: In normal operation, the differential comparator receives the data.
- CAN_L comparator: Single-ended comparator to monitor CAN_L only.

- CAN_H comparator: Single-ended comparator to monitor CAN_H only.
- CAN_L Vbatt Comp: Comparator to monitor high voltage on CAN_L>7 V.
- CAN_H Vbatt Comp: Comparator to monitor high voltage on CAN_H>7 V.

2.6.1.4 Differential Receiver

The differential receiver has a high common-mode range (−2–6V). This guarantees an EMC and ground shift robust communication. This comparator is active if there is no short on the bus.

2.6.1.5 CAN_L, CAN_H Single-Ended Comparator

These comparators monitor one bus CAN_H or CAN_L only. In case of a short of one bus wire, the corresponding comparator on the short-free wire receives the data. The robustness against EMC and ground shift is low compared with the performance of the differential mode.

2.6.1.6 CAN_H, CAN_L Vbat Comparator

These comparators monitor the voltage on CAN_H and CAN_L to detect overvoltage on these wires (V>7 V).

If the voltage on CAN_H or CAN_L is higher than 7 V, this indicates a failure on the bus wire (for example, short to Vbatt) in sleep mode and CAN_L will be terminated to Vbatt. In this mode, a high voltage is a normal situation. The Vbatt comparator will be disabled in this mode to reduce current consumption and to avoid a failure report.

2.6.1.7 Bus Failure Detection

To detect the kind of failure on the bus all comparators are used. The combination of the information allows the failure detection logic (Fig. 2.56) to find out which kind of failure is on the bus. Depending on the type of failure, the hidden output stage can be switched off to protect them.

2.6.1.8 Application Circuit

Figure 2.57 shows the typical application circuit of a low-speed CAN application. Sometimes CAN coils are used to improve the emission and EMC robustness. The termination resistors are different. Minimum on two nodes 500 Ω are set. The value for the termination resistors on the other nodes depends on the number of node in the network. In total, the value of all resistors must be above 100 Ω. For example, if

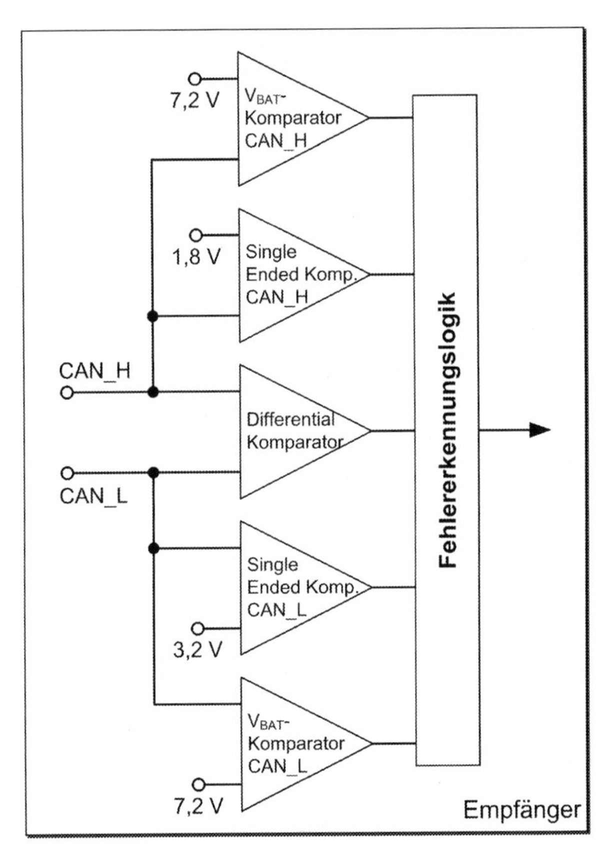

Fig. 2.56 Receiver comparators to receive the data and detect bus failures`

ten nodes are in the network and two nodes have 500 Ω, the value for the termination resistors of the other nodes is 1,400 Ω.

RxD, TxD, the mode pins NSTB, ENT and the failure flag pin NERR are connected to the microcontroller.

The local wake pin should be connected to the switch via a 10-kΩ resistor to increase the robustness against disturbance. A pull-up resistor is necessary; otherwise, the node floats if the switch is open. The value on these resistors depends on the required wetting current. The typical value is 1 kΩ, if the local wake pin is not used. This pin can be connected to the supply pin VS.

The pin INH can be connected to a voltage regulator or DC/DC converter. The transceiver controls now the power supply. In case of sleep mode, the power supply is switched off via the INH pin and in case of an:

- Power up
- Local wake event on pin WK
- Remote bus wake

The power supply will be switched on again.

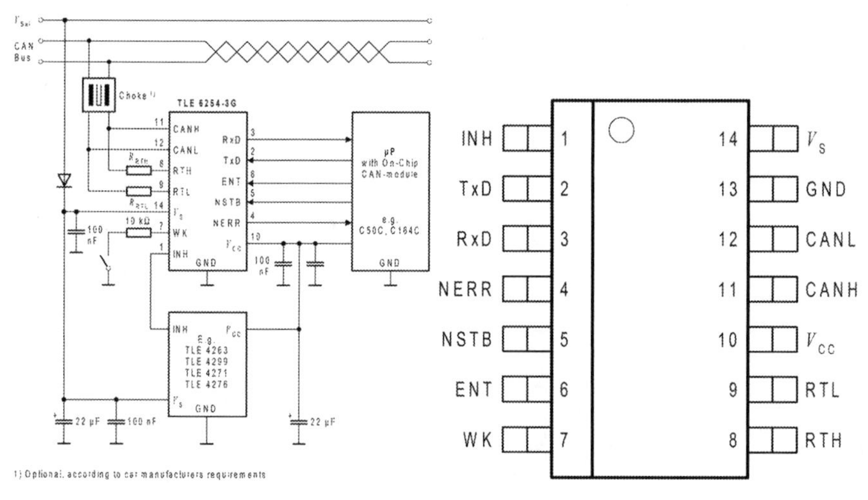

Fig. 2.57 Application circuit and pinning of TLE6254-3G

Normally, the ESD and EMC robustness of the TLE 6254-3G is very high. However, to use this performance, the board layout must be able to handle high currents in case of EMC and ESD events with a low voltage drop. In addition, the coupling between signals with fast transients on the board and the CAN_H and CAN_L wires must be very low. The blocking capacitors on VCC and VIO must be close to the pins to develop the best performance.

2.6.2 *Implementation Example TLE 6250 G (ISO 11898-2)*

TLE 6250 G from Infineon technologies is a high-speed CAN transceiver according to ISO 11898-2. Excellent EMC performance and high ESD robustness are the key features of this transceiver. The block diagram in Fig. 2.58 shows the main blocks of this device:

- A transmitter to transmit data
- A receiver to receive data from the CAN bus
- A mode control to control the device
- A signal shaper to optimize the output signals to reduce the emission on the bus

2.6.2.1 Pin Description

- TxD: Transmit data. Logic input pin. The data on pin TxD will be transmitted on the bus.
- GND: Ground.
- VCC: 5 V supply for the transceiver.

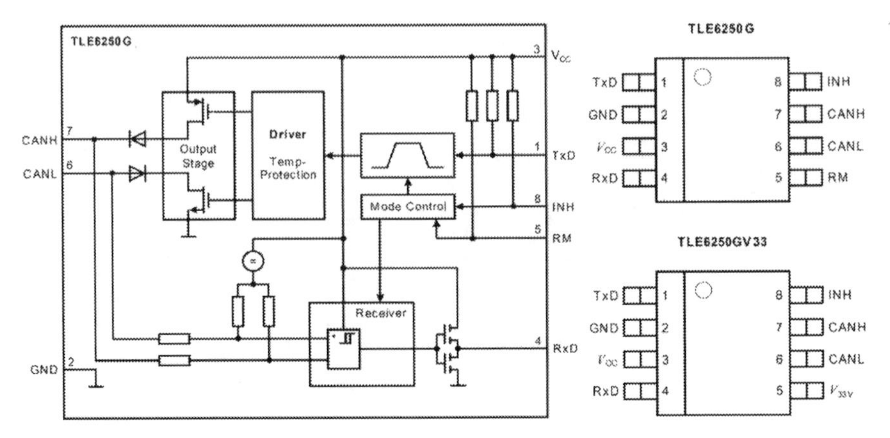

Fig. 2.58 Block diagram and pinning of TLE6250-G and TLE6250-GV33

- RxD: Receive data. Transform the differential signals from the bus into a local signal for the microcontroller.
- RM: Receive only mode. In this mode, the transmitter is switched off.
- INH: Logic input pin to control the modes (standby mode and normal mode).
- CAN_H, CAN_L: Bus pins. High-voltage pins to transmit and receive data.
- V33V: To support the I/O pin (TxD, RM, INH and RxD) with the microcontroller.

2.6.2.2 Transmitter

High voltage robustness, high EMC immunity and high ESD robustness are the requirements for the output stages CAN_H and CAN_L. At the same time, the CAN_H and CAN_L output signals should be shaped very well to reduce the emission. This is done by the signal shaper in the transmitter. The output stage CAN_H is a high-side switch and CAN_L is a low switch. Both stages switch on and off together at the same time.

2.6.2.3 Receiver Block

The receiver transforms the bus differential signal into a logical signal for the microcontroller (on pin RxD); the common-mode range of this receiver is between −20 V and +25 V to allow a very high immunity against disturbances from the bus. The high common-mode range also allows a high ground shift.

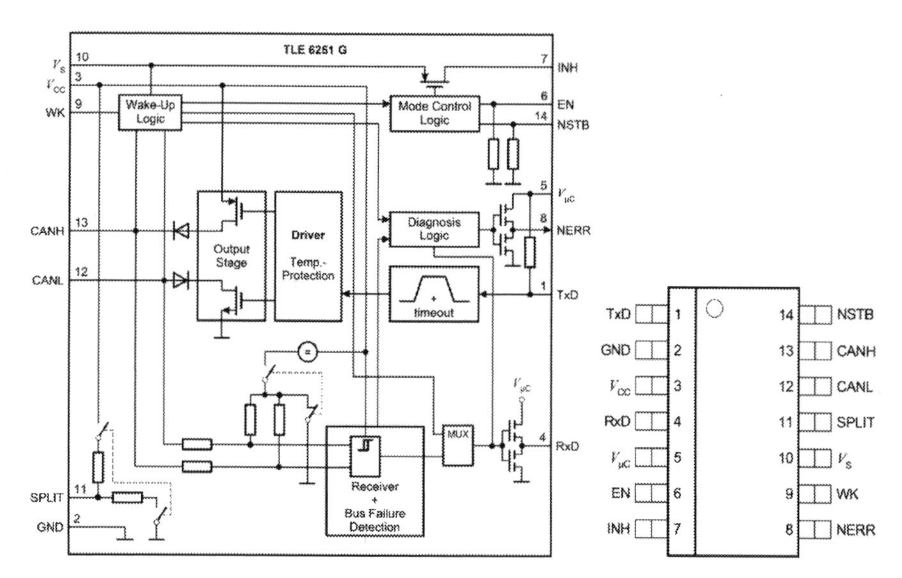

Fig. 2.59 Block diagram and pinning TLE6251-2G

2.6.2.4 Mode Pins: INH and RM

The INH pin controls the mode of the transceiver. In normal mode, the transceiver transmits and receives data. In standby mode, the transceiver current consumption is reduced to a minimum and the transmitter and receiver are blocked. The pin RM blocks the transmitter if the CAN controller is not working correctly or TxD is shorted to ground. The receiver works correctly in this mode.

2.6.2.5 V33V-I/O Supply Voltage

In some applications, different supply voltages are used for the microcontroller (for example, 3.3 V) and the transceiver (5 V).To allow this separate voltages, the interface pins are supplied from the microcontroller. An internal level shifter shifts the levels from the V33 domain into the 5 V transceiver domain. In case of a missing V33 voltage, the transmitter is blocked.

2.6.2.6 TLE 7250G and TLE 7250GVIO

The TLE 6250G and TLE 6250GV33 have now followers called TLE 7250G and TLE 7250GVIO with higher ESD robustness and a TxD time-out function. In case of a dominant signal longer than 300 µs the transmitter will be switched off to release the bus. After high on TxD the latch will be released.

2.6.3 TLE 6251-2G (ISO 11898-5)

TLE 6251-2G is a high-speed CAN transceiver according to ISO 119898-5 with remote bus wake-up with excellent EMC and ESD performance. In addition to the standard high-speed CAN transceiver, a remote wake-up receiver with a very low current consumption is implemented. If there is no communication on the bus, the transceiver can be set into sleep mode and the remote wake-up receiver will be woken up, if CAN bus communication starts again. In block diagram in Fig. 2.59 the fundamental blocks are shown.

The transceiver contains the following:

- A transmitter to transmit data on the bus
- A receiver to receive data from the bus
- The mode-control block to control the transceiver unit
- A physical bus failure detection unit
- A signal shaper to optimize the waveform on CAN_H and CAN_L and reduce the emission

2.6.3.1 Pin Description

TLE 6251-2G pins:

- WK: Local wake-up pin to wake up the transceiver via a switch.
- INH: High-voltage high-side switch to control a voltage regulator.
- NSTB and EN: Logical input mode pins to control the transceiver modes.
- NERR: Logic output pin; output is set to low in case of physical bus failure.
- TxD: Transmit data. Logical input pin to transmit data on CAN_H and CAN_L.
- RxD: Receive data. Transforms the differential signal on the bus into logical signal on RxD.
- CAN_H, CAN_L: Bus pins. To transmit and receive data.
- SPLIT: 2.5 V output buffer to improve signal integrity.
- VS: Battery supply pin (via a reverse polarity diode).
- VCC: 5 V supply pin for transmitter and receiver.
- $V\mu_C$: Supply pin for the logic I/Os.
- GND: Ground.

2.6.3.2 Transmitter

High voltage robustness, high EMC immunity and high ESD robustness are the requirements for the output stages CAN_H and CAN_L. At the same time, the CAN_H and CAN_L output signals should be shaped very well to reduce the emission. This is done by the signal shaper in the transmitter. The output stage CAN_H is a high-side switch and CAN_L is a low switch. Both stages switch on and off together at the same time.

The transceiver is very robust against ESD disturbance. According to ISO 61000-4-2 the device withstand ESD pulses higher than 6 kV. Current peaks up to 20 A have to be handled in the device for a short time.

A current limitation is implemented to protect the device against shot circuits to ground and Vbatt. In case of a short, the CAN_H and CAN_L output current will be regulated to 120 mA. In case of overheating, the transmitter will be switched off. A high level on TxD unlocks the transmitter (if the temperature is cooled down).

2.6.3.3 Receiver

The receiver transforms the bus differential signal in a logical signal for the microcontroller (on pin RxD).

The common-mode range of this receiver is between −40 V and +40 V to allow a very high immunity. The high common-mode range allows also a high ground shift

2.6.3.4 Bus Failure Detection

Bus failure detection in differential voltage systems with a high common-mode range is very difficult. In this kind of concepts, the bus can only be analysed when the transmitter sends data. In the TLE 6251-2G, the CAN_H and CAN_L output currents are measured and the difference will be detected as a bus failure. These concepts support:

- Short circuit CAN_H to ground. In this mode, the CAN_H current is much higher than in CAN_L.
- Short circuit CAN_L to supply voltage. In this mode, the CAN_L current is much higher than in CAN_H.
- Short circuit CAN_H to VCC or battery. In this mode, the CAN_L current is much higher than in CAN_H.
- Short circuit CAN_L to ground. In this mode, the CAN_L current is much smaller than in CAN_H.

The failure CAN short with CAN_L will be analysed with bit compare. If the transmitter transmits data, the receiver must receive these data. If not a short between CAN_H and CAN_L is possible.

2.6.3.5 $V\mu_C$-I/O Supply

Microcontroller has a wide range of supply voltage. To guarantee a reliable communication between the microcontroller and the transceiver the microcontroller supply voltage is used to support the transceiver interface pins. The $V_{\mu C}$ undervoltage detection unit blocks the transmitter and receiver function in case of undervoltage on pin $V_{\mu C}$ and set the device to sleep mode.

2.6.3.6 Split Pin

2.5 V (VCC/2) with an input resistance of 600 Ω will be provided on this pin. This voltage can be used to stabilize the recessive level on the bus. The centre pin of two termination resistors (2 times 60 Ω) should be connected to this pin. The voltage source is active in normal mode. In sleep mode and standby mode, the voltage source is disabled and high ohmic. This function is not offered in all transceivers with remote wake-up function.

Chapter 3
Data Link Layer Implementation

Wolfhard Lawrenz, Florian Hartwich, Ursula Kelling, Vamsi Krishna, Roland Lieder and Peter Riekert

3.1 M_CAN—Modular CAN Controller

The modular controller area network (M_CAN) module was developed to expand Bosch's well-known family of CAN modules (e.g. the C_CAN module, which is found in many microcontrollers) and support standardized (Automotive Open System Architecture, AUTOSAR) software drivers in particular, as well as applications with multiple CAN channels. The M_CAN module's internal partitioning in CAN core, Tx Handler, and Rx Handler provides flexibility for easy adaptations to future requirements. CAN messages are stored in a separate memory, the *Message RAM*,

W. Lawrenz (✉)
Waldweg 1,
38302, Wolfenbuettel, Germany
e-mail: W.Lawrenz@gmx.net

F. Hartwich
Robert Bosch GmbH, Tuebinger Strasse 123,
72703 Reutlingen, Germany
e-mail: Florian.Hartwich@de.bosch.com

U. Kelling
Infineon Technologies AG, Am Campeon 1-12,
85579 Neubiberg, Germany
e-mail: ursula.kelling@infineon.com

V. Krishna
Xilinx India Technology Services Pvt. Ltd.,
Cyber Pearl, Hi-tec City, Madhapur, Hyderabad 500 081, India

R. Lieder
Renesas Electronics Europe GmbH, Arcadiastrasse 10,
40472 Duesseldorf, Germany

P. Riekert
Ingenieurbüro für IC-Technologie, Kleiner Weg 3,
97877 Wertheim, Germany
e-mail: ifi@ifi-pld.de

W. Lawrenz (ed.), *CAN System Engineering,* DOI 10.1007/978-1-4471-5613-0_3,

© Springer-Verlag London 2013

not inside the M_CAN. The M_CAN module is compliant with CAN protocol 2.0 A, B, and *ISO 11898-1*. Figure 3.1 shows its internal structure.

All functions specified in the CAN protocol, such as CAN protocol controller state machines as well as the shift registers for transmission and reception, are implemented in the CAN core. This protocol unit has been adopted from earlier CAN modules and is part of a direct line of development that begins with the introduction of the CAN protocol. The CAN core's interface signals are connected to the rest of the M_CAN via a synchronization logic. This makes it possible to supply the CAN core with a dedicated clock for CAN communication, whereas the rest of the module is in the same clock domain as the host central processing unit (CPU). For example, the CAN core might be operated with an 8 MHz crystal clock while the CPU is supplied with a phase-locked loop (PLL) clock of significantly higher frequency which—to limit noise emission—may also be modulated.

The Tx Handler controls the transmission of messages. The host CPU may set transmission requests for several messages; transmit cancellation is also supported. The Tx Handler then transfers the messages—according to the priority of their identifiers—from the *Message RAM* to the CAN core's shift register. Up to 32 dedicated transmit buffers are available. They may—partially or completely—be combined to operate as a transmit first-in-first-out (FIFO) or as a transmit queue. Status information regarding the requested transmissions, including a 16-bit transmit time stamp, may be logged into the optional Tx Event FIFO.

Dedicated receive buffers and up to two receive FIFOs may be configured for the reception of messages, under the control of the Rx Handler. The Rx Handler performs acceptance filtering and transfers received messages into the *Message RAM*. The following filter types are available for the acceptance filtering:

- Range filter: Matches for identifiers in the range from start identifier to end identifier.
- Bit masking: Matches for a specific identifier while some identifier bits may be masked.
- Dual filter: Matches for two specific identifiers.
- Dedicated Rx: Matches for the identifier of a dedicated receive buffer.

The filters can each be used as acceptance or as rejection filter; they also decide where accepted messages are to be stored. In total, up to 128 filter elements may be configured for *11-bit identifiers* and up to 64 for *29-bit identifiers*. This may be combined with a global mask for *29-bit identifiers*, in support of J1939 applications. The various filter options allow a targeted filtering of received messages ensuring that only messages which are relevant for the particular node are stored in the *Message RAM*; others are rejected. The reception time, a 16-bit time stamp, is optionally stored with the message.

The M_CAN module combines both qualities of the "Full CAN" concept and of the "Basic CAN" concept. Received messages are stored in dedicated receive buffers as well as in FIFOs; no software acceptance filtering is needed. The transmit messages may be—depending on the application—stored in dedicated transmit buffers or managed in a transmit FIFO or in a dynamic transmit queue.

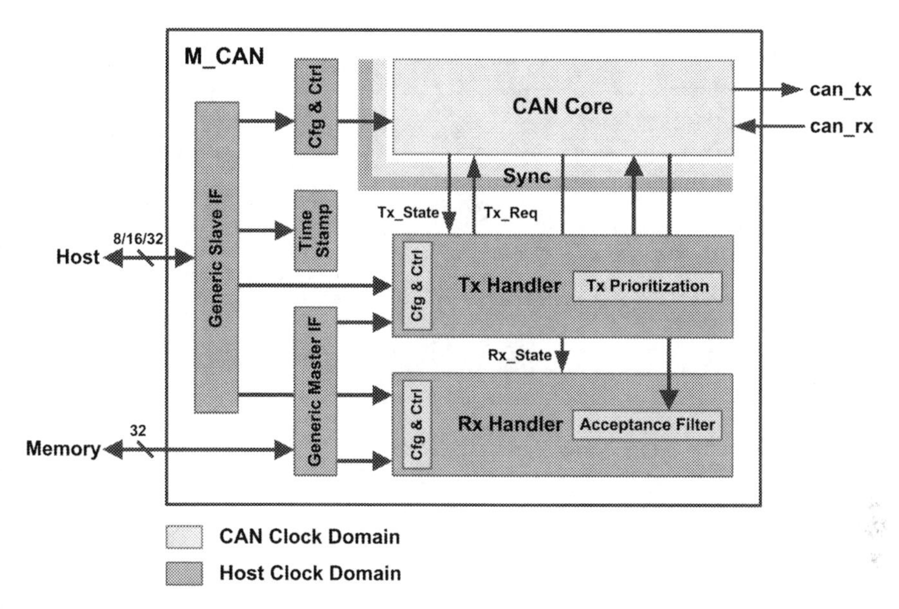

Fig. 3.1 M_CAN block diagram

A 16-bit timer counter is available to check for timeout conditions in the handling of the receive FIFOs and the Tx Event FIFO. Both the timer counter and the time-stamp generation are clocked, via a prescaler, with multiples (1–16) of the CAN bit time.

The configuration and control of the M_CAN module is done by the host CPU, via the Generic Slave interface. Through this interface, the CPU also reads status information from the CAN core, the Rx Handler, and the Tx Handler. The Generic Slave interface may be connected to 8/16/32-bit CPUs.

The Generic Master interface is used to access the 32-bit-wide *Message RAM* (single or dual channel). The CPU also has direct access to the *Message RAM*. The transmit buffers, the Tx Event FIFO, the dedicated receive buffers, the receive FIFOs, and the acceptance filter elements are stored in the *Message RAM*, outside of the module. The partitioning of the *Message RAM* can be configured flexibly (see Fig. 3.2). A maximum of 1,216 (32-bit-wide) words can be used per M_CAN module; the minimum size of the RAM is determined by the application.

Gateway (GW) configurations consisting of several M_CAN modules sharing one *Message RAM* (see Fig. 3.3) can easily be set up. Access conflicts between the M_CANs and the CPU are resolved by the attached RAM Arbiter state machine. No modifications to the M_CAN module are required for their use in a GW. It is also possible to connect several M_CAN modules to the same CAN bus, for example, to enlarge the number of message buffers for that channel.

The interrupt flags of the M_CAN module signal status or error conditions of CAN core, Tx Handler, and Rx Handler. The interrupt flags may be evaluated by polling, or they may be assigned (individually) to one of two interrupt lines that are connected to the host CPU.

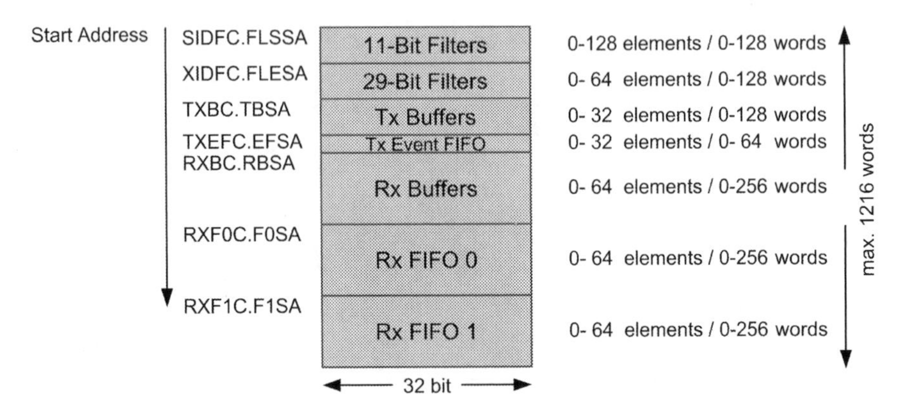

Fig. 3.2 Message RAM configuration

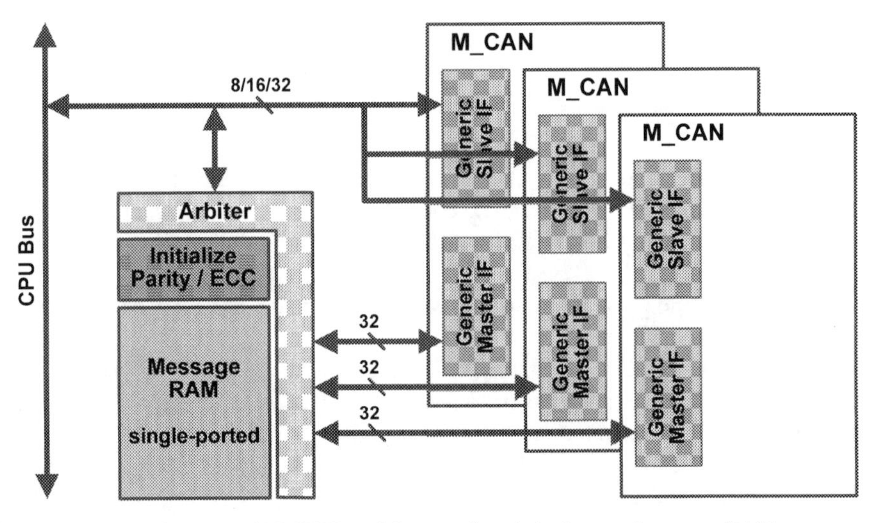

Fig. 3.3 Connecting several M_CAN modules to a shared single-ported message RAM

In addition to the normal operating mode, the M_CAN module provides several test modes such as the bus monitoring mode to silently observe the CAN communication or the loop back mode, in which the M_CAN treats its transmitted messages as received messages. Self-test of the internal transmit and receive path is possible without disturbing the communication on the CAN bus. A power-down support (sleep mode) completes the feature list.

The M_CAN's modular structure makes it easy to add new functions, such as new communication features as CAN with Flexible Data Rate (CAN FD; see Sect. 3.3). Another configuration of the M_CAN, with an additional frame synchronization entity that supports Time-Triggered CAN (TTCAN; see Sect. 3.2), is also available.

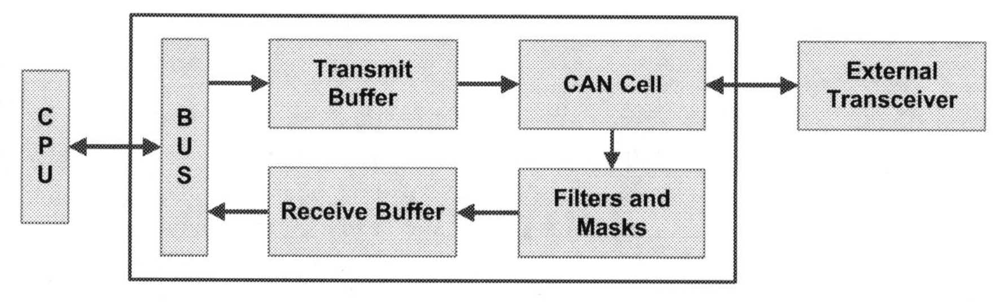

Fig. 3.4 IFI Advanced CAN block diagram

More information on Bosch's Internet Protocol controller area network (CAN-IP) modules can be found at the following URL: http://www.semiconductors.bosch.de/en/ipmodules/can/can.asp

3.2 IFI Advanced CAN

Prior to designing a CAN communication system, some decisions must be made. Is the intention to apply a standard CAN controller, which performs the communication tasks together with a standard CPU, or is it intended to make use of a CPU containing an integrated CAN controller or is it planned to integrate the CAN interface into a field-programmable gate array (FPGA), perhaps in conjunction with a CPU? In any case, however, an important criterion is to plan carefully how the CAN controller shall be operated in order to avoid the CPU to be loaded with unnecessary tasks. Normally operation of all CAN controllers is based on the same basic principle; some more or less mailboxes are installed along with a couple of filters and masks. IFI Advanced CAN, however, takes another route.

The background for the development of the IFI Advanced CAN is not to launch the n+ 1st variation of a standard controller as an application-specific integrated circuit (ASIC) but to combine the flexibility of an FPGA with a high-performing and resources-saving integration. For that purpose, this controller provides multiple parameters, which, in the compilation run for the FPGA code, are chosen in such a way that only those resources of the FPGA are allocated which are really necessary for the application.

This concept starts with the interface between CPU and controller (Fig. 3.4). As this Internet Protocol (IP) core is designed for Altera FPGAs, the interface is a so-called Avalon bus. This is a synchronous bus, allowing a dynamic adaptation of the bus width. When the controller is intended to be used in conjunction with the CPU in an FPGA, it is recommended to apply an Altera NIOSII CPU, which accesses the core with a data width of 32 bits. On the application of an external CPU, a parameter selects the desired data bus width of 8, 16, or 32 bits. This simplifies the connectivity for the designers. The following will not go into details on the architecture of the Avalon bus interface, but the structure of the controller will be enlarged, offering quite some specialties.

3.2.1 Transmit Buffer

For offloading the CPU, a sufficiently large buffer memory is required, which intermediately stores the CAN messages ready for transmission until they are finally transmitted by the CAN cell. For that purpose, most of the CAN cell implementations provide mailboxes into which the messages ready for transmission must be written. If on top the transmission of messages is priority controlled by the controller, a lot of software control is needed to know at what time actually a message had been transmitted. IFI Advanced CAN does not make use of this method, but applies a FIFO instead. The transmission of messages in the same sequence as the CPU generates the messages is not only desired by many communication tasks but also even much easier to handle. The number of messages which can be stored in that FIFO can be controlled by a parameter between 30 and 254. Theoretically, there is a case that due to a very busy CAN bus, a message never may be transmitted because of always losing arbitration. A possible solution to that problem would be to clear the FIFO and rewrite it with a different sequence. However, this would imply that the user always had to know which messages are still waiting in the queue. Even for this case, IFI Advanced CAN provides another way out. Any message can be written into the FIFO in the normal way or a priority identification may be assigned to the messages. Those messages are not written into the back of the FIFO but into the front of the FIFO, while passing the queue. This architecture avoids clearing the FIFO and memorizing the history by software. Therefore, three functions are implemented:

- Removal of a message from the CAN cell, if it is not currently in the transmission process. That is to say, the bus is busy transmitting another message and that message is waiting to be transmitted or that message had been interrupted by an error frame and must be retransmitted. In both cases, that message can be removed without corrupting any frames.
- The removed message must be written into the FIFO again as a not-yet-transmitted message in order to guarantee that this message is not lost.
- The most important message is handed over to the CAN cell for transmission.

In order to make sure that this method is working smoothly even for more than one message only, the FIFO buffer is switched into a last-in-first-out structure and identifications are assigned to all messages which are contained in the buffer. The identification enables the controller to recognize which messages still must be transmitted. This concept enables the CPU to insert easily the messages into the buffer while still maintaining control on the transmission. Each message ready for transmission is written into the FIFO as a sequence of four addresses each 32 bits long containing the following information:

- The addresses 2 and 3 reserve space for 8 data bytes.
- The address 1 contains the standard or the extended ID.
- The address 0 contains the data length code and the remote transmit bit as well as an optional frame number. By this number, the controller knows whether

transmission of this message must be controlled. In order to activate this function any number greater than 0 must be inserted. After a successful transmission and independently of any filter conditions, the controller writes this message into the receive buffer together with this number and a 32-bit-long time stamp, if desired. Writing into address 0 automatically enables transmission of the message.

Removing a message from a CAN cell can also be done without inserting another message (Remove Pending Message). In case the node is only alone at the bus, because the communication lines were disconnected, the node would continuously retry to transmit the message. After reconnecting the node, the node would retransmit the message again. A removal of the actual message together with a reset of the transmit FIFO pointer enables once again the setup of the messages to be transmitted. The number of message within the FIFO can also be read back.

3.2.2 Masks and Filters

In order to ease the evaluation of received CAN messages, this controller provides 256 pairs of masks and filters. There is an object number assigned to each of the filter pairs, which is written into the receive buffer after a successful check. Because programming is done in two steps, each of the masks and each of the filters provide an additional bit. When applied, both of the bits must be set, in order to avoid the controller to use non-valid combinations for comparison. This makes sure that masks and filters can be reprogrammed even though the system is running. As only those messages are written into the receive buffer which have passed the filter condition, a filter can be applied as such to pass all messages. Setting a 1 in bit position x of the mask defines that the value of the filter bit and the received Identifier (ID) bit in position x must match. A 0 indicates that the comparison is switched off and the comparator always would indicate a match.

3.2.3 Receive Buffer

The receive buffer is FIFO organized. It can be parameterized to be 32–256 messages long. Each message is stored as a sequence of four messages.

- The address 0 contains the 8-bit-long frame number for messages which had been transmitted by this controller itself, the object number as an identification which of the filters had passed this message, the remote frame bit, as well as the data length code.
- The address 1 contains the standard and extended ID.
- The addresses 2 and 3 contain the received data bytes.
- An additional address allows the receive time stamp of the message to be read.

In contrast to the transmit buffer, the FIFO pointer is only set to the next message after confirmation of the read process by writing into a dedicated address. This technique allows multiple read of each message as well as free choice of the sequence without the risk to lose data. Reading and clearing of the pointer is provided in order to give any kind of control support to the applicant besides information on interrupts.

3.2.4 Time Stamp

The FIFO implementation of transmit and receive buffers safeguards the chronological sequence of the messages. But at any time for any message the generation of a time stamp can be activated or a time stamp be read back respectively, in case the application requests more precise information. Not all systems operate under the same constraints; therefore, the time base is supplied by an external signal, which internally is fed into a 32-bit counter. The actual value of the counter is stored and assigned to the message when the acknowledge bit is recognized. This counter can be separately read and reset for synchronization purposes.

3.2.5 Conclusion

Depending on the application, either the standard component or the FPGA solution may be better suited. Nevertheless, there are more and more arguments arguing for an FPGA implementation. One of the pro-FPGA implementation arguments may be the cancellation of standard components; another one is the continuously decreasing costs of new FPGA product families. Furthermore, FPGAs provide an enormous flexibility with respect to the number of required interfaces. If an application requires more than one CAN node, just implement the number of CAN-IP cores as needed. Furthermore, the increasing complexity of FPGAs offers new possibilities up to the implementation of a complete system on one chip. In order to reduce development time to a reasonable degree while complexity is rising, the application of IP cores becomes a major issue.

The purchase of an IP core should not be based only on trust but on the option to check the functionality before buying. Altera supports this option with its OpenCorePlus concept. The desired IP core can be applied and tested without any restrictions as long as the FPGA is connected to the programming device. After cutting this connection, the controller will still continue to operate for another hour until it automatically stops. Another disadvantage of a standard component is that it only provides exactly the functionality as specified in the data sheet. The flexibility of the FPGA enables the designer to react on customer requests and to implement new functions if required.

3.3 Renesas RS-CAN

As part of its most recent generation of microcontroller devices, Renesas is introducing a new kind of CAN controller function. In contrast to previous implementations, the RS-CAN module supports shared memory among several channels, flexible sizes of memory areas used, and consequent assignment of FIFO structures.

The RS-CAN module contains a proprietary CAN transfer layer from Renesas which fulfils all the requirements of the ISO 11898, SAE J1939, and CAN 2.0B standards.

Besides its capabilities for the full support of "Full-CAN" or "Basic-CAN" applications, there are interesting new ways the RS-CAN module can be used. Several FIFO structures for reception and transmission allow streamed data processing, and by combining this with the AFL (acceptance filter list), a very efficient CAN controller hardware for GW applications is created.

The RS-CAN naturally also supports the conventional method of message processing via message boxes in both the receive and transmit directions. Here, RS-CAN can handle queued messages (with prioritized sending) concurrently.

The shared memory for all associated CAN channels allows easy transfer of messages and signals from one channel to another. The RS-CAN hardware has a built-in mirroring engine, which can perform this job on the message level without any CPU interaction.

If the shared memory is used consistently, it is possible to assign individual sizes of FIFO memories and filtering lists to the different channels, in order to tailor the amount of memory resources available to each channel. In this way, a channel that needs more data and filter resources can take advantage of another channel needing less of these resources (Fig. 3.5).

3.3.1 Properties of RS-CAN

One RS-CAN module supports up to eight CAN channels. The most popular implementation includes three channels and its characteristics are described in detail below (see Fig. 3.5).

- CAN protocol according to ISO 11898 (2.0B active), full functionality for extended identifiers and remote frames.
- Maximum baud rate: 1 Mbit/s. This baud rate can be achieved using a module clock at 22 MHz and a transfer layer clock at 8 MHz, if the bit timing is set to 8 tq per bit. The transfer layer clock can be derived from a separate clock source, by using a PLL bypass, for example.
- Identical hardware structure for all derivatives and channel configurations, which allows easy porting of software. Compatible with AUTOSAR requirements.

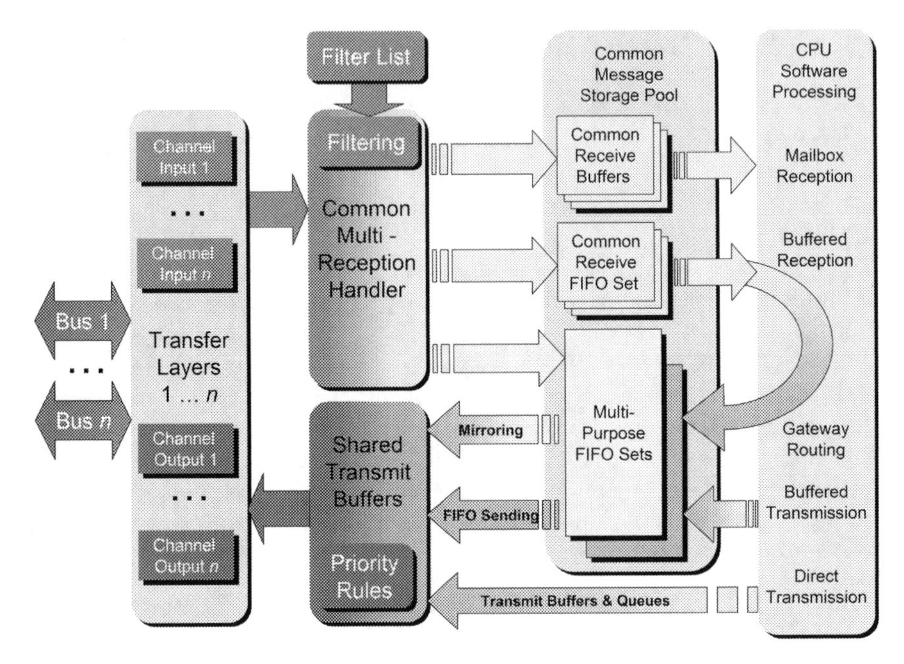

Fig. 3.5 RS-CAN architecture

- 192 receive objects, shared flexibly by user configuration between
  - up to 48 commonly shared receive message buffers for mailbox reception
  - up to eight commonly shared receive FIFO units, variable in depth up to 128 messages per FIFO
  - up to three multi-purpose FIFO units per channel (nine in total), variable in depth up to 128 messages per FIFO.
- 192 applicable acceptance filtering rules, including 29-bit identifier masking, masking for remote and extended frames, data length control (DLC) filtering, and GW hardware routing rules. Up to eight-way reception is possible, so that a received message can be stored into up to eight different locations in parallel, including an additional software identifier (Hardware Receive Handle (HRH) number of AUTOSAR COM stack processing).
- 16 transmit message buffers per channel (in total 48 buffers), assigned flexibly by user configuration to either
  - one transmission queue per channel (three in total), using a variable subset of the transmit message buffers
  - up to three multi-purpose FIFO units per channel (nine in total), variable in depth up to 128 messages per FIFO
  - up to 16 standard (direct) transmit message buffers per channel (48 in total).
- 16 transmit history list (THL) entries per channel.
- Many interrupt sources, including:

  - global error interrupt for DLC errors and lost messages
  - reception interrupt for each receive FIFO unit (fill level of FIFO is adjustable individually)
  - reception interrupt for each multi-purpose FIFO unit (fill level of FIFO is adjustable individually)
  - transmit interrupt for each multi-purpose FIFO unit (adjustable either on every message, or on the last sent message)
  - transmit interrupt for each transmission queue (adjustable either on every message, or on the last sent message)
  - transmit interrupt for every channel, where the message buffer is not assigned to a multi-purpose FIFO nor to a transmission queue
  - transmit abortion interrupt for every channel
  - THL interrupt for every channel (adjustable either on every new entry or on fill level)
  - error interrupt for every channel (adjustable on various and multiple error sources).
- Time stamp of reception.
- Transmission delay timers.
- Individual activation and deactivation of channels.
- Diagnostic capability: automatic routing of received messages from selectable or all channels to be output on another (diagnostic) channel.
- Diagnostic mirroring capability: automatic routing of received and sent messages from selectable or all channels to be output on another (diagnostic) channel.
- Self-test modes with internal and external (including transceiver) loop, to fulfil ISO safety requirements.
- Listen-only mode for bus analysis purposes.

3.3.2 Initialization of RS-CAN

3.3.2.1 Operation Modes

The RS-CAN module is able to communicate with several CAN channels, where each channel may have its individual configuration. For this reason, besides the global operation mode, there are operation modes for each channel. After a hard reset, the RS-CAN module is globally disabled, which means that all operation modes are set to *sleep mode*. As a general rule, the operation modes of the channels always follow the global operation mode in the direction of shutdown or stopping, but the channels can only be moved into the activation direction if the global operation mode already has this state.

Sleep Mode After its entry upon a hard reset, the RS-CAN module automatically initializes its local RAM, where messages, lists, queues, and configurations are stored. Consequently, all settings are well defined to start up values. There is no need to clear any memory with software. The completion of the initialization process is indicated by a flag.

The sleep mode disables the channels' clocking in order to save power effectively.

Reset Mode In this mode, the configuration can be changed. Global reconfiguration covers the definition of the memory usage and global behaviour, such as setting the AFL.

Within the channels, the reset mode allows users to set communication parameters such as bit timing.

Operation Mode A channel can be put into operation mode if this mode has already been set globally. At this point, the CAN channel starts its communication on the CAN bus.

3.3.2.2 Test Modes

Within the operation mode of a channel, several test modes are available besides regular operation.

Listen-Only Mode All transmit functionality is disabled. This is also effective for the *bus acknowledgement* and *error/overload reporting* on the CAN bus by the transfer layer. The CAN channel behaves as a listener on the CAN bus, but it cannot be seen by other bus participants. It is possible to use this mode to detect a valid baud rate among a known selection.

Self-Test Modes RS-CAN distinguishes between external and internal loops within the self-test modes. In general, the self-test modes are used to verify the functionality and safety of the RS-CAN with software.

The internal loop modes allow internal communication to enable internal transmitted messages to be received either in the same channel (using emulated *bus acknowledgement*) or by other internal channels. In internal loop mode, the CAN transceiver is not included, and the test messages are invisible for the other CAN bus participants.

In external loop mode, the CAN transceiver is also included in the test loop.

In all self-test modes, the transfer layer is fully included in the test path.

3.3.3 Transmission of Messages

RS-CAN includes four methods of sending messages: the classical use of message buffers, sending from a transmit queue, streamed sending through a FIFO, and auto-

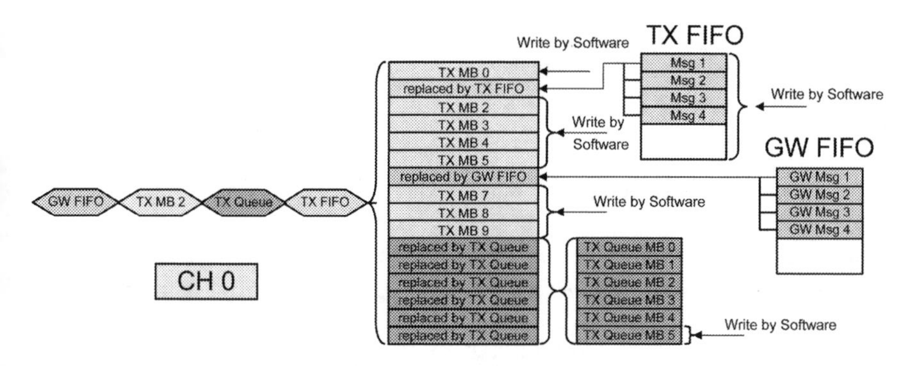

Fig. 3.6 Transmission of messages

mated routing by hardware (diagnostic and mirroring function). There are 16 transmit message buffers available for each channel, and these are used and shared for all these methods of sending messages. By means of sharing, all four sending methods can be enabled at the same time for a channel. Thus, by sharing a channel between several software applications, each application can use its favourite method. The priority of transmission is evaluated on message-buffer level; this means that all four methods of sending are in competition with each other according to the priority rules, but every method may have its internal rules as well.

Figure 3.6 shows an example of a valid usage of the message buffers and how they can be shared among the different sending methods.

3.3.3.1 Sending from Message Buffers

Every message buffer that has not been assigned to a transmit (TX) queue, a FIFO or to automated GW routing can be used in this way. When sending from a message buffer, this single message competes in priority with the remaining 15 message buffers. The message buffer stores all the information required to generate a valid and complete frame on the CAN bus.

In addition, there is a flag to enable the generation of an entry in the THL after successful transmission, and there is an optional pointer value (usable as *AUTOSAR HTH*), which will appear in the THL, too. This allows tracing and defined processing of transmit objects.

3.3.3.2 Sending from Transmit Queues

Several message buffers can be grouped to form a *transmit queue*, always starting with the uppermost buffer. Within this type of transmit queue, software simply writes all its messages into one single message buffer (the uppermost one), and the

RS-CAN hardware performs the sending according to priority rules. This method has the advantage that the software does not need to check for a free message buffer. RS-CAN indicates the fill state of the transmit queue.

3.3.3.3 Sending from Multi-Purpose FIFO

A *multi-purpose FIFO in transmit mode (TX)* has its own memory area to queue up messages. Again, software writes into a single location to feed in the messages, and the RS-CAN hardware takes care of sending and fill-level indication. The difference between this and the *transmit queue* is that within a FIFO, the sequence of messages will be kept, ignoring any priority rule (message identifier). One transmit message buffer must be assigned to a FIFO.

3.3.3.4 Sending from GW FIFO

If a *multi-purpose FIFO* is operated in *GW mode*, it can be assigned to be a reception target for other CAN channels, so that selected messages from them can be routed to it. In this configuration, the RS-CAN hardware performs all the tasks for this message routing without any necessary software interaction. Again, one transmit message buffer must be assigned to every multi-purpose FIFO.

The key prerequisites for smooth operation of the GW are that incoming messages are well selected from the channels by using the appropriate *access filtering list (AFL)* settings, and that the bus transport capacity of the output CAN channel is sufficient.

There is only one task that remains for the software, and this is the supervision of the FIFO overflow. If the configuration of the CAN channels allows (at least temporarily) more data to be routed through the hardware GW than the output channel can transmit, the FIFO may run into overflow, so that messages are lost.

3.3.3.5 Transmit History List

Every sent message can be recorded in the *THL*. The THL represents the confirmation for the software that a message has been sent and *acknowledged* by another CAN bus participant. The THL can generate interrupts on new entries or on fill level.

3.3.3.6 Transmission Intervals

When using the FIFO methods for sending messages, the minimum interval between two subsequently sent messages can be defined by an internal timer in RS-CAN. This functionality is required to fulfil *Transport Protocol* requirements of ISO 15765-2 and to avoid a full bus load caused by one node.

3.3.4 Reception of Messages

The reception of messages in RS-CAN is possible using several methods, but every method begins in the filtering section of the *multi-reception handler*. Here, the *AFL* determines where a received message will go. For each received message, the AFL is parsed for a match. If a match is found, the associated AFL entry contains up to eight storage targets, which can be loaded in parallel with the message. Valid storage targets are reception FIFO units, multi-purpose FIFO units, and a selectable receive message buffer.

An AFL entry is shown in Fig. 3.7.

The AFL entry stores *identifier* values (ID) for the standard and extended identifier frame formats of CAN, and associated *mask* flags, where the relevance of each identifier bit (including remote flag bit *RTR* and extended flag *IDE*) can be masked. Masked bits will be set as "don't care" for the filtering; this works like a so-called wildcard.

Reference *DLC* values can be entered for DLC checking. If a received message matches the ID, but does not have enough data bytes as specified in the DLC specification of the AFL entry, it will not pass the filter.

Furthermore, the AFL entry contains several pointers:

- A flat RX direction pointer with its associated enable flag (FE). If FE is set, the message will be stored in the receive message buffer with the number of the pointer value.
- One or several FIFO direction pointers. Here, each bit represents one of the available receive or multi-purpose FIFO units where the message can be stored.
- An additional pointer (PTR) value which is a freely configurable value. This value will be attached as a property with the message, so that an identification of the message is possible. This functionality corresponds with the HRH values of AUTOSAR communication stacks.

3.3.4.1 Reception into a Receive Message Buffer

The receive message buffer stores the whole message including the *PTR* value and a reception time stamp. Old data within the buffer are overwritten.

The method of storing a message in a receive message buffer is designed to be used in conjunction with polled message reception, i.e. for non-interruptive information, which is read and checked by software at certain intervals.

3.3.4.2 Reception into a FIFO Unit

This kind of reception is used for streamed and interruptive data processing. Sorting into different FIFO units makes it possible to distinguish between higher and lower priority messages. Every FIFO unit can be configured when its interrupt is

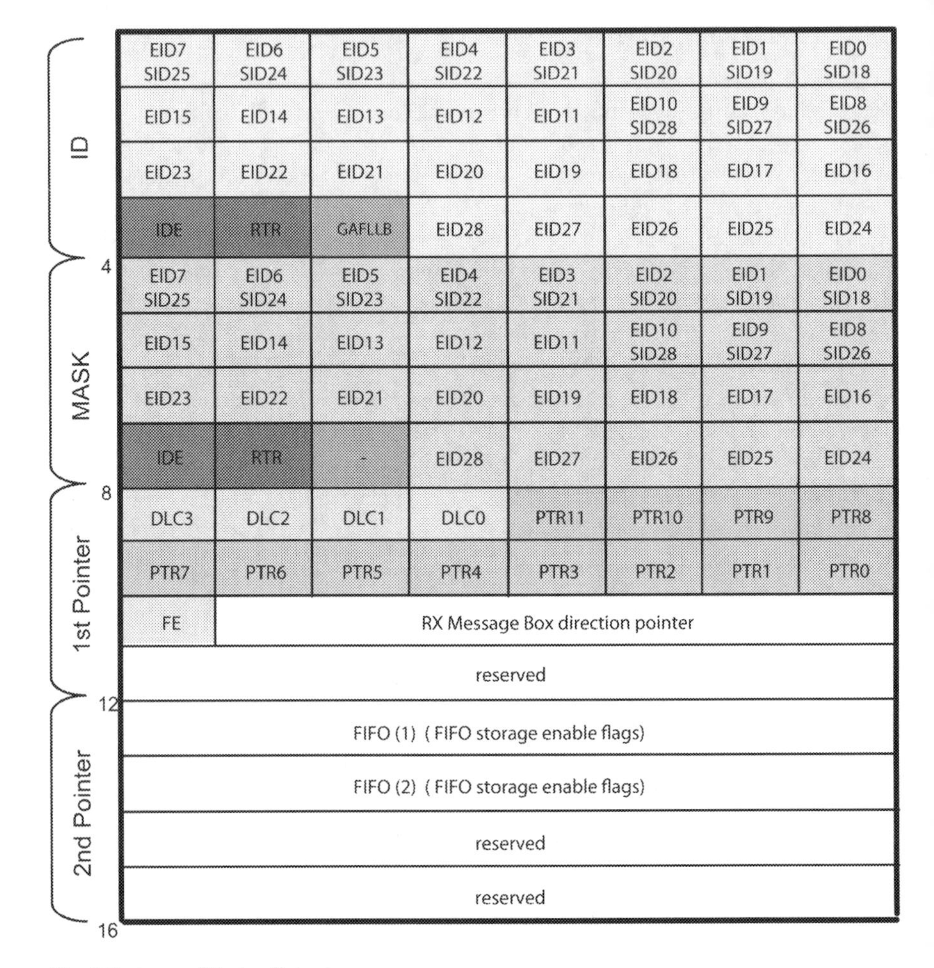

Fig. 3.7 Access filtering list entry

generated, depending on its fill level, even on every reception. Overflow of a FIFO is indicated, too.

Like the message buffers, a FIFO entry also contains the *PTR* value and the reception time stamp of the received message.

When receiving into a multi-purpose FIFO unit, hardware GW operations are possible. If enabled within a multi-purpose FIFO, its received messages can be transmitted by another channel.

- In diagnostic mode, a multi-purpose FIFO can collect received messages from one or several channels and send them to another channel.

- In loop back and mirroring mode, a multi-purpose FIFO collects received and transmitted messages from one (loop back) or several (mirror) channels, and sends them to another channel.
- At the same time, while performing the hardware GW functions, the messages can be copied into a standard receive message buffer, so that software can monitor all messages processed by the GW.

3.3.5 Summary

The RS-CAN module handles the increased complexity of current requirements for CAN controllers. Streamed data processing support for GWs with high efficiency is combined with flexibility in usage. Greatly enhanced filtering methods and shared resources for all channels allow RS-CAN to be adapted to most application needs. At the same time, the new structure of the RS-CAN hardware is smaller than those of its predecessors.

3.4 Infineon's CAN Modules of the XC16x- and XC2000/XE16x family

In this section, a short introduction to the two actual CAN modules of Infineon will be given. Both standards comply to "CAN 2.0B active". At the end of the section, an example how to do a GW application with an XC2000/XE16x family device (a 16/32-bit-microcontroller family with up to CAN nodes) is shown.

3.4.1 TwinCAN and MultiCAN from Infineon

The CAN modules of the current Infineon microcontroller families from Infineon are defined in a scalable approach. The CAN modules have a message control block and separated nodes, building one module. There it is possible to append the message objects to the node, wherever they are needed. It is possible to build FIFO message buffers as well as an automatic rerouting of messages, the so-called GW function. The GW does not cost any CPU performance.

The implementation of the TwinCAN module can be found on the XC16x microcontroller family. The MultiCAN module is available on the XC2000 family as well as on the 32-bit TriCore controllers and last but not least on the 8051-based family, the XC8xx microcontrollers. Therefore, a porting of CAN code among the families is given.

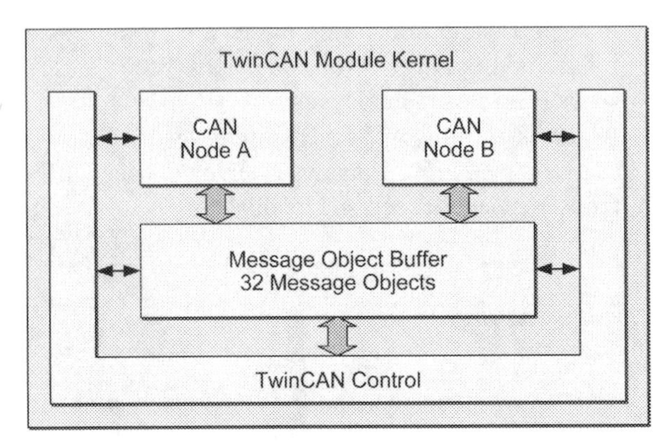

Fig. 3.8 Basic block diagram of the TwinCAN module

3.4.2 TwinCAN

The TwinCAN module can handle standard as well as extended identifiers. It is able to receive and transmit all identifier types.

- All 32 message objects:
  - All message objects can be assigned to one of the two CAN nodes.
  - All message objects can be used to receive or transmit.
  - All message objects can be part of a FIFO structure with a size of power of two.
  - All message objects do have a local acceptance mask.
  - All message objects do support frame counters for bus analysis (for example, statistics).
  - All message objects can be part of a GW.
  - All message objects can be used for remote monitoring in the GW use case.
- Up to eight interrupt nodes can be assigned to interrupt events.
- All nodes support the analyser function (listen mode).

First, the structure of the module is shown. The TwinCAN module is having a block of 32 message objects and two independent CAN nodes. Additional control logic makes these three blocks to act as one module. Figure 3.8 shows the basic block diagram of the TwinCAN module.

We start with a brief description of the message object function.

3.4.2.1 Message Objects

Each message object has a local acceptance mask. Therefore, it is possible to receive a group of identifiers. The acceptance mask is ANDed to the identifier. A 0 on a bit position means "don't care" for the TwinCAN module. Message objects can be part of a FIFO, a GW or one message object can build also a so-called Shared GW.

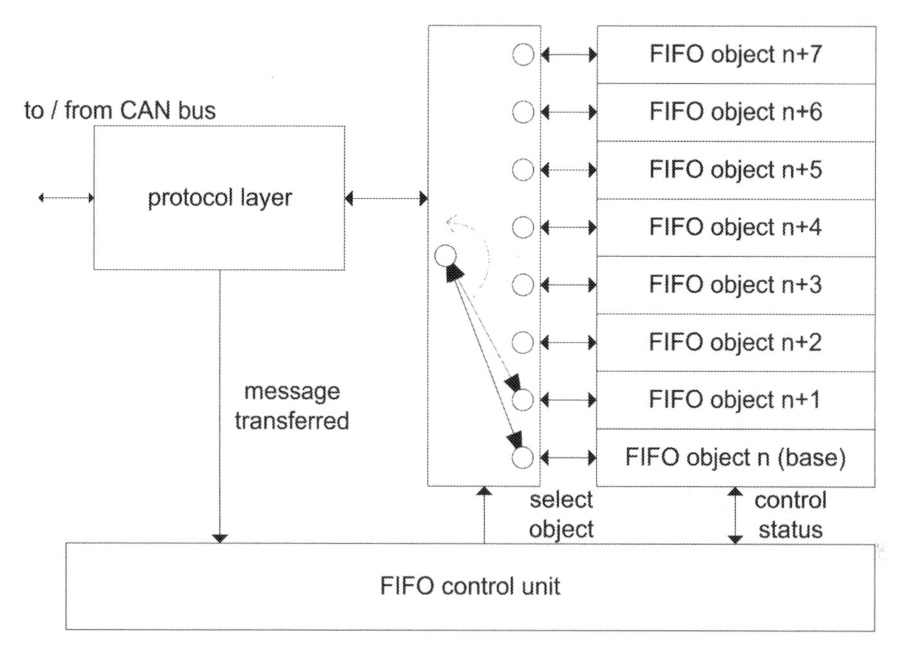

Fig. 3.9 FIFO block diagram

It is also possible to combine the FIFO and the GW feature. Each message object can trigger a receive or a transmit interrupt.

Each message object can be part of a FIFO, which is described here.

3.4.2.2 FIFO

With the TwinCAN module, the FIFO can consist of 2, 4, 8, 16, or 32 message objects. In a system, where specific messages or message groups are coming in, in a high frequency, the FIFO gives the possibility to buffer these messages until the CPU is ready to read out these messages. The probability of overwriting messages can be reduced. During reception, the CPU is not used, until the interrupt is triggered, at a predefined level. The FIFO is also available for transmission, so that the CPU can write all messages to be sent within one block and the TwinCAN module will take care of the transmission to the bus.

In addition to the FIFO, a rerouting function is available, the so-called GW mode (Fig. 3.9).

3.4.2.3 Automatic GW

The GW function of the TwinCAN module allows interconnecting two different bus systems. These two bus systems are allowed to run on different baud rates. No

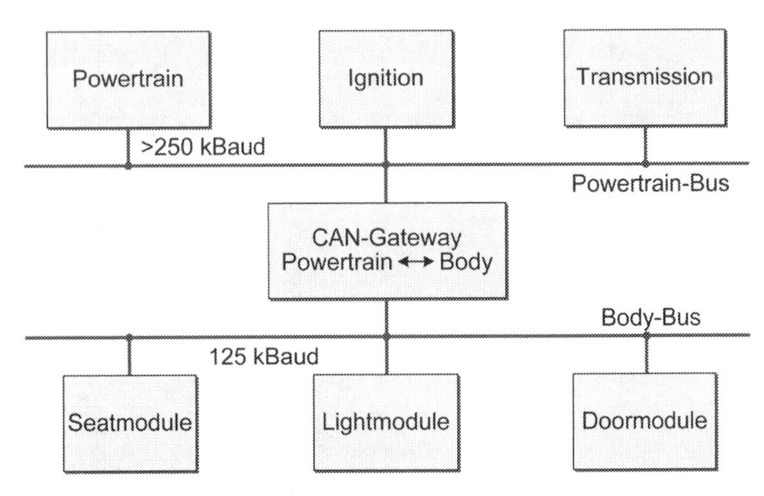

Figure 3.10 Example: CAN gateway for automotive

CPU load will be generated. A message received on bus A will be copied to bus B and depending on the settings, the transmission request can be set automatically; the amount of data bytes can be reduced and/or the identifier can be changed automatically. This feature is quite useful also in combination with the GW feature. The FIFO/GW combination allows to reroute messages automatically between two buses running at different speeds, without overwriting message objects, or to buffer their contents via software.

Figure 3.10 shows an example of the speed; information can be automatically forwarded from the Powertrain module to the door module, to lock the door at speeds greater than 30 km/h automatically.

In the following example, a message block of four messages having the same identifier shall be routed to a bus running at a lower baud rate.

3.4.2.4 FIFO/GW Combination

To route the messages from bus A to bus B (Fig. 3.11) in an optimal way, a message object assigned to bus A has to be configured a source GW object and to point to a four-state FIFO, assigned to bus B. This combination allows rerouting the messages automatically and it does not cost any CPU performance after initialization. The received messages are copied to the FIFO and the message will be sent according to CAN prioritization rules. Software activities are only needed, in case the data bytes are changed. No software interaction is needed, in case the data length core or the identifier is changed. Be aware that in case the data length code is increased, the new data byte will include a 0×0, if not changed by software.

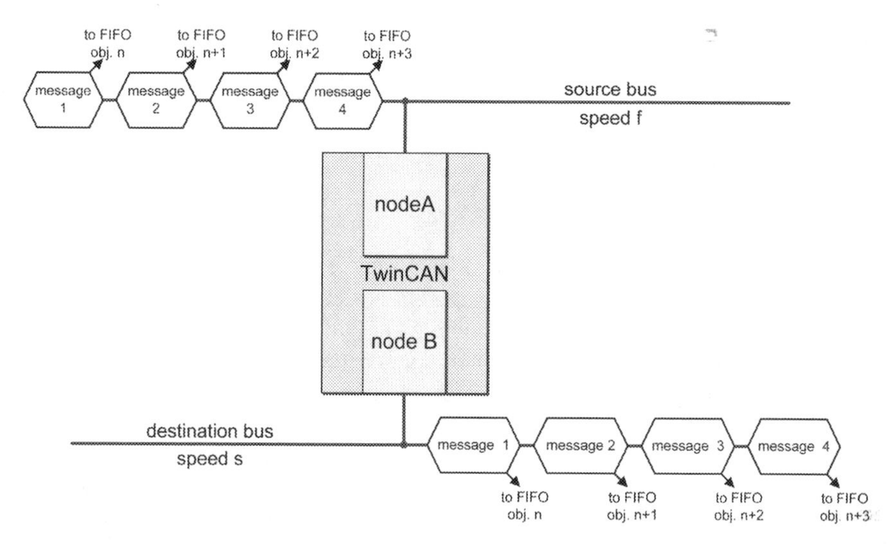

Fig. 3.11 FIFO/gateway running at different speeds

3.4.2.5 Shared-GW-Modus

In case of sporadic messages, which shall be rerouted to the other bus, it is possible to configure a single message object to be a GW. This is the so-called Shared GW Mode. If a message is received, the message object switches to transmission on the second CAN node. The message is sent according to CAN prioritization rules. Depending on the settings, it switches back automatically. This feature only exists on TwinCAN but not on its successor, the MultiCAN module.

Another feature of the TwinCAN module is the so-called analyser mode.

3.4.2.6 Analyser Mode

Starting with TwinCAN, a so-called analyser mode has been introduced to Infineon-microcontrollers. It is comparable to a listen-only mode. This feature gives the possibility to switch a CAN node silent. The CAN node will listen to bus and save the understood messages with the corresponding identifiers into the message objects, but is not taking part actively on the bus. The node will not send any acknowledgement or any error signalling to the bus. If a baud rate detection is implemented, this mode can be used to switch the node silent during the detection phase and not to spam the bus with error messages. If the CAN module becomes active, the analyser feature needs to be disabled, but not the complete node needs to be reconfigured.

With this feature, it is also possible to have both nodes on the very same bus, but only one node is active. With the help of the reception on the second node, a software comparison can be done. This is useful in case of safety application (higher levels) to guarantee the correct reception. The analyser feature can also be applied to disconnect a failing node which permanently disturbs communication from the bus in order to enable the remaining still functional system to continue communication.

3.4.2.7 The Interrupt System

TwinCAN has 72 interrupt sources; 32 interrupt sources are for reception and another 32 for transmission, as the very same message object can trigger a receive as well as a transmit interrupt. (In case of a remote frame) For each CAN node, four status and error interrupts are existing. The interrupt generation of the TwinCAN module allows to have up to eight independent interrupt routines as eight interrupt nodes can be assigned. Therefore, it is quite easy to have a well-defined prioritization among the events. Thus, it is possible to assign a Peripheral Event Controller (PEC) or Direct Memory Access (DMA) to a special message object, without having a complete interrupt routine and having a fast copy process from the TwinCAN to the needed memories.

The successor of the TwinCAN module is the MultiCAN module. The MultiCAN module can be found on the XC2000/XE16x family, the XC8xx family, and on all devices of the Audo Next Generation, Audo Future, and Audo MAX family. By having the same CAN module over all families, the software compatibility of over 8-, 16-, and 32-bit microcontrollers is given to a high extent.

3.4.3 MultiCAN

Like for the TwinCAN module, first the core functions of the module will be explained briefly before discussing the most important point in very detail.

The MultiCAN module offers:

- CAN functionality, which is V2.0B active.
- A CAN bus analyser mode and baud rate detection mechanisms for each CAN node.
- Up to 256 (dependent on the device) message objects that
  - can be assigned to the CAN nodes
  - can be used for transmission or reception
  - offer the remote monitoring mode in case of GW
  - have “Frame Counter Monitoring”.
- Acceptance filtering:
  - Each message has its own local mask, which allows to receive a group of messages.
  - Each message object is able to receive and transmit messages in standard or extended format, and by masking also both types can be received via the very same message object.
  - It is possible to have different prioritization rules on the internal arbiter, running on the message objects.

- Each message object can be part of a FIFO of any size. The only limitation is the absolute amount of message objects. The message objects are part of a double-chained list and the FIFO is done in the very same structure. The double-chained lists can be changed during runtime.
- Each message object can be part of a GW, rerouting messages from one CAN bus to another.
- The list/a part of list can be rerouted to another CAN node at any time.
- Up to 16 interrupt nodes can be assigned to interrupt sources within the module.

The MultiCAN module can include up to 256 message objects and eight CAN nodes. Depending on the implementation, the amount of CAN nodes and message objects can be different from one controller to another. For example, a XC878 included two CAN nodes and 32 message objects, a TC1167/TC1767 has two CAN nodes and 64 message objects, the TC1797 has four CAN nodes and 128 message objects and most family members of the XC2000/XE166 family are available with six CAN nodes and 256 message objects. All members of the Audo MAX family do include the MultiCAN module. Like the TwinCAN module, the module is split between node logic and message objects. Node logic and the message objects are combined and made on module by the control logic. The message objects can be freely assigned to any of the nodes. However, they are not part of a static structure, which is controlling the message objects, but they are part of a list structure. Therefore, it is possible to reassign message objects during runtime, but also to have FIFOs of any size. Figure 3.12 shows a block diagram of the module.

Like on TwinCAN, each message object has a local mask, giving the possibility to receive a group of message identifiers. The flexibility of the FIFO has been increased by using the list structure within the module. Therefore, all message objects (against TwinCAN being limited) can be part of a FIFO, GW, or FIFO/GW combination. The message objects can be scattered over the RAM and do not need to be behind each other.

Figure 3.13 shows an example, with a FIFO consisting of message objects 5, 16, and number 3. Message object 5 is the source object. The source object does not need to be part of the FIFO. Message object 3 is the end of the list and, therefore, pointing to itself. The current position within the FIFO is shown by the pointer CUR(rent). The terminology in Fig. 3.13 is identical to that in the User's Manual.

With the help of the list structure, a GW can be built. In contrast to TwinCAN, the number of the message object is no longer relevant to build such a GW.

As on TwinCAN, the FIFO and the GW feature may be combined to a GW/FIFO feature. By using the list structure, it is possible to change the FIFO size during the runtime to react flexibly on different busloads. For example, in case that a diagnosis is activated, a different set of message identifiers becomes relevant and the software is able to react to this change.

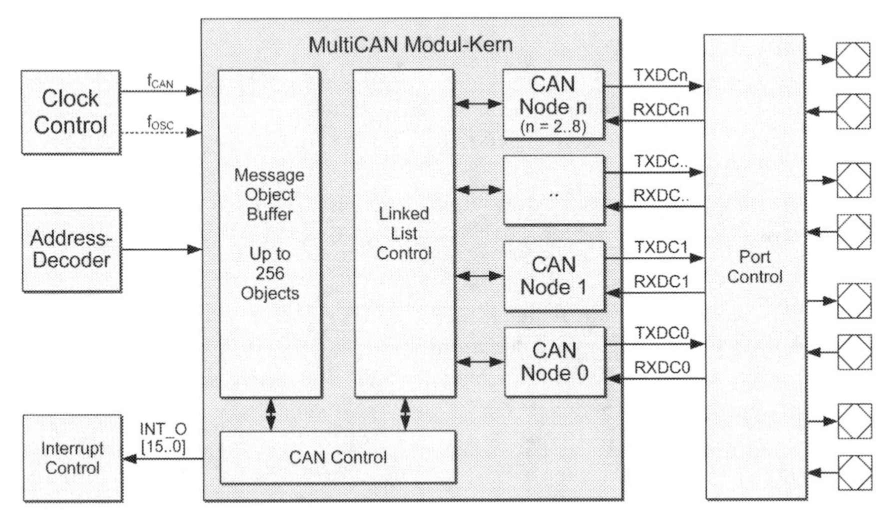

Fig. 3.12 Block diagram of the MultiCAN module

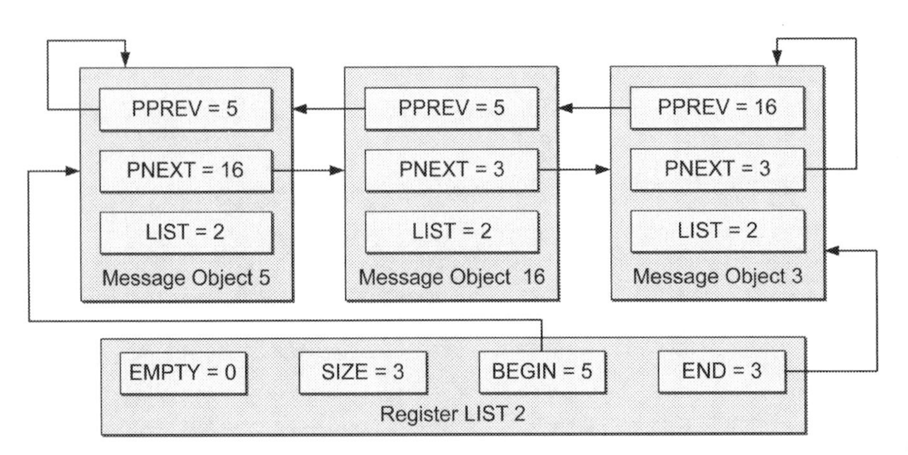

Fig. 3.13 List structure of the MultiCAN

3.4.3.1 Advantages of the MultiCAN

Besides the already explained list structure, the MultiCAN module is also having an analyser mode. This allows, for example, auto-baud detection or supervision of incoming traffic for safety applications with a second node. A module internal counter also allows calculating the amount of time quanta within one bit. By having such a function, the automatic baud rate detection becomes easier.

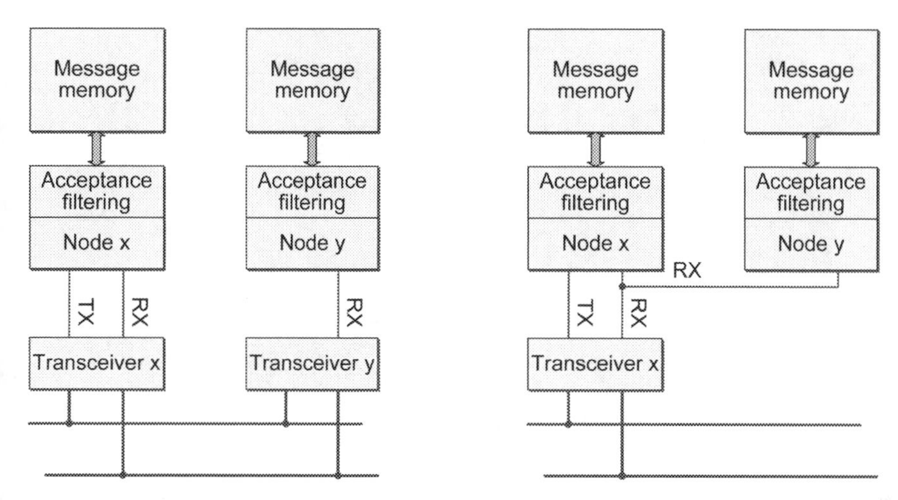

Fig. 3.14 System with two redundant CAN nodes connected with two transceivers (*left*) as well as system with only one transceiver (*right*)

On MultiCAN, a feature has been introduced to measure the actual transceiver delay and to check if the bus termination is properly done. By being able to measure the time between a falling edge being transmitted by the CAN node, until this falling edge is received, the actual transceiver delay can be measured.

3.4.3.2 MultiCAN Supports CAN-Debugging

For safety critical applications, MultiCAN is giving some possibilities to support fault tolerant implementations. An example is shown within this section.

If two CAN nodes are connected to the very same bus, we do have the possibility as shown in Fig. 3.14.

Two CAN nodes, via two CAN transceivers, are connected to the very same bus (Fig. 3.14, left). The advantage is at the same time the disadvantage of the concept. Two transceivers are used, which means higher cost. As one module shall only listen to bus, the transmit line does not need to be connected. The comparison between the messages is done via software. Due to the fact that two transceivers are used, also transceiver errors can be found, as two transceivers are used.

In the right part of Fig. 3.14, a similar approach is shown, having only one CAN transceiver. The advantage is on the cost side, having only one transceiver. The disadvantage remains that here transceiver errors cannot be found.

In case of having two separate modules, small disturbances can be seen on both nodes at the very same time; therefore, such an issue remains undetected. The MultiCAN module helps to overcome the absolute synchronism.

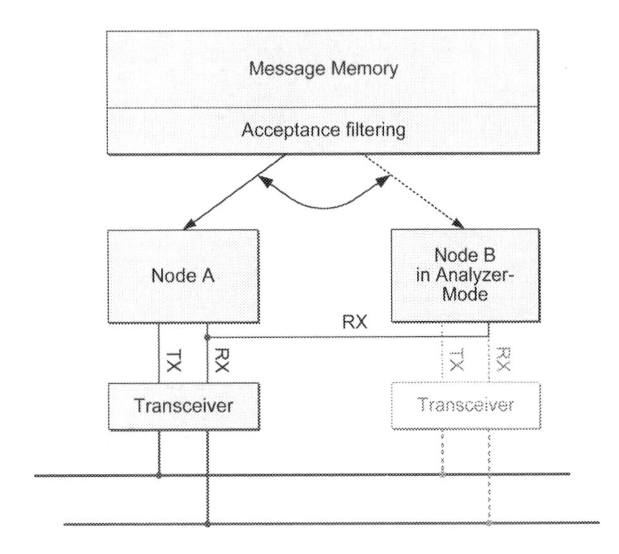

Fig. 3.15 MultiCAN in analyser mode

3.4.3.3 MultiCAN in Analyser Mode

As already described, the CAN node is only listening to the bus, if it is switched to the analyser mode. It is not actively participating in bus traffic. The message objects are assigned to the nodes. If the application needs a higher safety level and, therefore, has to test if messages are properly received, two nodes have to be on the same bus, receiving the same messages, and the application has to compare the received information. As the nodes are transferring the information not exactly at the very same time, short time disturbances can be detected. Therefore, an erratic behaviour between the protocol machine and the nodes can be detected. The amount of unknown errors, relevant for high-level safety applications can be decreased (only necessary in case of Safety Integrity Level 3, SIL3, or Automotive Safety Integrity Level C, ASIL C, and higher). This concept can be implemented with one or two transceivers, depending on the safety level. A possible solution is shown in Fig. 3.15; the grey line shows the solution with two transceivers. Node A is taking part in the bus traffic, whereas node B is in the analyser mode. The messages are saved with a small offset; therefore, short time disturbances become detectable. If further redundancy is needed, both CAN nodes can be connected, so that in case of a failure on node A, node B can take over and signal the situation to the network and, in case required, take over the role of the active node. These concepts are especially interesting for safety applications, for example, with the TC1387 designed for ASIL C and SIL 3 applications.

In addition, the MultiCAN can be used to detect errors within bus termination. The time between the first dominant edge and the sample point can be measured (Fig. 3.16) with the frame counter, so bus extensions are made easier.

The analyser mode and the measurement of bus termination help to detect faulty transceiver circuitries on a bus system.

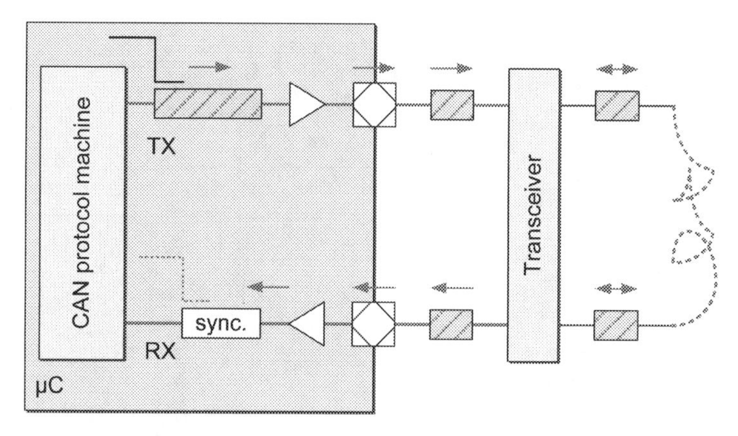

Fig. 3.16 Tx to Rx measurement

3.4.3.4 MultiCAN: Flexible Interrupts

MultiCAN has up to 16 interrupt nodes, called Service Request Nodes (SRN), which can be assigned to interrupt events on the MultiCAN module. Each message object can cause an interrupt on receive, transmit, or both. The frame counter can be used to get time and message information. It is also having an overflow interrupt, to enable a better monitoring of the CAN traffic. Therefore, the initialization is assigning interrupt events (Fig. 3.17) to interrupt nodes. All used interrupt sources are mapped to an interrupt node (an interrupt node can handle more than one interrupt source). This is done via bit-field setting. It is also possible to keep some interrupt nodes on polling, but to handle others within an interrupt routine.

By having this variety of interrupt sources and interrupt nodes, it is possible to have, for example, status information on a lower interrupt level or to poll these, but, for example, error situations like a bus-off (Alert, ALRT) can be handled on higher priority levels. Therefore, an appropriate reaction to an event can be assigned.

3.4.4 The XC2000 Family in GW Applications—An Application Example Using MultiCAN

Microcontrollers of the XC2000/XE16x family do have the "C166SV2 Core" with MAC (multiplier-accumulator, set of digital signal processor (DSP) functions) unit. The very high-end one even includes a cache. They are the logical progression of the previous C16x and XC16x family. The CPU and the peripherals can run up to 128 MHz. The family members do have different MultiCAN modules, with different amount of nodes and message objects. Most of the family members, besides the very low-end one, do have six nodes and 256 message objects. In addition, depending on the device, up to ten Local Interconnect Network (LIN) buses can

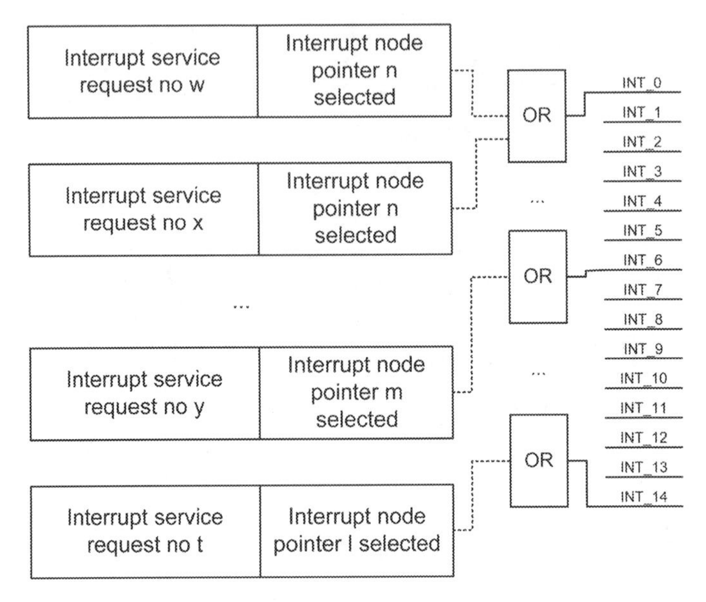

Fig. 3.17 Assignment of interrupts

be handled, in the very high-end one of the XC2000 family, in which a FlexRay module is integrated.

In this example, the CAN module is having six nodes and 256 message objects. The XC2000 family is a scalable microcontroller family, scalable as the usage of a different package will not cause a change in software.

The following example also works with the TriCore family and, in a restricted way, also with the XC8xx family.

3.4.4.1 GW Between Two CAN Bus Systems

In a GW application, different ways of message rerouting are needed. On the one hand, there are messages, which simply need to be transferred to another bus; these messages can be rerouted by the automatic GW feature. The GW mode is configured to copy the complete message and to transfer it directly afterwards. Here no interaction with software is necessary; all actions can be handled within the hardware. Other messages have to be changed in some of the bytes. In this case, the GW mode needs to be configured in such a way that the transfer to the destination message object takes place, but there is no automatic sending, as the software still needs to do the change. Nonetheless, a software transfer is not necessary. The transfer can be, for example, signalled by an interrupt. The corresponding bytes are changed, a new message transmission is requested, and the message object will take part in the internal arbitration process. Other messages need to be sent with a different identifier; if it is always the very same, this change has to be done in the hardware, by using

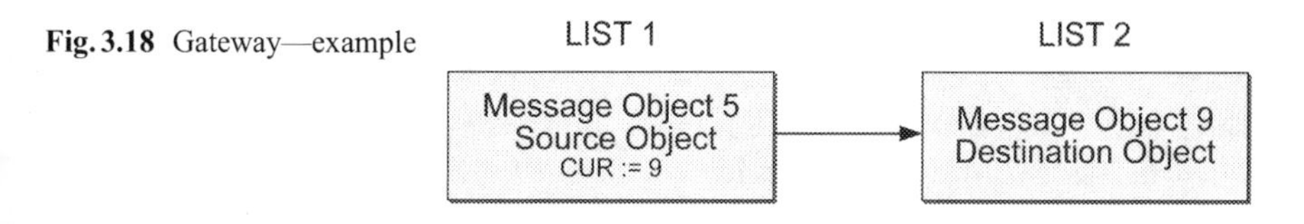

Fig. 3.18 Gateway—example

the feature appropriately. In case the identifier needs to be changed depending on the contents of the message, a software interaction is necessary. The basic actions are comparable to the change of data bytes.

In Fig. 3.18, a GW consisting of two message objects is shown. Message object 5 is defined as the source object and number 9 as the destination object by the usage of CUR. The different GW transfer features can be used as described above; they can be found within the register MOFCRn (here n = 5).

3.4.4.2 Gateway Between CAN- and LIN-Bus

A LIN-compliant node on the XC2000 is implemented on the USIC (Universal Serial Interface Controller) module. The USIC includes a module internal counter, which eases the baud rate measurement during BREAK signal. The module includes a FIFO structure, which allows in combination with the collision detection feature, to have only a single interrupt for the header and one for the frame response. For frames copied from one bus system (here CAN) to another or signal groups, it also eases the usage of writing to the module, as the information can be written at one time. The application does not need to reserve the occupied message objects for longer times or to buffer the messages within RAM in this case, as they can directly be written to the FIFO of the USIC. A CAN message can be copied in only for transactions from the CAN module to the USIC and then be sent via LIN protocol. The received amount of interrupts can be reduced as the FIFO structure allows not to receive the information byte-wise. The speed difference between CAN bus and LIN is normally quite high in an automotive environment, as the LIN bus has a maximum baud rate of shortly below 20 kBaud. Therefore, buffering LIN messages has a high impact on the application.

3.4.4.3 Gateway from CAN to FlexRay

Depending on the XC2000 family device, it is possible either to use a device internal FlexRay module or to use the CIC-310 as extension device, to enable FlexRay communication. The different baud rates need some buffering, especially transferring data from the time-driven high-speed FlexRay bus to the (compared to FlexRay) low-speed event-driven CAN bus. Here the CAN module internal FIFO structures ease handling the traffic coming from the FlexRay module and have to be sent via CAN bus.

3.4.4.4 Outlook

The MultiCAN module includes several useful extensions against the TwinCAN module. The list structure grants a high degree of flexibility to user and more freedom for CAN applications. The double-chained list-structure enables a highly flexible FIFO structure, as the FIFO elements can be collected among the CAN module. Therefore, the reception on faster CAN bus can be buffered in-between for the transmission on a low-speed bus. The USIC on the XC2000 family rounds out this flexibility to the LIN bus. The external bus controller (EBC) allows attaching an external FlexRay device, if this is not included internally. Therefore, the currently existing serial bus systems widely used in automotive environment are available on the device. The FIFOs within CAN and USIC, which is used for LIN, ease the implementations and reduce the overall CPU load for transferring data from one bus system to another. However, these features become useful not only for automotive applications but also for industrial applications. The MultiCAN is used on several members of the industrial microcontroller families, for example, the TC11xx, XE16x, and XC8xx. These devices are optimized for automation and industrial drivers.

3.5 Xilinx CAN-Controller LogiCORE™ IP

The features of Xilinx CAN Controller are that it:

- Conforms to ISO 11898-1, CAN 2.0A, and CAN 2.0B standards
- Supports Industrial (I) and Extended Temperature Range (Q)
- Supports standard frames (11-bit identifier) as well as extended frames (29-bit identifier)
- Supports bit rates up to 1 Mbps
- Transmits message FIFO with a user-configurable depth of up to 64 messages
- Prioritized message transmission through High-Priority Transmit Buffer
- Automatic re-transmission on errors or lost arbitration
- Receive message FIFO with a user-configurable depth of up to 64 messages
- Acceptance filtering by (a user-configurable number of) up to four acceptance filters
- Sleep mode with automatic wakeup
- Loop back mode for diagnostic applications
- Maskable error and status interrupts
- Has readable error counters (Fig. 3.19).

3.5.1 User Interface

The external interface of the CAN controller is a subset of the Xilinx Intellectual Property Inter-connect (IPIC) signalling. This enables the CAN controller to be

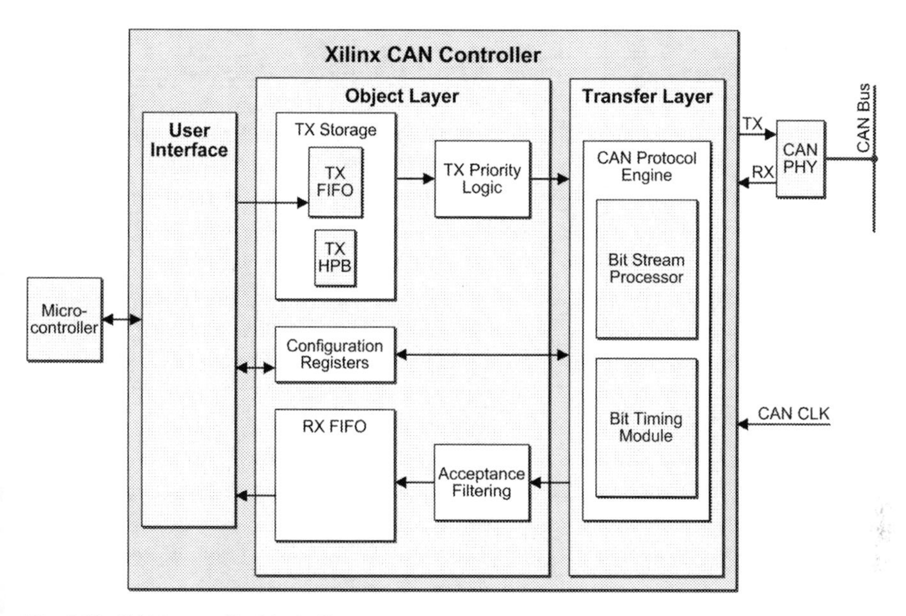

Fig. 3.19 CAN controller block diagram

interfaced to any microcontroller in a stand-alone mode. When coupled with an on-chip peripheral bus/processor local bus (OPB/PLB) Intellectual Property Interface (IPIF), which attaches to the core through the IPIC interface, the core can be connected to the MicroBlaze. This allows the core to be used in an Embedded Development Kit (EDK) environment. Table 3.1 describes the interface signalling of the CAN controller.

3.5.1.1 Interface Description

The CAN controller supports the following two modes of transfers

- Single read
- Single write

3.5.1.2 Single Read Transaction

For a read operation, when the transfer is enabled (Bus2IP_CS='1' and Bus2IP_RNW='1'), the core samples the address on the Bus2IP_Addr pins and returns the corresponding read data on the IP2Bus_Data pins. Read data are returned on a successive clock rising edge, after a wait time. IP2Bus_Ack is asserted when the data are ready on the IP2Bus_Data pins. For a read operation, it should be noted that address is assumed to be valid on the Bus2IP_Addr pins when Bus2IP_CS is asserted and the core samples the address on the next rising edge of SYS_CLK.

Table 3.1 External I/Os

#	IPIC name	I/O	Default value	Description
1	Bus2IP_Reset	Input	0	Active high reset
2	Bus2IP_Data(0:31)	Input	X"00000000"	Write Data bus
3	Bus2IP_Addr(0:7)	Input	"00000000"	Address Bus
4	Bus2IP_RNW	Input	1	Read or Write signalling '1' for a Read Transaction '0' for a Write Transaction
5	Bus2IP_CS	Input	0	Active high CS
6	IP2Bus_Data(0:31)	Output	X"00000000"	Read Data bus
7	IP2Bus_Ack	Output	0	R/W data acknowledgement
8	IP2Bus_IntrEvent	Output	0	Active high interrupt line. See Note 1.
9	IP2Bus_Error	Output	0	Active high R/W error signal. Reserved for future use.
10	CAN_PHY_TX	Output	1	CAN bus transmit signal to PHY
11	CAN_PHY_RX	Input	1	CAN bus receive signal from PHY
12	CAN_CLK	Input		24 MHz oscillator clock input
13	SYS_CLK	Input		Input interface clock

Note 1: The Interrupt line is an edge-sensitive interrupt. Interrupts are indicated via the transition of the interrupt line from logic '0' to logic '1'

IP2Bus_Ack is asserted for all read transactions, irrespective of whether the transaction is valid or not. Successive read operations require that the Bus2IP_CS be de-asserted and reasserted. The timing diagram for a single read transaction is shown in Fig. 3.20.

It should be noted that

- read transactions from address locations defined as reserved return all '0's on the IP2Bus_Data bus,
- read transactions from write-only address locations return all '0's on the IP2Bus_Data bus,
- read transactions from the AFR register when C_CAN_NUM_ACF=0 return all '0's on the IP2Bus_Data bus,
- read transactions on the Acceptance Filter ID Register (AFIR) and Acceptance Filter Mask Register (AFMR) address locations when C_CAN_NUM_ACF=0 return all '0's on the IP2Bus_Data bus,
- read transactions on any or all of the AFIR and AFMR address locations when C_CAN_NUM_ACF>0 return the data that were written to these locations, and
- read transactions on an empty RX FIFO return invalid data.

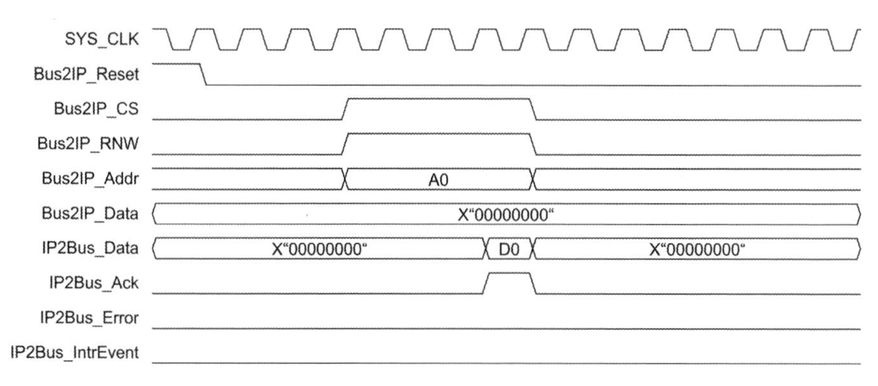

Fig. 3.20 Single read transaction

3.5.1.3 Single Write Transaction

For a write operation, when the transfer is enabled (Bus2IP_CS='1' and Bus2IP_RNW='0'), the core samples both address and data from the Bus2IP_Addr and Bus2IP_Data pins, respectively, and IP2Bus_Ack is asserted on a successive clock rising edge. For a write operation, it should be noted that address on the Bus2IP_Addr bus and data and Bus2IP_Data bus are assumed to be valid when Bus2IP_CS is asserted.

IP2Bus_Ack is asserted for all write transactions, irrespective of whether the transaction is valid or not. Successive write operations require that Bus2IP_CS be de-asserted and reasserted. The timing diagram for a single write transaction is shown in Fig. 3.21.

3.5.2 Object Layer

3.5.2.1 Transmit and Receive Messages

Separate storage buffers exist for transmit (TX FIFO) and receive (RX FIFO) messages through a FIFO structure. The depth of each buffer is individually configurable up to a maximum of 64 messages.

3.5.2.2 TX High-Priority Buffer

The Transfer High-Priority Buffer (TX HPB) provides storage for one transmit message. Messages written on this buffer have maximum transmit priority. They are queued for transmission immediately after the current transmission is complete, pre-empting any message in the TX FIFO.

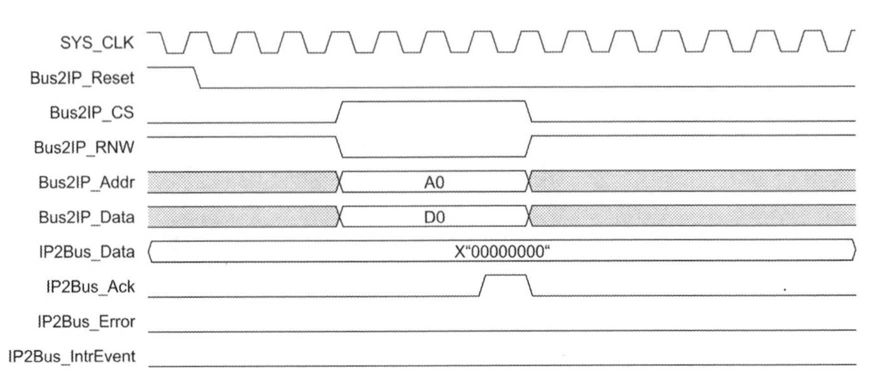

Fig. 3.21 Single write transaction

3.5.2.3 Acceptance Filters

Acceptance filters sort incoming messages with the user-defined acceptance mask and ID registers to determine whether to store messages in the RX FIFO, or to acknowledge and discard them. The number of acceptance filters can be configured from 0 to 4. Messages passed through acceptance filters are stored in the RX FIFO.

3.5.2.4 Configuration Registers

This module provides access to the registers through the external microcontroller interface.

Table 3.2 defines the CAN controller configuration registers. Each of these registers is 32-bit wide and is represented in big endian format. Any read operations to reserved bits or bits that are not used return '0'. A '0' should be written to reserved bits and bit fields not used. Writes to reserved locations are ignored.

3.5.3 Transfer Layer

3.5.3.1 Bit Timing Module

The primary functions of the Bit Timing Logic (BTL) module include:

- Synchronizing the CAN controller to CAN traffic on the bus
- Sampling the bus and extracting the data stream from the bus during reception
- Inserting the transmit bit stream onto the bus during transmission
- Generating a sampling clock for the Bit Stream Processor (BSP) module state machine

Table 3.2 Configuration registers

Register Name	Address	Access
Control Registers		
Software Reset Register (SRR)	0 × 000	Read/Write
Mode Select Register (MSR)	0 × 004	Read/Write
Transfer Layer Configuration Registers		
Baud Rate Presale Register (BRPR)	0 × 008	Read/Write
Bit Timing Register (BTR)	0 × 00C	Read/Write
Error Indication Registers		
Error Counter Register (ECR)	0 × 010	Read
Error Status Register (ESR)	0 × 014	Read/Write to Clear
CAN Status Registers		
Status Register (SR)	0 × 018	Read
Interrupt Registers		
Interrupt Status Register (ISR)	0 × 01C	Read
Interrupt Enable Register (IER)	0 × 020	Read/Write
Interrupt Clear Register (ICR)	0 × 024	Write
Reserved		
Reserved Locations	0 × 028 to 0 × 02C	Reads Return 0/Write has no effect
Messages		
Transmit Message FIFO (TX FIFO)		
ID	0 × 030	Write
DLC	0 × 034	Write
Data Word 1	0 × 038	Write
Data Word 2	0 × 03C	Write
Transmit High-Priority Buffer (TX HPB)		
ID	0 × 040	Write
DLC	0 × 044	Write
Data Word 1	0 × 048	Write
Data Word 2	0 × 04C	Write
Receive Message FIFO (RX FIFO)		
ID	0 × 050	Read
DLC	0 × 054	Read
Data Word 1	0 × 058	Read
Data Word 2	0 × 05C	Read
Acceptance Filtering		
Acceptance Filter Register (AFR)	0 × 060	Read/Write
Acceptance Filter Mask Register 1 (AFMR1)	0 × 064	Read/Write
Acceptance Filter ID Register 1 (AFIR1)	0 × 068	Read/Write
Acceptance Filter Mask Register 2(AFMR2)	0 × 06C	Read/Write
Acceptance Filter ID Register 2 (AFIR2)	0 × 070	Read/Write
Acceptance Filter Mask Register 3(AFMR3)	0 × 074	Read/Write
Acceptance Filter ID Register 3 (AFIR3)	0 × 078	Read/Write
Acceptance Filter Mask Register 4(AFMR4)	0 × 07C	Read/Write
Acceptance Filter ID Register 4 (AFIR4)	0 × 080	Read/Write
Reserved		
Reserved Locations	0 × 084 to 0 × 0FC	Reads Return 0/Write has no effect

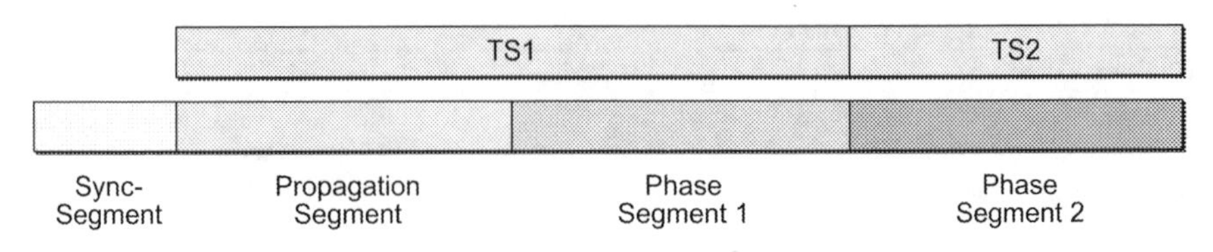

Fig. 3.22 CAN bit timing

Figure 3.22 illustrates the CAN bit time divided into four parts:

- Sync segment
- Propagation segment
- Phase segment 1
- Phase segment 2

The four bit time parts are comprised of a number of smaller segments of equal length called *time quanta* (tq). The length of each time quantum is equal to the quantum clock time period (period=tq). The quantum clock is generated internally by dividing the incoming oscillator clock by the baud rate prescaler. The prescaler value is passed to the BTL module through the Baud Rate Presale (BRPR) register. The propagation segment and phase segment 1 are joined together and called 'time segment1' (TS1), while phase segment 2 is called 'time segment2' (TS2). The number of time quanta in TS1 and TS2 vary with different networks and are specified in the Bit Timing Register (BTR), which is passed to the BTL module.

The Sync segment is always 1-tq long. The BTL state machine runs on the quantum clock. During the start-of-frame (SOF) bit of every CAN frame, the state machine is instructed by the BSP module to perform a hard sync, forcing the recessive (r) to dominant edge (d) to lie in the sync segment. During the rest of the recessive-to-dominant edges in the CAN frame, the BTL is prompted to perform re-synchronization.

During re-synchronization, the BTL waits for a recessive-to-dominant edge. After this is over, it calculates the time difference (number of tqs) between the edge and the nearest sync segment. To compensate for this time difference, and to force the sampling point to occur at the correct instant in the CAN bit time, the BTL modifies the length of phase segment 1 or phase segment 2.

The maximum amount by which the phase segments can be modified is dictated by the Synchronization Jump Width (SJW) parameter, which is also passed to the BTL through the BTR. The length of the bit time of subsequent CAN bits is unaffected by this process. This synchronization process corrects for propagation delays and oscillator mismatches between the transmitting and receiving nodes. After the controller is synchronized to the bus, the state machine waits for a time period of TS1 and then samples the bus, generating a digital '0' or '1'. This is passed on to the BSP module for higher level tasks.

3.5.4 Bit Stream Processor

The BSP module performs several MAC/logical link control (LLC) functions during reception (RX) and transmission (TX) of CAN messages. The BSP receives a message for transmission from either the TX FIFO or the TX HPB and performs the following functions before passing the bit stream to BTL:

- Serializing the message
- Inserting stuff bits, cyclic redundancy check (CRC) bits, and other protocol-defined fields during transmission

During transmission, the BSP simultaneously monitors RX data and performs bus arbitration tasks. It then transmits the complete frame when arbitration is won, and retrying when arbitration is lost. During reception, the BSP removes stuff bits, CRC bits, and other protocol fields from the received bit stream. The BSP state machine also analyses bus traffic during transmission and reception for Form, CRC, ACK, Stuff, and Bit violations. The state machine then performs error signalling and error confinement tasks. The CAN controller will not voluntarily generate overload frames but will respond to overload flags detected on the bus. This module determines the error state of the CAN controller: Error Active, Error Passive, or Bus-off. When TX or RX errors are observed on the bus, the BSP updates the transmit and receive error counters according to the rules defined in the CAN 2.0 A, CAN 2.0 B, and ISO 11898-1 standards. Based on the values of these counters, the error state of the CAN controller is updated by the BSP.

3.5.5 Configuring the CAN Controller

This section covers the various configuration steps that need to be performed to program the CAN core for operation.

The following are some of the key configuration steps:

- Choose the mode of operation of the CAN core.
- Program the configuration registers to bring up the core.
- Write messages to the TX FIFO/TX HPB.
- Read messages from the RX FIFO.

3.5.5.1 Programming the Configuration Registers

The following steps are to be performed to configure the core when the core is powered on or after system reset or software reset.

1. Choose the mode of operation
   - Write a '1' to the LBACK bit in the Mode Select Register (MSR) and '0' to the SLEEP bit in the MSR to choose loop back mode.

   - Write a '1' to the SLEEP bit in the MSR and '0' to the LBACK bit in the MSR to choose sleep mode.
   - Write '0's to the LBACK and SLEEP bits in the MSR to choose normal mode.

2. Configure the Transfer Layer Configuration Registers
   - Program the Baud Rate Priscilla Register and the BTR to correspond to the network timing parameters and the network characteristics of the system.

3. Configure the AFRs

The number of AFMR and AFIR pairs that are used is chosen at build time. To configure these registers, the following steps should be taken:

- Write a '0' to the UAF bit in the AFR register corresponding to the AFMR and AFIR pair to be configured.
- Wait till the ACFBSY bit in the Status Register (SR) is '0'.
- Write the appropriate mask information to the AFMR.
- Write the appropriate ID information to the AFIR.
- Write a '1' to the UAF bit corresponding to the AFMR and AFIR pair.
- Repeat the steps mentioned above for each AFMR and AFIR pair.

4. Write to the Interrupt Enable Register (IER) to choose the bits in the Interrupt Status Register (ISR) that can generate an interrupt.
5. Enable the CAN controller by writing a '1' to the CEN bit in the Software Reset Register (SRR).

3.5.5.2 Transmitting a Message

A message to be transmitted can be written to either the TX FIFO or the TX HPB. A message in the TX HPB gets priority over the messages in the TX FIFO. The TXOK bit in the ISR is set after the CAN core successfully transmits a message.

1. Writing a Message to the TX FIFO
   - Poll the TXFLL bit in the SR. The message can be written into the TX FIFO when the TXFLL bit is '0'.
   - Write the ID of the message to the TX FIFO ID memory location (C_BASEADDR + 0 × 030).
   - Write the DLC of the message to the TX FIFO DLC memory location (C_BASEADDR + 0 × 034).
   - Write the Data Word 1 of the message to the TX FIFO DW1 memory location (C_BASEADDR + 0 × 038).
   - Write the Data Word 2 of the message to the TX FIFO DW2 memory location (C_BASEADDR + 0 × 03C).

Messages can be continuously written to the TX FIFO until the TX FIFO is full. When the TX FIFO is full, the TXFLL bit in the ISR and the TXFLL bit in the SR are set. If polling, the TXFLL bit in the SR should be polled after each write. If using interrupt mode, writes can continue until the TXFLL bit in the ISR generates an interrupt.

2. Writing a Message to the TX HPB
   - Poll the TXBFLL bit in the SR. The message can be written into the TX HPB when the TXBFLL bit is '0'.
   - Write the ID of the message to the TX HPB ID memory location (C_BASEADDR+0×040).
   - Write the DLC of the message to the TX HPB DLC memory location (C_BASEADDR+0×044).
   - Write the Data Word 1 of the message to the TX HPB DW1 memory location (C_BASEADDR+0×048).
   - Write the Data Word 2 of the message to the TX HPB DW2 memory location (C_BASEADDR+0×04C).

After each write to the TX HPB, the TXBFLL bit in the SR and the TXBFLL bit in the ISR are set.

3.5.5.3 Receiving a Message

Whenever a new message is successfully received and written into the RX FIFO, the RXNEMP bit and the RXOK bits in the ISR are set. In case of a read operation on an empty RX FIFO, the RXNEMP bit in the ISR is set.

1. Reading a Message from the RX FIFO

The RXOK or RXNEMP bits in the ISR can be polled. In interrupt mode, the reads can occur after the RXOK or RXNEMP bits in the ISR generate an interrupt.

- Read from the RX FIFO memory locations. All the locations must be read regardless of the number of data bytes in the message.
- Read from the RX FIFO ID location (C_BASE_ADDR+0×050).
- Read from the RX FIFO DLC location (C_BASE_ADDR+0×054).
- Read from the RX FIFO DW1 location (C_BASE_ADDR+0×058).
- Read from the RX FIFO DW2 location (C_BASE_ADDR+0×05C).

After performing the read, if there are one or more messages in the RX FIFO, the RXNEMP bit in the ISR is set. This bit can either be polled or generate an interrupt. The process mentioned above should be repeated till the FIFO is empty.

3.5.5.4 CAN Graphical User Interface

The CAN graphical user interface (GUI) provides a single screen for configuring the CAN core. Parameter C_BASEADDR defaults to X"00000000" in the GUI, while the parameter C_HIGHADDR does not exist (Fig. 3.23).

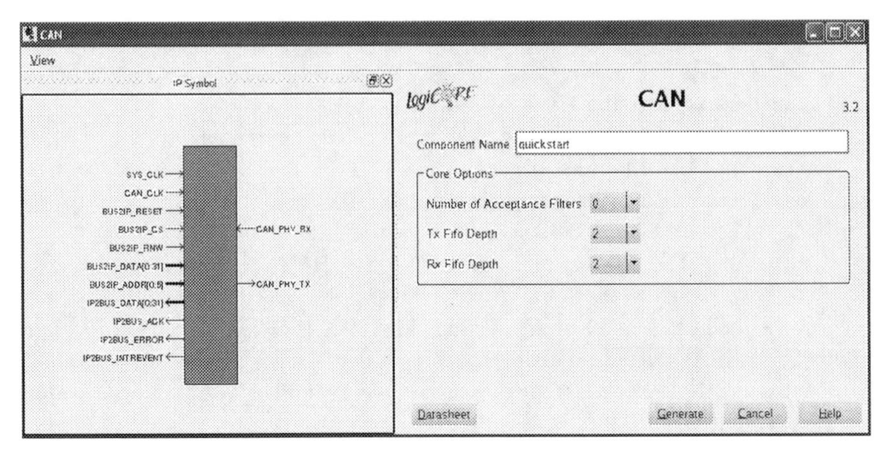

Fig. 3.23 LogiCORE

3.5.5.5 Component Name

Base name of the output files generated for this core. The name must begin with a letter and be composed of the following characters: a–z, A–Z, 0–9, and "-".

3.5.5.6 Xilinx CAN Controller Design Parameters

To obtain a CAN controller that is uniquely tailored to the minimum system requirements, certain features can be parameterized. This results in a design that utilizes only the resources required and gives the best possible performance. The features that can be parameterized in the CAN controller are shown in Table 3.3. The interface parameters C_BASEADDR and C_HIGHADDR need to be specified only when the core is interfaced to the OPB IPIF. For the core generated by CoreGen, C_BASEADDR defaults to X"00000000". C_HIGHADDR parameter does not exist for CoreGen cores.

Number of Acceptance Filters Valid range is from 0 to 4. This specifies the number of acceptance filter pairs used by the CAN controller. Each acceptance filter pair consists of a mask register and an ID register. These registers can be configured so that a specific Identifier or a range of Identifiers can be received. This determines the value of C_CAN_NUM_ACF.

TX FIFO Depth Valid values: 2, 4, 8, 16, 32, 64. This configures the depth of the TX FIFO.

The TX FIFO depth is measured in terms of the number of CAN packets. For example, TX FIFO with a depth of 2 can hold at most two CAN packets. This determines the value of C_CAN_TX_DPTH.

Table 3.3 Xilinx CAN Controller design parameters

Features	Feature/description	Parameter name	Allowable values	Default value
CAN controllerfeatures	Depth of the RX FIFO	C_CAN_RX_DPTH	2,4,8,16,32,64	2
	Depth of the TX FIFO	C_CAN_TX_DPTH	2,4,8,16,32,64	2
	Number of acceptance filters used	C_CAN_NUM_ACF	0 to 4	0
Interface	Base address	C_BASEADDR	32 bit address	None
	High address	C_HIGHADDR	32 bit address	None

RX FIFO Depth Valid values: 2, 4, 8, 16, 32, 64. This configures the depth of the RX FIFO. The RX FIFO depth is measured in terms of the number of CAN packets. For example, RX FIFO with a depth of 2 can hold at most two CAN packets. This determines the value of C_CAN_RX_DPTH.

3.5.6 Ordering the CAN Controller

A free evaluation version of the CAN core is provided with the Xilinx CORE Generator, which lets you assess the core functionality and demonstrates the various interfaces of the core in simulation. After purchase, the core may be downloaded from the Xilinx IP Center for use with the CORE Generator v9.2i and higher. The CORE Generator is bundled with ISE Foundation v9.2i software at no additional charge. Contact your local Xilinx sales representative for pricing and availability about the CAN LogiCORE module or go to the CAN product page www.xilinx.com/systemio/can/index.html for additional information.

Chapter 4
Higher Level Protocols

Gangolf Feiter, Lars-Berno Fredriksson, Karsten Hoffmeister, Joakim Pauli and Holger Zeltwanger

4.1 CANopen

CANopen is a standardized communication system that specifies communication profiles for the two lowest International Organization for Standardization (ISO)/ Open Systems Interconnect (OSI) layers as well as protocols for the ISO/OSI transport and application layers. Furthermore, CANopen provides device and application profiles in which transmitted process and configuration parameters as well as diagnostic information are specified.

The basic idea of CANopen was developed within an Elite Sport Performance Research in Training (ESPRIT) project. The registered association *CAN in Automation* (CiA) took over the first specification in 1994. Since then, the members of this international users and manufacturers association extend and maintain the CANopen specifications. Some of these specifications are freely available on the Internet, while the remaining specifications are only available for CiA members. Within the scope of this book, not all CANopen functionalities can be described. Further in depth going special literature is available.

G. Feiter (✉)
Concepts & Services Consulting, Alte Landstrasse 34,
52525, Heinsberg, Germany
e-mail: gangolf.feiter.csc@online.de

L.-B. Fredriksson
Kvaser AB, Aminogatan 25 A,
43153, Mölndal, Sweden

K. Hoffmeister
Elektrobit Automotive GmbH, Max-Stromeyer-Strasse 172,
78467, Konstanz, Germany

J. Pauli
Volvo Powertrain Corporation, Gropegårdsgatan,
SE-405 08, Göteborg, Sweden

H. Zeltwanger
CAN in Automation (CiA) GmbH, Kontumazgarten 3,
90429, Nuremberg, Germany

W. Lawrenz (ed.), *CAN System Engineering,* DOI 10.1007/978-1-4471-5613-0_4,
© Springer-Verlag London 2013

Table 4.1 Location of sampling point for several bit rates

Bit rate	Nominal bit time t_b (μs)	Valid range for location of sample point (%)	Recommendation location of sample point (%)
1 Mbit/s	1	75–90	87.5
800 kbit/s	1.25	75–90	87.5
500 kbit/s	2	85–90	87.5
250 kbit/s	4	85–90	87.5
125 kbit/s	8	85–90	87.5
50 kbit/s	20	85–90	87.5
20 kbit/s	50	85–90	87.5
10 kbit/s	100	85–90	87.5

Table 4.2 Maximal recommended bus and stub length

Bit rate	Bus length (m)	Stub length (max; m)	Accumulated stub length (max; m)
1 Mbit/s	25	1.5	7.5
800 kbit/s	50	2.5	12.5
500 kbit/s	100	5.5	27.5
250 kbit/s	250	11	55
125 kbit/s	500	22	110
50 kbit/s	1,000	55	275
20 kbit/s	2,500	137.5	687.5
10 kbit/s	5,000	275	1,375

Originally, CANopen was considered for usage in "embedded" networks in machine control systems. Meanwhile CANopen is used in a variety of industrial sectors as an "embedded" communication system. This also includes particular special purpose vehicles, trains, lifts and medical devices (e.g. computed tomography (CT) scanners) as well as ships.

4.1.1 *Profiles for the Lower Layers*

Currently, CANopen uses mainly the physical transmission according to ISO 11898-2. Alternative physical interfaces (ISO 11898-3 and Powerline) are being developed for specific applications. However, the bit timing is not explicitly specified in ISO 11898-1 and ISO 11898-2. Out of this reason, the CANopen communication profile (EN 50325-4 respectively CiA 301) defines sample points for several bit rates. Table 4.1 shows an overview of the bit timing parameters. Other bit rates are not allowed. Bit rates less than 50 kbit/s are not supported by all Controller Area Network (CAN) transceivers.

Table 4.2 shows the achievable bus length, on the basis of the recommended sample points, if direct current (DC) and alternating current (AC) voltage parameters for cables and connectors compliant to ISO 11898-2 are used. The actual

Table 4.3 CANopen pinning for the nine-pin DIN-Sub-D connector

Pin	Signal	Description
1	–	Reserved
2	CAN_L	CAN_L bus line (dominant low)
3	CAN_GND	CAN ground
4	–	Reserved
5	(CAN_SHLD)	Optional CAN shield
6	(GND)	Optional ground
7	CAN_H	CAN_H bus line (dominant high)
8	–	Reserved
9	(CAN_V+)	Optional CAN external positive supply (dedicated for supply of transceiver and optocouplers, if galvanic isolation of the bus node applies)

achievable bus length depends also on the internal delay times of the CAN controller and the CAN transceiver. For the exact design of the physical transmission, the formulae for line theory or appropriate empirical formulae should be used.

Basically, a linear bus topology with a 120-Ω termination resistor is required. Longer bus lengths may require the usage of a higher termination resistor. Not terminated bus lines are tolerable only to a certain degree, due to the reflections that could lead to a falsification of the bit value.

There are no detailed regulations on which bus cables or connectors should be used. However, there are recommendations for the pin assignments for a variety of connectors available. One of the widely used connectors is the nine-pin DIN-Sub-D connector (DIN41652). Table 4.3 shows the pin assignment given in the recommendation CiA 303-1.

CANopen uses the CAN protocol as described in ISO 11898-1. Although usage of remote frames is not recommended, they are basically allowed in some CANopen protocols. The various implementations of remote frames in CAN controllers could lead to problems when using CANopen protocols.

Some CANopen protocols use the base frame format (11-bit identifier) exclusively for data and remote frames, whereas other CANopen protocols allow the usage of the extended frame format (29-bit identifier). Error and overload frames are transparent for CANopen protocols. They are automatically processed by the CAN controller and lead, in case of an error frame, to an automatic retransmission of the interrupted frame or to a shutdown of the device (bus-off). Under rare conditions, it is to be kept in mind that a frame can be sent twice. Because of this situational condition, relative data or toggle commands shall not be transmitted.

4.1.2 Device Model

A CANopen-compatible field device features, from the hardware point of view, at least one CAN driver module and a CAN controller which implements the CAN protocol. Every CANopen device has to have its own object dictionary.

Table 4.4 Structure of CANopen object dictionary

Index range	Object
0000_h	Not used
0001_h–$001F_h$	Static data types
0020_h–$003F_h$	Complex data types
0040_h–$005F_h$	Manufacturer-specific complex data types
0060_h–$025F_h$	Device profile-specific data types
0260_h–$03FF_h$	Reserved
0400_h–$0FFF_h$	Reserved
1000_h–$1FFF_h$	Communication profile area
2000_h–$5FFF_h$	Manufacturer-specific profile area
6000_h–$67FF_h$	Standardized profile area 1st logical device
6800_h–$6FFF_h$	Standardized profile area 2nd logical device
7000_h–$77FF_h$	Standardized profile area 3rd logical device
7800_h–$7FFF_h$	Standardized profile area 4th logical device
8000_h–$87FF_h$	Standardized profile area 5th logical device
8800_h–$8FFF_h$	Standardized profile area 6th logical device
9000_h–$97FF_h$	Standardized profile area 7th logical device
9800_h–$9FFF_h$	Standardized profile area 8th logical device
$A000_h$–$AFFF_h$	Standardized network variable area
$B000_h$–$BFFF_h$	Standardized system variable area
$C000_h$–$FFFF_h$	Reserved

Table 4.5 Structure of the communication parameters

Index range	Object
1000_h–1029_h	General communication parameters
1200_h–$12FF_h$	SDO parameters
1300_h–$13FF_h$	CANopen safety parameters
1400_h–$1BFF_h$	PDO parameters
$1F00_h$–$1F11_h$	SDO manager objects
$1F20_h$–$1F27_h$	Configuration manager parameters
$1F50_h$–$1F58_h$	Parameters for program control
$1F80_h$–$1F91_h$	NMT parameters

Table 4.4 shows the structure of this parameter list. Table 4.5 shows the structure of the communication profile parameters. The single parameters of the object dictionary are addressable through a 16-bit index and an 8-bit sub-index.

If a field device implements several CANopen devices, it contains several object dictionaries as well. Additionally, it can provide bridge or gateway functionality. For example, the specification CiA 302-7 describes a CANopen-to-CANopen gateway with up to 32 CANopen devices, each having its own interface.

Three of the general communication parameters are mandatory:

- Device type (Object 1000_h)
- Error register (Object 1001_h)
- Identity object (Object 1018_h)

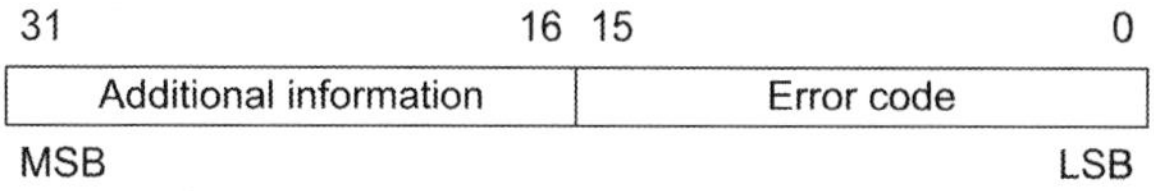

31	16	15	0
Additional information		Error code	
MSB			LSB

Fig. 4.1 Structure of the device-type parameter

Only the vendor ID has to be implemented in the identity object; all other parameters (product code, revision number and serial number) are optional. CiA uniquely assigns the vendor ID. Together with the other manufacturer-managed identity sub-parameters, every CANopen device can be addressed uniquely. The error register provides information if an error occurred in the device. The device type indicates which device or application profile is supported by the interface. The device-type parameter is a 32-bit value with the structure illustrated in Fig. 4.1. If the device does not support a standardized device or application profile, the profile number field shows a zero. If the CANopen device implements a standardized profile from CiA, it is shown through a number in the profile number field.

The value $FFFF_h$ in the additional information field indicates that this CANopen device supports several profiles. All other values are interpreted as profile specific.

The object dictionary provides space for eight logical devices (see Table 4.4). In case of several logical devices, the device type of the first logical device is implemented in index $67FE_h$. Each logical device has an 800_h address section available for profile parameters. The device-type object of the second logical device has the index $6FFE_h$.

Every logical device could contain several virtual devices. A minimum implementation of a virtual device may exist out of only one process date. More complex virtual devices comprise various process data and provides (if necessary) configuration data. The internal structure of a field device with several CANopen devices is illustrated in Fig. 4.2.

Every CANopen device must be assigned by the system developer with a unique 7-bit node ID. Because the node ID zero is reserved, a maximum of 127 devices is addressable in a CANopen network. The CAN identifiers, used in a variety of communication profiles to transmit and receive data frames, derive out of these node IDs. The setting and allocation of the node ID is not generally standardized in CANopen. The device manufacturer could, e.g. use a dual inline package (DIP) switch or provide an additional configuration interface. Another possibility is the coding of the connector: The device receives its node ID through an additional plug-in connection. In the CANopen profile for building door control systems, the patented node ID claiming procedure is used. When the CANopen interface is exclusively available, the Layer Setting Service (LSS), described in the specification CiA 305, can be used.

Every CANopen device must implement a communication state machine also referred to as network management (NMT) finite state automation (FSA) machine. Additional device-specific state machines could be necessary in the virtual devices. The NMT-FSA is part of the NMT, which is explained in the next subsection.

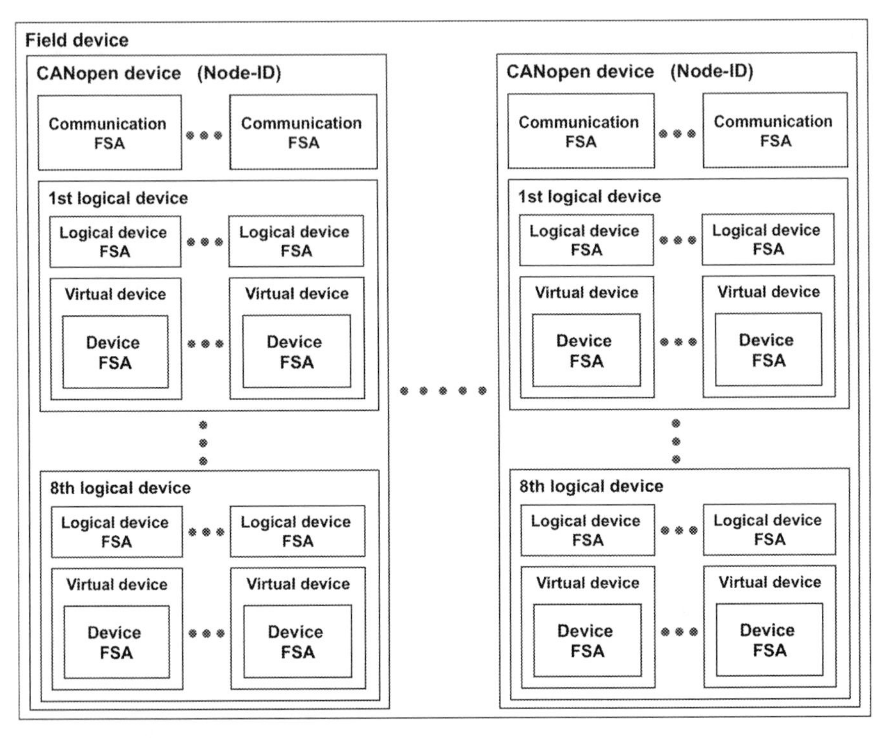

Fig. 4.2 Field device model

4.1.3 Network Management

The NMT is based on the master/slave principle. Only one active NMT master exists per CANopen network. The master controls all NMT-FSAs (referred to as NMT slave state machine) of the NMT slave devices. This also includes the FSA of its own CANopen NMT slave. To avoid misunderstanding, even the CANopen manager, which contains the NMT master, has to have an NMT slave state machine and an object dictionary. The so-called CANopen masters that do not have their own object dictionary are by definition not a CANopen device!

The NMT slave state machine supports four states (see Fig. 4.3): the volatile state "initialization" as well as the "pre-operational", "operational" and "stopped" states. The state transition from initialization to pre-operational takes place automatically. When the device reaches the pre-operational state, it starts its boot-up protocol, i.e. it sends its boot-up message, a single CAN data frame with a CAN identifier built-up out of the 4-bit function code (most significant CAN-ID bits) 1110_b as well as the 7-bit node ID (least significant CAN-ID bits). The boot-up message has a 1-byte data field that contains a zero. It is used to introduce the device after power-on or reset to the other members in the network. The same data frame is used in pre-operational and operational states to send the periodical *heartbeat protocol*. It contains

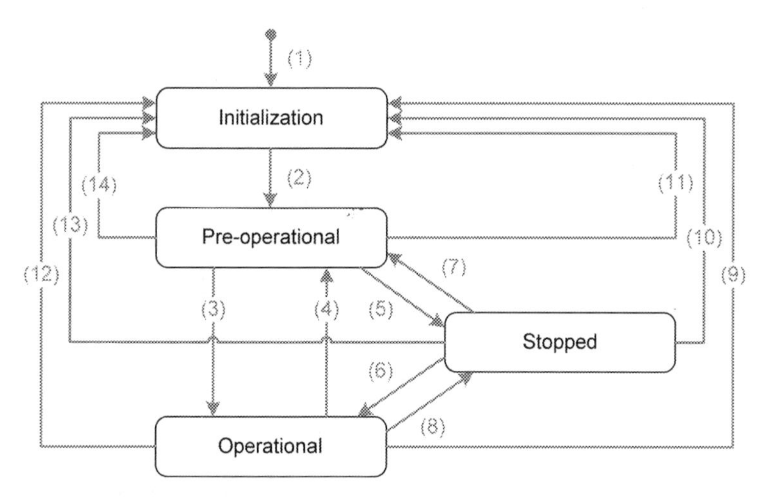

Fig. 4.3 CANopen NMT slave state

in 6 bits of the 1-byte data field the NMT state in which the device is situated at the moment (see Fig. 4.4). The frequency of the heartbeat could be configured in the *producer-Heartbeatheartbeat-time* (1017_h) parameter. If a CANopen device shall receive a heartbeat from another device, it has to be configured in the parameter set *consumer-heartbeat-time* (1016_h). Of course, the consumer time always has to be bigger than the corresponding producer time; otherwise, the device is always considered to be "lost".

If the consumer heartbeat time expires without reception of the corresponding heartbeat message, it will lead to an event (heartbeat event) in the heartbeat application of the device. The action is taken because this event is device or profile specific.

The state transitions are normally commanded by the CANopen device with NMT master functionality. For this purpose, the NMT master sends the CAN data frame with the highest prior identifier (CAN ID=0). In this 2-byte message, the first byte contains the command (command specifier) and the second byte contains the node ID of the device that shall perform the state transition. In case of an error, the devices are able to change their states autonomously. The value 0 is interpreted as broadcast command, meaning all nodes have to perform this command. The executive state transition could be configured in the *error-behaviour* (1029_h) parameter.

For applications where no master/slave NMT is allowed due to safety issues, flying master solutions, as specified in CiA 302, are available. The protocols, specified in this standard, allow that if a device with NMT master functionality has a failure another device activates its NMT master functionality. In fact, it is possible that more than one device with "sleeping" NMT masters are located in the network. If the NMT master with the highest priority returns into a functional state, after a temporary failure, it could reclaim the NMT with the help of the specified protocols.

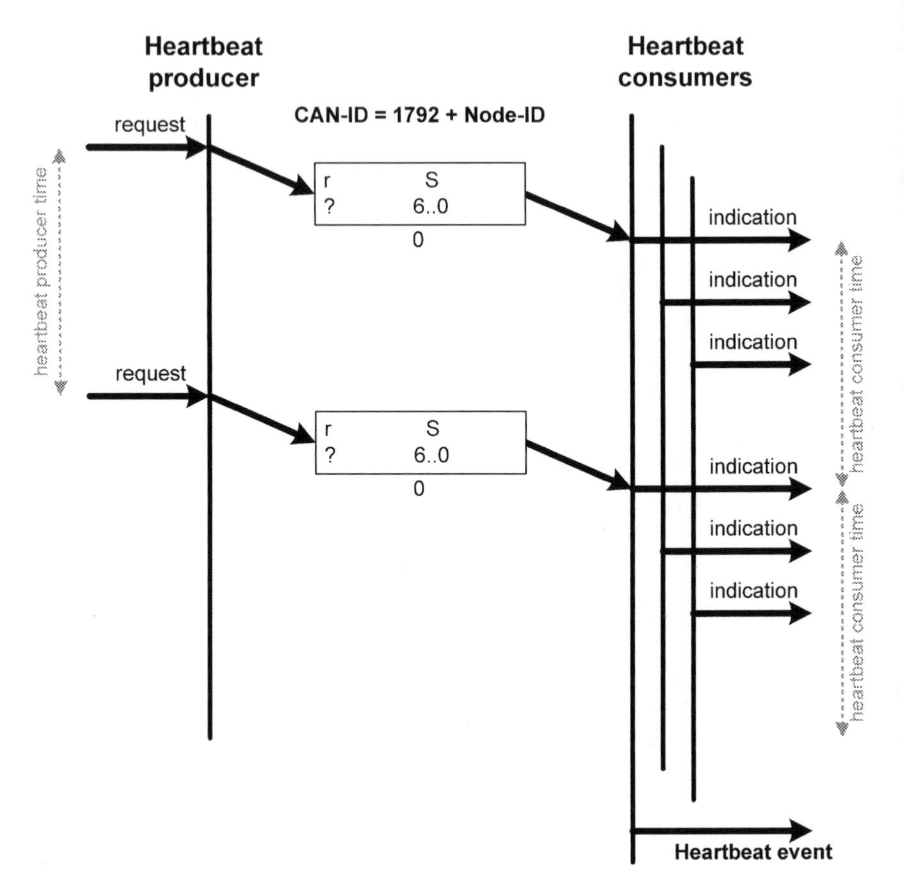

Fig. 4.4 Heartbeat protocol

4.1.4 Transport Protocols

The CANopen specification (EN 50325-4 respectively CiA 301) does not explicitly describe transport protocols. However, implicitly the service data object (SDO) protocols could be regarded as transport protocols in terms of the ISO/OSI reference model. The SDO protocols always work after the client/server model. The communication initiative always comes from the client and the server reacts to the "customers' wishes". SDO communication is allowed in pre-operational and in operational state. With SDO protocols, it is possible to write or read an object dictionary entry. Addressing takes place through the 16-bit index and the 8-bit sub-index. The two 8-byte data frames making an SDO are needed for: one, that the client could send his commands (command specifier) and if necessary the parameter data, and, the second one, so that the server could reply if the command was executed and could transmit in case of a read access the desired parameter data.

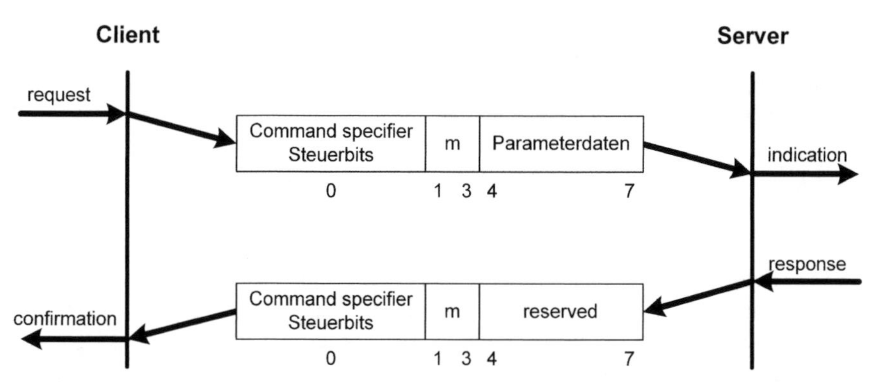

Fig. 4.5 SDO download protocol for not segmented data (m = index and sub-index)

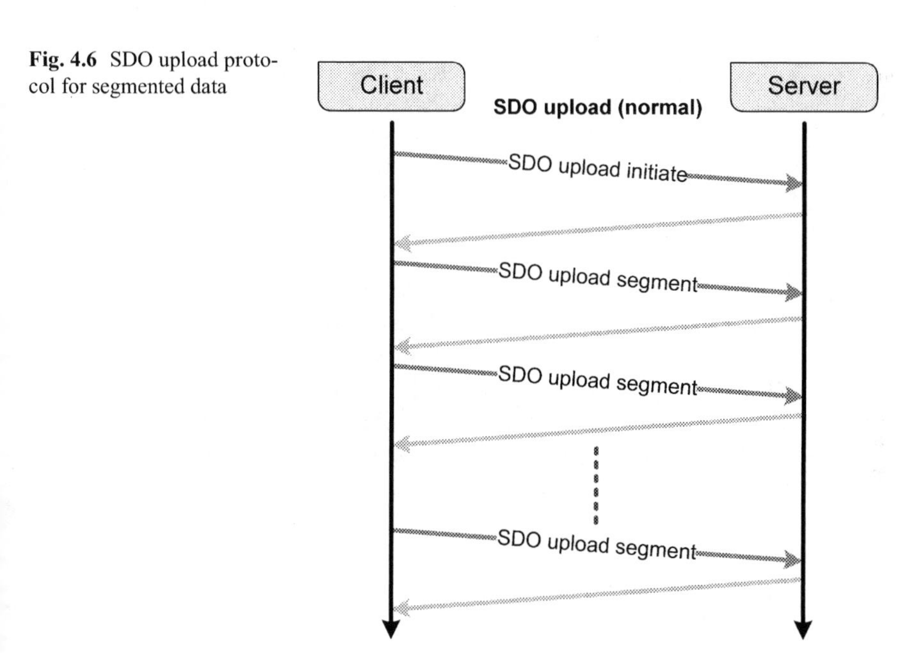

Fig. 4.6 SDO upload protocol for segmented data

Figure 4.5 illustrates the protocol for parameter with a maximal length of 4 bytes (expedited SDO). The 8-byte data field consists of 1-byte command specifier, 3-byte index/sub-index and up to 4 bytes of parameter data.

If data, to be read or written, are larger than 4 bytes, they are segmented and transmitted with the same CAN data frames (segmented SDO) one after another (see Fig. 4.6). By doing so, every segment is confirmed. The first segment contains 4-byte parameter data. The following segments consist of up to 7-byte parameter data. Principally, the parameter data can have any length. An end-identification is sent in the command specifier of the last transmitted segment.

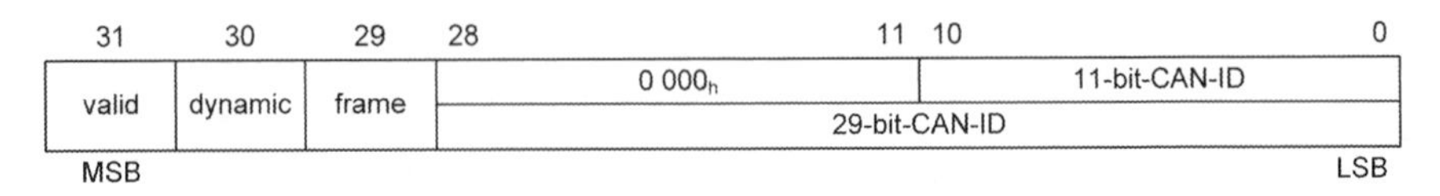

Fig. 4.7 Structure of the SDO COB-ID sub-parameter

To recognize if a segment was sent twice, every segment consists of a toggle bit. If a segment is sent twice, the server responds with an abort code. Both client and server are able to send an abort code in their SDO messages that aborts the current SDO communication, when problems occur.

Every CANopen device has to implement at least one SDO server to assure that at least one other CANopen device (normally the device with NMT functionality) is allowed to access (write or read) the object dictionary. The two needed CAN identifiers depend on the allocated node IDs: 1100_b+node ID for the to-be-received SDO message and 1011_b+node ID for the responded message. Every device is able to build up an SDO client and an SDO server relationship with all other devices in the CANopen network. Because there are no existing predefined CAN identifiers for these additional SDO connections, the connections have to be configured. Corresponding server and client SDO parameter sets are provided in the object dictionary (1200_h–$127F_h$ or 1280_h–$12FF_h$). Figure 4.7 illustrates the format of the configuration parameters for the CAN identifier. The 32-bit value consists of a valid byte for generating and deleting the data frame. The dyn-bit denotes if it is a static or a dynamic SDO connection. The frame-bit defines the frame format to be used and the CAN-ID bits with which the message shall be transmitted or received.

To improve the data throughput, the SDO block transfer exists. By using the SDO block transfer, only a configurable amount of segments, instead of all segments, are confirmed.

4.1.5 Application Protocols

Besides the NMT, boot-up, heartbeat and SDO protocols, CANopen also specifies protocols that uniquely assign the application layer of the ISO/OSI reference model. These include the process data object (*PDO*) *protocol* used for the transmission of time critical process data, the *emergency protocol* used for the notification of device and application failures, the synchronization (*SYNC*) *protocol* used for the synchronization of data capturing and data actuation as well as the *time protocol* used to set the system time.

Every CANopen device can transmit up to 512 PDOs and can receive up to 512 PDOs. PDOs are only allowed to be transmitted and evaluated during the operational state. PDOs use a CAN data frame and are sent according to the producer/consumer principle, which means there is always exactly one producer and one or more consumers of the CAN data frame. The complete 8 bytes of the data field

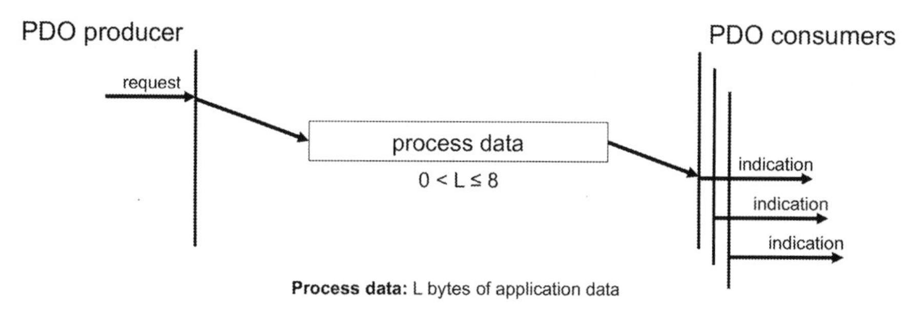

Fig. 4.8 PDO protocol

are available for the, to be transmitted, process data (see Fig. 4.8). Due to this the PDO does not need an overhead, in contrast to the CAN protocol, and therefore no additional bus bandwidth is necessary. The PDOs are configured with the help of two parameter sets: PDO communication parameter and PDO mapping parameter. The communication object identifier (COB ID), the transmission type, the inhibit time, the event timer and the SYNC start value belong to the PDO communication parameters. The 32-bit parameter COB ID consists of besides a few control bits, similar to the corresponding SDO parameter (see Fig. 4.7), the CAN identifier used for transmitting or receiving. The parameter transmission type defines how the PDO is to be transmitted or sent. The transmit PDO (TPDO) differs between a cyclic synchronous and an event-driven (asynchronous) transmission. When using the event-driven transmission, the event has to be defined in the device or application profile (transmission type: 255) or by the manufacturer (transmission type: 254). The event could be a temperature increase of 1°C or the change of a binary signal. The event could also be the expiration of the event timer. In this case, the PDO is sent periodically with the time (in milliseconds) configured in the event time parameter. If a signal-specific event and the event timer are defined, the PDO will be transmitted directly after the signal-specific event occurs. If no signal-specific event occurs, the PDO will be transmitted after the event timer has been elapsed. The event timer is restarted after every PDO transmission.

When a PDO gets transmitted cyclic synchronously (transmission type: 1), the process data are updated and transmitted after receiving a SYNC message. In order not to unnecessarily increase the busload with low-frequency signals, the PDO can be sent with only every second or 240th SYNC message (transmission type: 2–240). With the parameter SYNC start value, it is possible to configure the counter-value of the SYNC message (*sync counter*) to determine the first transmission. A possibility of reducing the busload lies in the usage of an acyclic synchronous transmission (transmission type: 0) in which the SYNC message has to be received and additionally a defined event has to occur. For receive-PDOs (RPDOs), a differentiation is only made with asynchronous or synchronous reception. With asynchronous reception the process data are immediately handled, whereas with synchronous reception the device waits for the next SYNC message to actuate the received data.

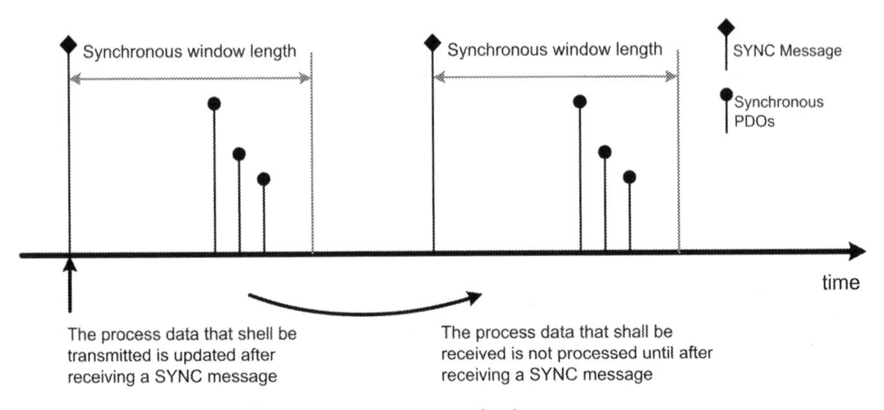

Fig. 4.9 Principle of the synchronous PDO communication

Figure 4.9 shows the synchronous transmission of PDOs. Synchronous PDO communication is especially common in the electric and hydraulic drive technology and in data acquisition of time-dependent linked values. The frequency of the SYNC message and the CAN identifier, on which the message is received or transmitted, are configurable via SDO.

Requesting PDOs via remote frames is possible in CANopen (transmission type: 252 and 253) but not recommended. Besides the additional busload, the variety of different implementations in the CAN controllers is problematic.

The inhibit time is the communication parameter in which the transmission of a PDO can be delayed for a certain time. With this method, the system developer can prevent the PDO from using the entire bus bandwidth. During the time period in which the highest priority PDO is not allowed to be sent, the second highest priority PDO becomes the highest priority PDO. With the usage of inhibit times, it is possible to achieve a completely deterministic behaviour of the PDO communication.

PDO mapping parameters determine what process data are to be received or transmitted via PDO. The PDO mapping parameters are defined by the device manufacturer or they are specified in the CANopen profiles. The entries in the mapping parameters contain pointers (index and sub-index) and determine which object dictionary entries are mapped into a PDO. Due to the fact that index and sub-index are local device addresses, the TPDO could contain pointers other than the corresponding RPDOs. The process data can be locally stored to different addresses in the object dictionary when more than one device receives the same PDO. Principally, the 8-byte data field of a PDO could be organized bitwise: This is the reason why a maximum of 64 mapping parameters is provided per PDO. If four 16-bit values are packed into a PDO, only four mapping parameters are needed. Simple CANopen devices only support static PDO mapping, i.e. the mapping set by the manufacturer is not configurable. When using variable mapping, the user is able to change the process data transmitted in the PDO, when the device is in pre-operational state. As a consequence, an optimization of the PDO transfer is possible. The user is able to group the needed process data for his/her application in one PDO. If the content of PDOs is changed when the device is in operational

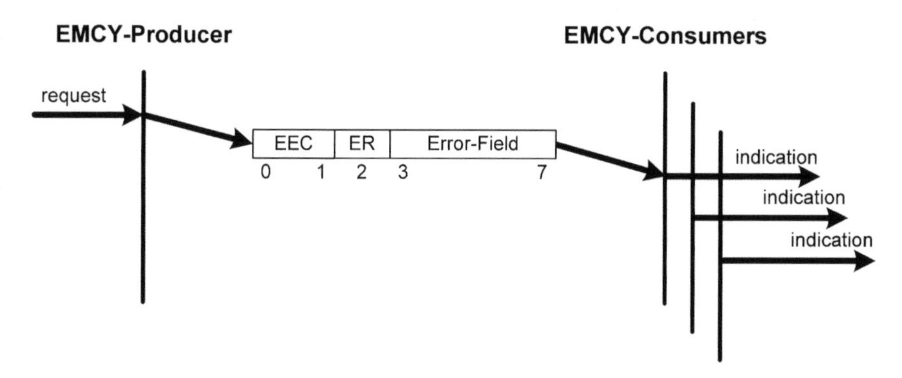

Fig. 4.10 Emergency protocol and structure of the message (EEC = emergency error code and ER = error register)

state (dynamic PDO mapping), the user has to pay attention to the data integrity of the PDO producer and consumer.

If the 512 TPDOs and the 512 RPDOs per device are not enough, it is possible to define a PDO as a multiplexed-PDO (MPDO). Similar to an SDO, it is only possible to transmit one process data in an MPDO. An MPDO contains the source or destination address of the object dictionary entry. Thus, a protocol overhead exists, but no confirmation about the reception of the MPDO is sent.

The emergency protocol is a special PDO that could be received by every other CANopen device if the emergency consumer parameter (1028_h) is configured. The emergency message consists of 8 bytes in which a standardized 2-byte emergency error code, the 1-byte error register and the manufacturer- or profile-specific 5-byte error field are transmitted (see Fig. 4.10). The CAN identifier, with which the emergency message is sent, is described in the emergency COB ID parameter (1014_h) and could be reconfigured if desired. The default value of the CAN identifier is 0001_b + node ID. It is reasonable to configure the emergency inhibit time (1015_h) to prevent an overload of emergency messages on the bus. In fact, it could happen that an error triggers a further error in another device, which could lead to a domino effect whereby many devices could start transmitting error messages over and over again.

The SYNC message is also transmitted after the producer/consumer principle. The SYNC message has no data field or a 1-byte data field containing the SYNC counter. The user can configure the counter (1019_h): 0 means that the counter-byte is not transmitted, and 2–240 represent the highest value that the counter supports. The other values are reserved. The SYNC message can be configured regarding the used CAN identifier, for transmitting and receiving (COB ID: 1005_h) as well as for the communication cycle period (1006_h) and the synchronous window length (1007_h) in microseconds. The synchronous window length serves as a measure for the producer and the consumer to indicate if a synchronous PDO was transmitted or received in the expected time. If the producer is not able to send the PDO (e.g. because bus access was denied), he could send an emergency message instead of the PDO. Same applies for the consumer: If a synchronous PDO is not received in

Table 4.6 Generic and application-specific CANopen device profiles

Number	Description	Status
CiA 401	Generic I/O modules	Free available
CiA 402	Drives and motion control	Only for members
CiA 404	Measuring devices and closed-loop controllers	Free available
CiA 406	Encoders (rotating and linear)	Free available
CiA 408	Fluid power technology	Free available
CiA 410	Inclinometer	Free available
CiA 412	Medical devices	Only for members
CiA 413	J1939-to-CANopen gateways	Only for members
CiA 414	Weaving machines	Free available
CiA 418	Battery modules	Free available
CiA 419	Battery charger	Free available
CiA 420	Extruder downstream devices	Free available
CiA 425	Medical add-on devices	Only for members
CiA 444	Crane add-on devices (e.g. spreader)	Only for members
CiA 445	RFID reader	Only for members
CiA 446	AS-Interface gateway	Only for members

the configured time, an emergency message is sent and the delayed reception will be ignored. The SYNC message is transmitted by default with the CAN identifier 128.

The time message is also based on the producer/consumer principle. The message contains a 6-byte value given in milliseconds after the 1 January 1984. The predefined CAN identifier is 256 and can be configured in the corresponding time COB ID (1012_h). With the time message, it is possible to do a network-wide time synchronization of the local timer units in the CANopen devices.

For transmission of safety-oriented information, special CANopen safety protocols, described in CiA 304, are available. The safety-related data object (SRDO) protocol uses two CAN identifiers that differ in at least 2 bits. Both frames are transmitted periodically, whereas the time interval between them is not allowed to exceed a certain value. The data of both frames are bitwise inverted and are double-checked by the consumers of the data. This concept of serial redundancy with integrated time expectation is able to detect all single faults. Due to this, the CANopen safety protocol is suitable for a safe data transmission up to Safety Integrity Level (SIL) 3 according to International Electrotechnical Commission (IEC) 61508.

4.1.6 Device Profiles

CANopen is one of the most standardized communication systems. This does not only apply to the communication protocols, specified in CiA 301 and CiA 302, but also apply to a variety of generic and application-specific device profiles (see Table 4.6). Device profiles specify the interface to the application program of a device in terms of process data and configuration parameter, which are normally writable and/or readable, in the range of 6000_h–$67FF_h$ in the object dictionary.

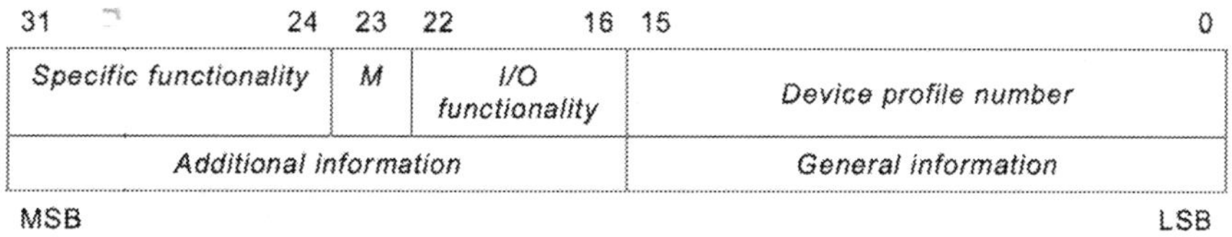

31 — 24	23	22 — 16	15 — 0
Specific functionality	M	I/O functionality	Device profile number
Additional information			General information
MSB			LSB

Fig. 4.11 Structure of a device-type parameter for CiA 401 modules

Table 4.7 I/O parameters for CiA 401 modules

Field name	Definition
Device profile number	401_d
I/O functionality—Bit 16	1^b=digital input(s) implemented 0^b=not implemented
I/O functionality—Bit 17	1^b=digital output(s) implemented 0^b=not implemented
I/O functionality—Bit 18	1^b=analog input(s) implemented 0^b=not implemented
I/O functionality—Bit 19	1^b=analog output(s) implemented 0^b=not implemented
I/O functionality—Bit 20 to Bit 22	Reserved
M(apping of PDOs)	1^b=device-specific PDO mapping is supported 0^b=predefined, generic PDO mapping is supported

Note: Any combination of digital/analog, inputs and outputs is allowed; one of the bits 16–19 shall be 1_b

The device-type object (1000_h) is described in detail in the device profiles. It does not only show which device profile is implemented but also show the available functions in the device. Figure 4.11 shows the structure of the device-type parameter for generic input/output (I/O) modules according to CiA 401 (Table 4.7).

The device profiles also determine the PDO communication and the PDO mapping parameters. The first four TPDOs as well as the first four RPDOs can each be assigned with a CAN identifier. The assignment, done in the device profiles, follows the same scheme as in other CANopen protocols. The four bits with the highest priority of the CAN identifier correspond to a PDO function (see Table 4.8), and the seven other identifier bits represent the node ID of the device.

This assignment guarantees that no CAN identifier is assigned twice. In other words, it is not possible that two CANopen devices send data frames with the same identifier leading to a not solvable bus access conflict. The CANopen device with the NMT master functionality is normally a programmable controller that has the corresponding RPDOs and TPDOs. Thus, per default there is only one master/slave relation concerning PDOs. If a system developer wants to realize PDO cross-communication or wants to implement more than one controller in the network, the configuration of PDO identifiers is necessary. This procedure is also known as PDO linking and is supported by software tools. In this way, it is even possible to realize distributed PLC systems.

Table 4.8 Default CAN identifiers for TPDOs and RPDOs

Name	Function code	Resulting CAN-IDs
TPDO1	0011_b	385 (181_h)–511 ($1FF_h$)
RPDO1	0100_b	513 (201_h)–639 ($27F_h$)
TPDO2	0101_b	641 (281_h)–767 ($2FF_h$)
RPDO2	0110_b	769 (301_h)–895 ($37F_h$)
TPDO3	0111_b	897 (381_h)–1023 ($3FF_h$)
RPDO3	1000_b	1025 (401_h)–1151 ($47F_h$)
TPDO4	1001_b	1153 (481_h)–1279 ($4FF_h$)
RPDO4	1010_b	1281 (501_h)–1407 ($57F_h$)

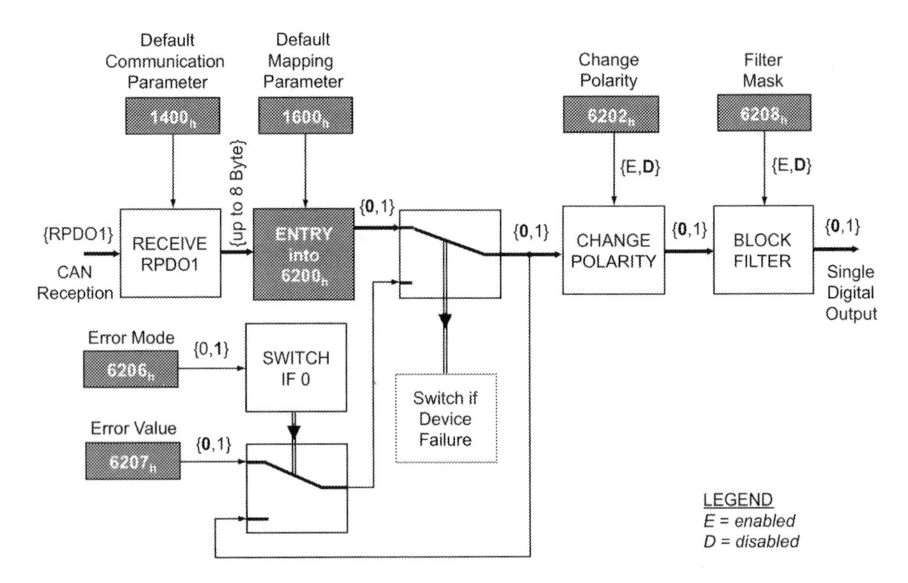

Fig. 4.12 Block diagram of a digital output module according to CiA 401

The most frequently implemented device profile is the profile for generic I/O modules (CiA 401). It supports digital and analog inputs and outputs. The user is able to parameterize the modules in a standardized way regarding the I/O functionality. With the parameter set "error value output (6207_h)", it is possible to set the value of the digital outputs that should be used, when an internal device error occurs. Another example is the parameter set "analog input interrupt trigger selection (6421_h)". By this parameter set conditions can be defined under which a PDO transmission is triggered by an analog input, which has been mapped into a TPDO. These conditions can be exceeding or falling below a certain threshold or the predefined change of a given value. It is also possible to set the limit and delta values under a fixed address (index and sub-index) in the object dictionary.

Figure 4.12 shows the block diagram of a digital output module, and Fig. 4.13 shows the block diagram of an analog input module. The device profile CiA 401

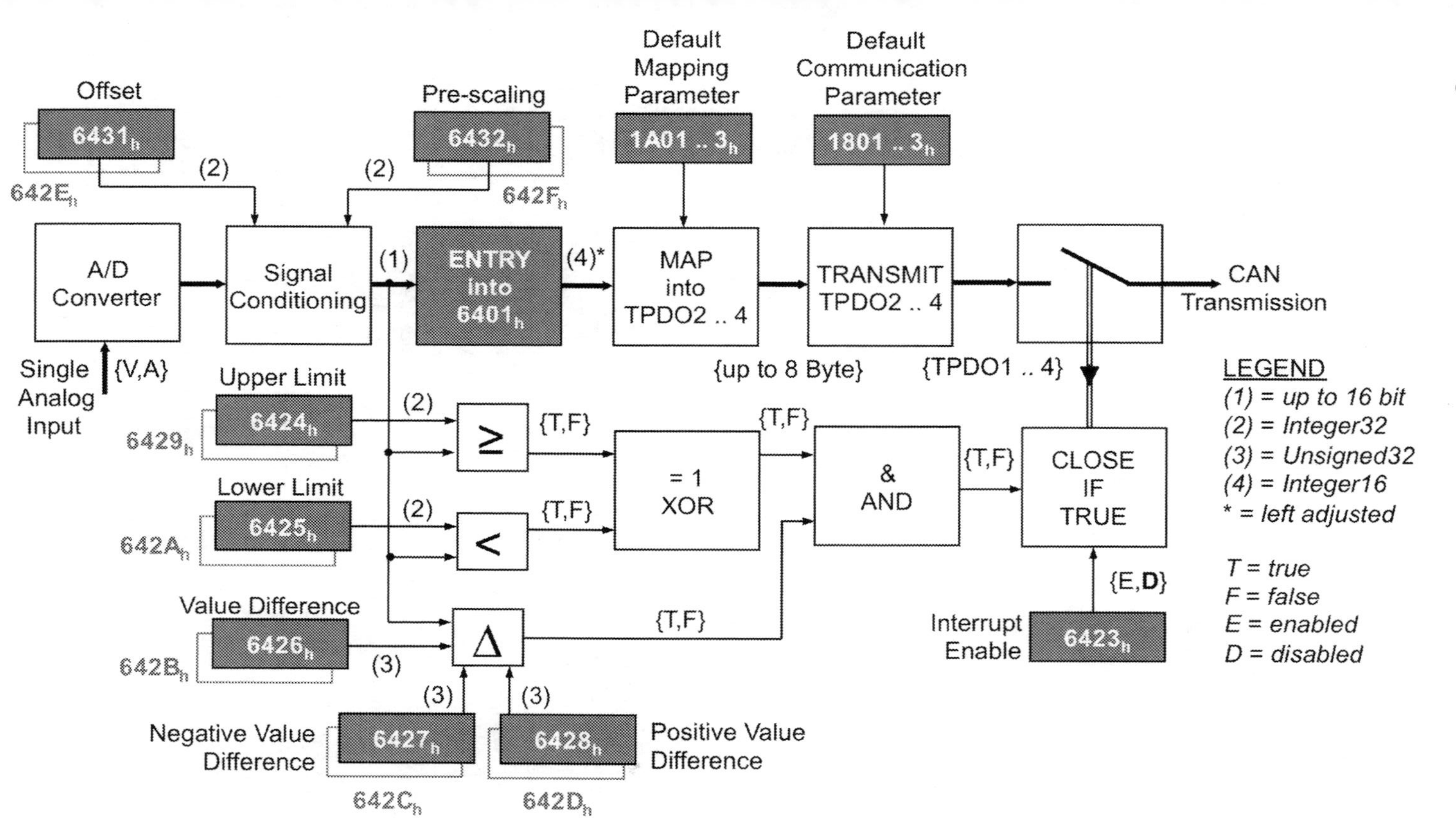

Fig. 4.13 Block diagram of a digital input module according to CiA 401

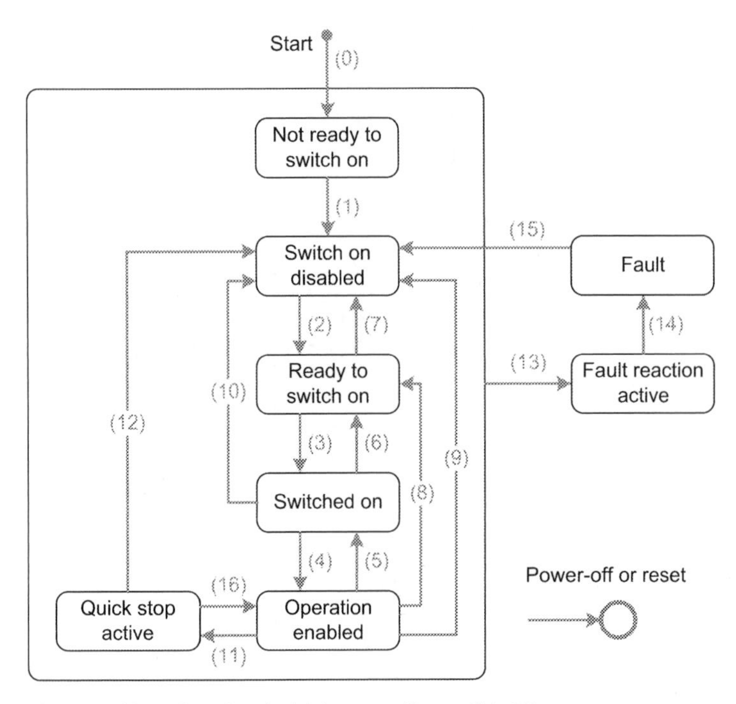

Fig. 4.14 State machine of an electrical drive according to CiA 402

also defines a default PDO mapping. According to the defined PDO mapping, 8 × 8 digital input signals are located in the first TPDO and 4 × 16-bit analog input values in the following three TPDOs. In the first RPDO 8 × 8 digital output signals and in the three following RPDOs 4 × 16-bit digital output values are located. Further digital and analog I/O signals could be placed in manufacturer-specific PDOs. All default PDOs are valid and have a predefined CAN identifier. They are transmitted event driven (transmission type: 255). The event is a configured trigger condition. Furthermore, all TPDOs that are not switched off generally are transmitted when a transition from pre-operational state into operational state occurs.

The device profile CiA 401 supports a digital granularity of 8 bytes per default. 1-bit, 16-bit and 32-bit accesses are optionally provided. Analog modules have a 16-bit resolution specified per default. Alternatively, analog 8-bit values, 32-bit values as well as floating-point arithmetic and manufacturer-specific analog formats can be implemented.

Some of the CANopen device profiles also specify application-specific state machines. The state machine, shown in Fig. 4.14, is described in the drives and motion control profile CiA 402 (see IEC 61800-7). It is controlled by a control word (6040_h) received via RPDO. After a state transition, the CANopen device transmits the status word (6041_h) to the host controller for purpose of acknowledgement on the application level.

In the drives and motion control profile, all necessary parameters for operation are listed in the object dictionary as defined addresses (index and sub-index). Even setpoints or actual values are transmitted via PDOs. The drives and motion control profile CiA 402 is suitable for simple frequency converters as well as for complex servo controller and stepper motor controls.

For some industries, specific device profiles were developed. Such profiles simplify in particular system integration in modular mechanical engineering. The system developer does not have to link generic digital and analog inputs and outputs together but is able to integrate a complete subsystem on a higher level. Typical examples are collimators and dosimeters in the field of medical devices, thread-feeding equipment in weaving machines or spreaders in crane systems. The members of CiA will keep publishing generic and branch-specific device profiles in the future.

4.1.7 Application Profiles

With the approach of the device profiles, the system developer is allowed to integrate devices from various manufacturers in a CANopen network with one or several programmable controllers. By doing so, the plug and play functionality is limited to a master/slave relation concerning the PDO communication. Although it is possible to configure a PDO cross-communication between any of the devices, cross-communication is not predefined. Out of this reason, branch-specific application profiles were made particularly to meet these requirements.

Application profiles describe all interfaces of a CANopen application in an object dictionary. The object dictionary entries have identical meaning in all devices. Application profiles specify functional units, also referred to as virtual devices. A CANopen device is able to integrate up to eight application profiles (each located in a logical device). Each logical device, on the other hand, contains a number of virtual devices. The information on what virtual devices are implemented could be found in the device-type parameter (1000_h) or in an application profile-specific parameter. With the concept of virtual devices, it is possible to describe transparent CANopen bridges/gateways in an easy way.

Furthermore, the concept of the application profiles opens the possibility of implementing, in an extreme case, only one virtual device in a single CANopen device and all other virtual devices in another CANopen device as well as the combination of both of these devices through a CANopen network. It is also possible to implement every virtual device in a CANopen device and let them communicate through CANopen. It is not possible to spread a virtual device over more than one CANopen device. Out of this reason, it is very important to ensure, when creating an application profile, that the granularity of the virtual devices is suitable for future requirements. Table 4.9 shows the already published application profiles and the application profiles that are still in development.

The application profile for lift control systems (CiA 417) specifies several virtual devices (see Table 4.10). The object dictionary (range 6000_h–$9FFF_h$) is divided into

Table 4.9 CANopen application profiles

Number	Description	Status
CiA 415	Road construction machine sensors	Only for members
CiA 416	Building door control systems	Only for members
CiA 417	Lift control systems	Free available
CiA 421	Train vehicle control networks	Only for members
CiA 422	Municipal vehicles	Only for members
CiA 423	Rail vehicle power drive systems	Only for members
CiA 424	Vehicle door control systems	Only for members
CiA 426	Exterior rail vehicle lighting	Only for members
CiA 430	Auxiliary rail vehicle lighting	Only for members
CiA 433	Interior rail vehicle lighting	Only for members
CiA 436	Construction machineries	Only for members
CiA 437	Grid-based photovoltaic systems	Only for members
CiA 447	Special-purpose car add-on devices	Only for members
CiA 455	Drilling machines	Only for members

Table 4.10 Virtual devices of the CANopen lift application profile

Virtual device	Function
Call controller	Receives all call requests
Input panel unit	In-car call panel or floor call panel
Output panel unit	In-car display panel or floor display panel
Car door controller	Transmits commands to the car door unit
Car door unit	Cabin doors
Light barrier unit	Light barrier in the cabin doors
Car position unit	Position measurement of the car (according to CiA 402)
Car drive controller	Transmits commands to the car drive unit
Car drive unit	Drive unit (according to CiA 406)
Load measuring unit	Measuring of current load of the car
Sensor unit	Glass breakage, smoke, pressure, temperature sensor, etc.

800_h segments. Every segment could represent a lift control, i.e. it is possible to describe a group control containing up to eight lift shafts with this profile. Every virtual device has several configuration parameters that are readable or writable via SDO as well as process data, that is transmitted or received, in PDOs.

Every CANopen device according to CiA 417 supports a Transmit-MPDO and up to 127 Receive-MPDOs as well as, depending on the implemented virtual device, further dedicated TPDOs and RPDOs. Due to the fact that all PDOs have an assigned CAN identifier (see Fig. 4.15), the system developer must not assign any CAN identifiers. However, if the devices support PDO linking, he/she is able to optimize the PDO communication in his/her lift application with regard to priority and data content.

The CANopen lift application profile supports up to 254 floors per lift control. In total, 32 times 254 digital inputs per lift control are addressable. When using eight lift controls, a total of 65,024 digital inputs are available.

PDO	Description
PDO1	not used
PDO2 to PDO128	MPDO
PDO129	not used
PDO130 to PDO256	Virtual input
PDO257 to PDO272	Lift 1
PDO273 to PDO288	Lift 2
⋮	
PDO369 to PDO384	Lift 8
PDO385	not used
PDO386 to PDO512	Sensor input

PDO	Description
PDO257	Virtual output
PDO258	not used
PDO259	Load measuring
PDO260	not used
PDO261 to PDO264	Car drive control and status
PDO265 to PDO268	Car position sensor 1 to 4
PDO269 to PDO270	Car door control and status
PDO271	Light barrier
PDO272	Car door position

Fig. 4.15 PDO assignment in a CANopen-lift network system compliant to CiA 417

It is possible to implement up to eight application profiles in a single CANopen device, if the object dictionary entries of a logical device are used for the application profile. This possibility could be used in the application profiles for train vehicle control systems. From a logical point of view, it is a hierarchically arranged virtual network that can be mapped one-to-one on the physical CANopen network. However, it is also possible to represent up to eight virtual networks on a CANopen interface. The implementation flexibility of the application profiles allows the possibility of using the same specification in a simple and in a complex system. The "bridges" needed for the implementation of several physically separate CANopen networks are PDO transparent, i.e. the PDOs are just reached further due to their system-wide validity. If it is wished to configure the entire system from one single point via SDO, it is necessary to implement CiA400, which enables remote SDO communication. It is therefore necessary to assign every network with a unique network ID.

4.2 AUTOSAR

4.2.1 Introduction

4.2.1.1 AUTOSAR Foundation

"AUTOSAR" is the abbreviation for "AUTomotive Open System Architecture". The organization of the AUTOSAR standard was founded in 2003 as a development partnership of international automobile manufacturers and supplier industry. The goal is to develop a standardized software architecture and standardized software interfaces for automotive electronic systems.

The development of such standards for software development in the automotive industry was long overdue. The software functionality within vehicles increases a lot and the complexity is growing. At the same time there is a high demand to deliver high-quality software, partly also due to the growing amount of functional safety-related systems. Some of the original equipment makers (OEMs) react to this global trend with own standardized software platforms, for example, the BMW *Standard Core*.[1] This software platform was provided by BMW to the Tier 1 already in 1998. Because of the existence of different OEM-specific automotive middleware, the Tier 1 always needs to adapt their application software to these quasi-standards. The integration costs of these adaptations are sometimes beyond the application development costs. In parallel to these technical requirements, there is a change in the market environment. The major sales regions are saturated. This forces the OEM to change their strategy for vehicle production. In addition, niche markets will be addressed with special vehicle models and the model change will be faster. The pressure for vehicle development, and hence also for automotive software development, increases.

The only way to develop more complex system in a shorter time—and still control them—is to standardize a horizontal basic software while reusing the applications more and more. A first step to the right direction is the AUTOSAR standard.

The following pages will give a brief overview about the AUTOSAR standard by focusing the CAN-based communication path through the layered software architecture.

4.2.1.2 AUTOSAR Concept

One of the main ideas of AUTOSAR is the clear separation of hardware-dependent software from hardware-independent software. Therefore, an abstraction layer is placed between the microcontroller hardware and the application software. The AUTOSAR runtime environment (RTE) is on top of the layered architecture. It is the only visible interface from the application point of view. Below the RTE several software layers exist, from services via operating system (OS) down to the lowest software layer, the microcontroller abstraction layer (MCAL). Figure 4.16 shows an AUTOSAR compliant software architecture.

The AUTOSAR RTE defines the interface from the application to the basic software. For data exchange, the RTE implements, e.g. a client–server and/or a sender/receiver communication model.

All interactions of application software, called "application software components" in the AUTOSAR language, run via the RTE. This results in a hardware-independent application. This enables the exchange of applications among the different electronic control units (ECUs). If there is a lack of resources on one ECU, e.g. no more random access memory (RAM) left to run a special RAM-intensive

---

[1] *Standard Core* is a typical name for a complete automotive middleware respectively for the software between application and hardware of an electronic control unit (ECU). Additional to the pure runtime software a Standard Core contains support functions like a generic make environment or complex configuration tools.

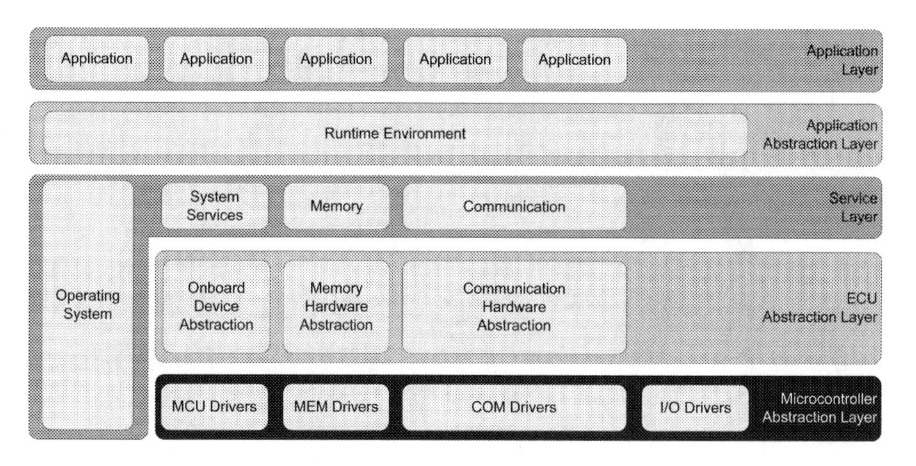

Fig. 4.16 The layered architecture of an AUTOSAR Standard Core. (Elektrobit Automotive GmbH, AUTOSAR—Getting Started)

application algorithm, it can be executed on another ECU if the input and output signals of these application software component are also available on the other ECU.[2] With this independence from hardware it is in addition possible to reuse the application software component on another hardware architecture, e.g. on the next generation of the ECU. Software will become also a "carry-over part" for the automotive development.

The basic software itself is also structured in hardware-dependent and hardware-independent software parts. The only real hardware-related software part is the OS and the MCAL—all other software can be reused, too. With this portability major parts of the basic software can become a carry-over part with all the benefits of reused standard software.

4.2.1.3 New Methodology

After the successful separation of application software and hardware-dependent software, new methodologies within the automotive vehicle development can be applied. Function-oriented development will replace ECU-oriented development. AUTOSAR defines a complete methodology for the function-oriented development. The methodology is not a complete process description, this was never the goal of AUTOSAR, but it gives a brief structure in the overall development process. In brief, the AUTOSAR methodology starts with the pure software functions, which are represented by the application software components. The inputs and outputs of these functions, the signals, are defined and the connections are known. All this information is collected in a so-called "System Configuration Description", which

[2] And the timing requirements to the signals are fulfilled (e.g., availability and, maximum jitter).

is an Extensible Markup Language (XML)-based data structure defined by the AUTOSAR meta-model. The system configuration description contains the description of all software components, the ECU resources and the system constraints for a complete vehicle. The next step, after the definition of many system parameters, the configuration of one ECU is extracted from this global vehicle description. This extract of the system configuration forms together with the configuration of the basic software the so-called ECU configuration description. Within this ECU-specific configuration, the exact definition, e.g. of OS tasks is done. System configuration and ECU configuration are two processes with high interaction and are done usually by two different developer groups.

With AUTOSAR, the role model of the automotive industry might change. The classic supplier pyramid will be replaced by a network of suppliers because the tasks, especially the software related, can be done in future a little bit more independently and in parallel. The software companies will play an important role; they provide the software platforms where all the applications will run on top. In addition, the same or different software suppliers can provide pure software functions as a product. If the standard is established, these software can be used in several cars, independent of the OEM.

4.2.2 *The AUTOSAR Platform*

4.2.2.1 Overview

The legacy software platforms from each OEM differ in interfaces and content. The goal of the AUTOSAR standardization is to form a universal standard which need not have to be adapted for each OEM.

The AUTOSAR specifications deal with many software modules, needed for an ECU development. Old standards, like the OSEK/VDX "Offene Systeme und deren Schnittstellen für die Elektronik im Kraftfahrzeug / Vehicle Distributed eXecutive" standard, comprised less functionality, e.g. the mentioned OSEK standard only defines an OS, a communication model and NMT. The implementation of these standard modules found their way into many cars of many OEM, but AUTOSAR is going beyond; Fig. 4.16 shows a much broader approach and shows much more standardized modules of the AUTOSAR-layered architecture.

In principle, every module can be assigned to one of the four layers: the application abstraction layer, service layer, ECU abstraction layer and MCAL. The application abstraction layer forms the only interface to the applications. The implementation of this layer is the RTE, and the RTE is the "glue code" which connects the applications to the lower layers. The service layer implements system services, memory access services, communication services and the OS services. The ECU abstraction layer is used to implement an abstraction of the peripheral I/O, memory (e.g. flash memory or electrically erasable programmable read-only memory (EEPROM)) and communication. The services are based on this abstraction layer and the abstraction layer itself connects, e.g. different communication channels by using the lowest layer, the MCAL. In this lowest layer, the drivers for accessing the hardware are implemented. This lowest layer has to be developed for each new microcontroller—it is the only piece of software which is not hardware independent.

4.2.2.2 Microcontroller Abstraction Layer

As mentioned above, the MCAL is the only layer which depends on the hardware. The above layers have no dependency any more, except the OS. This principle is the basis for an easy adaptation of the basic software to new hardware architecture or a new derivative; only this deepest layer will be exchanged. The layers above will be tested together with the exchanged lower layers in an integration test system. There are exceptions within the higher levels of software if special hardware features shall be used there, too. If these features are not able to be implemented only in the MCAL, a hardware dependency is placed in the higher level software—the supplier of this basic software has to check if this dependency is maintainable and/or if the costs for hardware dependency are less than the, e.g. speed advantage gained by using special hardware features.

The drivers, of the MCAL, can be subdivided into the following four groups:

- Microcontroller drivers—they contain software to control the microcontroller core (Micro-Controller Unit (MCU) driver), timer modules (General Purpose Timer (GPT) driver) and on-chip watchdog (watchdog driver).
- Memory drivers—these are the drivers to access EEPROM and FLASH memory.
- Communication drivers—these are used to connect bus systems like Serial Peripheral Interface (SPI) bus for inter-ECU communication, e.g. to connect an external SPI EEPROM device, and to access common used bus systems like CAN, Local Interconnect Network (LIN), FlexRay and also Ethernet.
- I/O drivers—these drivers connect the on-chip peripherals like digital I/O, analog digital converter (ADC), input capture unit, pulse width modulation (PWM) and the port driver for the configuration of each port pin functionality.

Within the AUTOSAR architecture, hardware access will be enabled by two software modules: the driver of the MCAL and the corresponding interface of the layer above—the ECU abstraction layer.

4.2.2.3 ECU Abstraction Layer

For almost each driver of the hardware-dependent layer exists an interface to encapsulate the lowest layer. The access to this interface layer is always identical for all layers above. For example, multiple CAN modules can be summed-up abstract by the interface layer, e.g. by controlling several CAN drivers. The AUTOSAR COM module in the service layer can access the CAN interface always in a same way, sending only a protocol data unit (PDU)[3] to the interface. The interface is

[3] PDU: protocol data unit.

configured to forward the PDU to the dedicated bus system—in this example, to a dedicated CAN bus. An exception to this driver/interface relationship is the group of the I/O driver. There exists no interface layer for the I/O drivers—instead there is an "I/O hardware abstraction" module to implement the bridge between the low-level hardware-dependent driver and the highest layer of the application abstraction layer. The specification of this I/O hardware abstraction is more like a guideline for the implementation than a real implementation with application program interface (API) specification, etc. The reason is the various possibilities of connecting I/O to a microcontroller; hence, the implementation of this I/O hardware abstraction is in many cases project specific and peripheral specific. There could be, for example, power switches which need already a combination of two I/O (ADC plus PWM) or an algorithm implemented in this I/O hardware abstraction (e.g. debouncing of digital input).

4.2.2.4 Service Layer

The service layer of an AUTOSAR-based software architecture provides many services which can be accessed from the applications via the RTE. The modules of this layer are complex state machines, e.g. the ECU state manager which controls the state of the ECU like OFF, RUN and SLEEP. The module groups of the service layer are as follows:

- State manager—it manages all the states of an ECU, communication and, e.g. of the Watchdog.
- Memory manager—it controls the access to persistent memory. The non-volatile RAM manager (NVRAM-Manager) is a core module of the memory services. This module manages the non-volatile memory blocks of an EEPROM, a FLASH–EEPROM emulation or other external memory devices. The tasks are, for example, write, read access, initialization of memory-mapped blocks and check if the cyclic redundancy check (CRC) is still valid.
- Vehicle communication services—services of this service module group enable the communication across the border of the ECU. The lower layers support the known protocols like CAN, LIN and FlexRay—in future also Ethernet. The jobs of these services are the abstraction of the access via standardized interfaces and data encapsulation via abstract PDUs instead of protocol-specific messages, control of the network by support of NMT services and access to diagnostic services.
- OS services—the OS provides several services. For example, the management of the central processing unit (CPU) time for tasks and interrupt service routines. There are event mechanisms and protection mechanisms for time and memory access if provided by hardware. The memory protection is only possible if the controller has a memory management unit (MMU) or memory protection unit (MPU). The AUTOSAR OS also provides mechanism for time synchronization—e.g. to synchronize the internal schedule with an external clock, provided by FlexRay.

4.2.2.5 Application Abstraction Layer

The highest layer of the AUTOSAR-layered architecture is the interface to the application. This application abstraction layer is the only interface of an AUTOSAR application to the AUTOSAR basic software. No AUTOSAR application shall access functions from the basic software direct. Out of the application perspective, there are no tasks or interrupts anymore and basic software interfaces do not exist. This highest layer will be implemented only by one software module: the AUTOSAR RTE.

The RTE implements the connection of all communication paths and the abstraction of the runtime management. An AUTOSAR compliant application—application software component—implements a software function. This application software component consists out of a formal description in XML, the Software Component Description (SWC-D) and the implementation in C source code. The RTE itself will be generated with XML input data of the ECU configuration and all the application software component descriptions. The API of the RTE depends on the software components, their signal names and internal structure. The RTE is the implementation of the virtual function bus (VFB), which is the core element of the AUTOSAR architecture.

The VFB is the communication medium for all functions and provides sender/receiver or client/server communication methods. In the world of the VFB, it is not relevant where the function is executed and how the data are transported. The "mapping" of functions to ECUs is part of the system design process—the result of this process is the system description. The path for the signals will be defined by the distribution of the application software components to real physical ECUs. For the application it is irrelevant how the data are transported; this will be done by the configured basic software and the RTE. A signal could be a direct result of a digital input line or it could be a part of a CAN message (Fig. 4.17).

4.2.2.6 AUTOSAR Applications

AUTOSAR compliant applications are connected via the RTE to the basic software platform. The API of the RTE provides many possibilities of designing an application. In principle, the communication is running via "ports" and the activation of runtime parts is done by the execution of the so-called "runnable entities". The communication ports can be of the type sender/receiver or client/server. The client/server type represents the typical function call, which is needed to call, e.g. library functions (e.g. mathematical functions). As already mentioned, the information from where the RTE gets the data is part of the RTE configuration (which is part of the ECU configuration). With this method it is possible to develop application software and define later the final communication path for each input or output signal. This is the basis for shifting application functions within the vehicle ECU topology—one of the goals of AUTOSAR. An ECU is in future AUTOSAR systems only one piece of the whole platform which host application software.

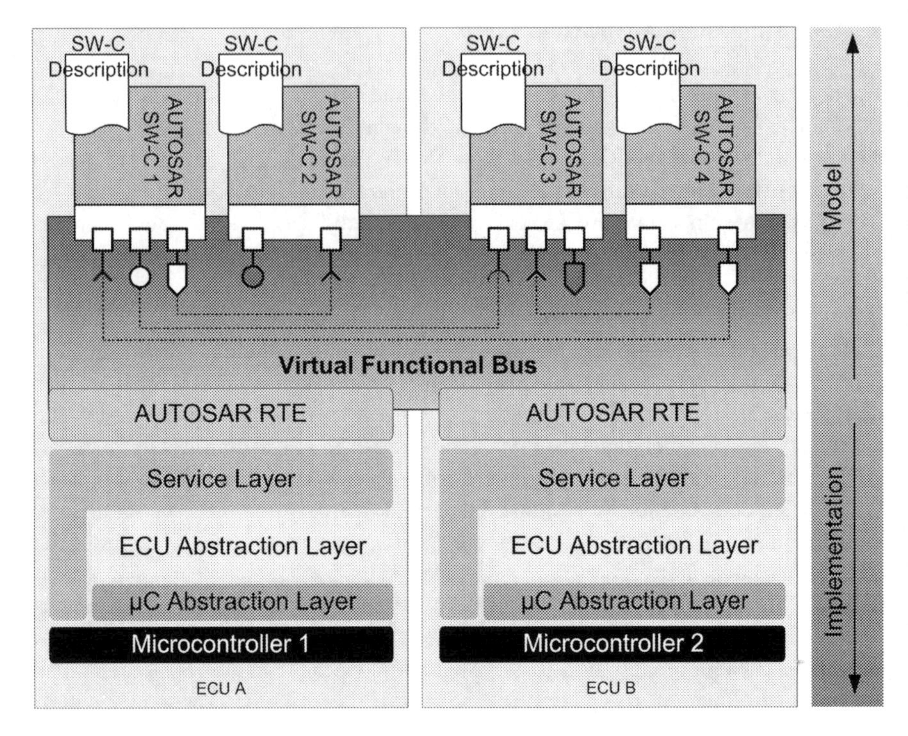

Fig. 4.17 From model to implementation. (Elektrobit Automotive GmbH, AUTOSAR—RTE User's Guide)

The requirements of the software components to their host environment are described in a standardized formal format. This format, the software component description, contains all information about the internal structure of the software component: Communication ports are described with their interfaces as well as the internal implementation details of the runtime parts. These runtime parts are called runnable entities and part of the description is how they are executed respectively by the event that triggers them. Trigger events can be set, e.g. by a cyclic clock or by the reception of a signal (data-received event).

The goal of the RTE is to hide all lower software layers. As such RTE hides the I/O driver layer and the communication layers as well as the access to the operating system. The application developer shall not write source code with direct access to counter, alarms and semaphores anymore. It is no longer necessary to define tasks and control respectively configure the task activation in the OS configuration. Instead, all runtime components of the applications will be represented by the runnable entities. It is the job of the RTE to manage all runnable entities, to map them into OS tasks and hence to predefine parts of the OS configuration. With this abstraction, it is possible to design, develop and test application software components

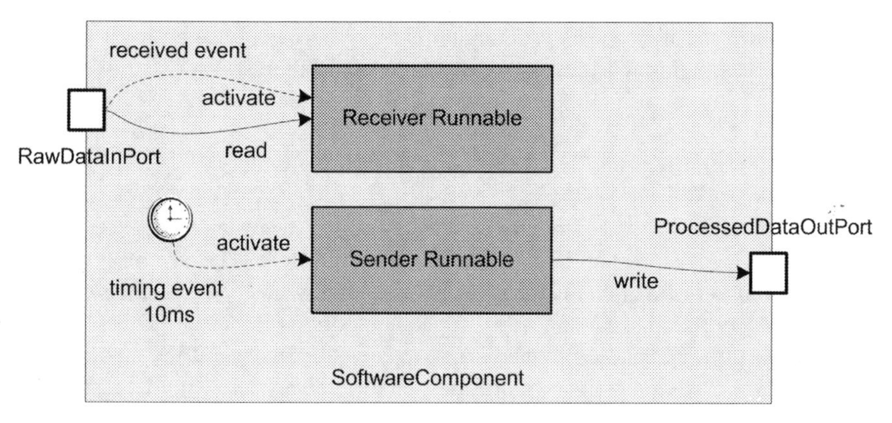

Fig. 4.18 Principle of a software component with ports and runnable entities (Elektrobit Automotive GmbH, AUTOSAR—RTE User's Guide)

independent of the final architecture. The configuration of the RTE, with all the architecture information, will be done later by the integration engineers (Fig. 4.18).

4.2.3 AUTOSAR Communication—An Example

4.2.3.1 Overview

The communication within an AUTOSAR architecture is running only across the ports of the application software components. The characteristic of a port is defined by its interface description. Available communication models are sender/receiver model for the transport of signals and client/server model for the execution of (remote) function calls. The receiver respectively the server can be on the same ECU (the so-called intra-ECU communication) or on different ECUs (the so-called inter-ECU communication). If the signals are shared across ECUs the data will be transported by the known automotive bus systems CAN, LIN and FlexRay.

With the abstraction of the communication channels, the bus protocols become invisible for the application developers. These developers do not need to know the details of the protocols or the register structure of the communication hardware module. The CAN module will be controlled by the CAN driver, and the "packaging" of the signals into CAN messages will be done by the COM module respectively by the CAN interface. These standard modules will be partly configured by the system configuration—there are communication parameters which have global validity. The CAN identifier and the position of signals in CAN frames (FlexRay, etc.) for all ECUs are the same and hence part of the system configuration. The message buffer configuration of the CAN hardware (e.g. used in first in, first out (FIFO) mode, ring-buffer or other) is part of the ECU configuration and will be defined by an integration expert who is integrating basic software with the application software.

4.2.3.2 Communication Path Through the Basic Software

AUTOSAR defines several modules which are taking part to send or receive a signal. On a first view, the modularity is complex and oversized. The advantages of the clear separation of functionality seem to be bought very costly—if at each module border, e.g. a buffer is needed to store the data temporarily. If each module is developed to be 100% exchangeable the buffers may be needed, and if the modules are bundled it is possible to apply optimizations across the borders with the result that resources are not wasted. This can be done also if the configuration tooling has a clear view to the overall configuration, knowing all the facts and hence be able to provide an optimized configuration.

The following example shows how a signal will run through the AUTOSAR-layered architecture from a sender to a receiver application software component. The example refers to the CAN stack.

4.2.3.3 AUTOSAR Application

Within the application, information will be represented as a signal. For example, the actual vehicle speed could be an 8-bit value from 0 to 255, representing the speed in km/h. The port of the application, through which the signal is sent, is defined by an interface containing an 8-bit data element. The interface and the data type are defined within the software component description (XML). The generation of the RTE will also generate a component header (C-code) where these data types are defined. The RTE has two modes for the generation: In the contract phase, all c-headers with all data definitions are generated which are needed by the application software components (data types, constants, interface structures, function prototypes of the RTE API, etc.). In the second phase, the RTE generation phase, also the implementation of the functions will be generated. This is only possible if the complete path of the signal is known, i.e. the receiver and the medium for transport are defined (another software component on the same ECU, an output port of the microcontroller or something on another ECU).

If the interface and data types are defined in the software configuration description, the port can be configured. The name of the port is later a part of the RTE-send-API within the application software component. A runnable entity with the name *MyRunnable* can call the send function of the port *pPortname* with the value *Value* and data element *Speed* like shown in the following code-fragment:

```
void MyRunnable (Rte_Instance self)
{
 static SpeedValueType Value=0;
/* Compute Value */
...
/* Send Value */
 Rte_Send_pPortname_Speed(self, Value)
}
```
[4]

The parameter *self* is the instance handle of the runnable entity which is calling this function. This is needed to implement a methodology to use multiple instantiation of one (code) identical application software component. It is similar to the "this" pointer of a C++ class.

Calling the RTE send function will forward the value—depending on the configuration—to the lower software layers of the AUTOSAR architecture.

4.2.3.4 AUTOSAR RTE

The RTE contract phase generates prototypes of the send function. The RTE generation phase will generate the content of the send function. The RTE generator knows now the path of the signal. In this example, we assume that the signal shall be sent to another software component on another ECU by using the CAN bus. Hence, the RTE will forward the signal to the AUTOSAR COM module, which is responsible for all external communication. At the "sender signal mapping" (part of the ECU configuration), the interface data element of the port is mapped to a signal of the COM module (in the example the signal with the name COM_SIGNAL_1). The names can be different, but the data types have to be identical.

```
/* Rte_Send API function for DemoComponent */
Rte_StatusType Rte_Send_pPortname_Speed (SpeedType value)
{
 Status=Com_SendSignal (COM_SIGNAL_1, (void*) &value);
/* error handling, etc. */
}
```

If the receiver of this Speed-Signal would be on the same ECU, the implementation of this function could write direct into the memory of the receiver component. If this signal would be sent to a physical output port of the microcontroller, the I/O hardware abstraction would be called. This makes visible how deep the implementation of the RTE depends on the configuration. The RTE is simply the C-code representation of the configuration and unique for each ECU. In this example, the path continues at the AUTOSAR COM module.

[4] All used code examples are taken from the EB tresos AutoCore—they are only fragments and not complete.

4.2.3.5 AUTOSAR COM

The COM module receives the signal from the RTE and executes the "packaging" of the signal into a PDU. A PDU bundles several signals and will be forwarded to the lower layers. In the opposite way, the COM receives PDUs, extracts the signals and triggers actions of the RTE (e.g. by calling COM-callback-functions which fire an RTE event). The COM module implements many additional functions like timeout monitoring, byte conversion, different notification mechanisms, different transfer mode and so on. The lower levels of the communication stack do not know "signals" any more—they only handle PDUs.

The C-code of the application function and the RTE was straightforward and simple. The *Com_SendSignal* function of the COM module is very complex and does not fit to one page of paper. The core functionality of this function is the correct packaging of the signals into the PUD and the final triggering of the send mechanism of lower layers. Depending on the signal type, the packaging is complex or simple. A bool signal is easy; the correct placement of a byte array is more complex. For all operations, it is important to keep the data consistency, and the access shall be atomic. Hardware-specific library functions provide this atomic operation by using microcontroller-specific atomic functions.

If the signals are placed at the correct position of the corresponding PDU, two scenarios exist: If the PDU is send cyclic, it is sufficient to update the data—the function returns without further actions. In another context of the COM module all cyclic send operations will be processed independent, and the PDU will be sent to the configured time slot respectively with the configured cycle time. If the user defined, within the configuration, an instant sending of the signal or PDU, the lower layers will be called direct. From the COM module the path continues to the PDU router; the *PduId* for the message to be sent and a pointer to a complex data structure will be given to the transmit function:

```
PDU RouterPduR_ComTransmit(PduId, &pduinfo);
```

4.2.3.6 PDU Router

The job of the PDU router is to abstract the physical communication bus systems from the higher software layers. By using static routing tables the PDU router transfers PDUs from higher layers (e.g. the COM module) to lower layers (e.g. the CAN interface) and vice versa. The static routing tables, e.g. for the Tx-PDUs, contain for each PDU a target send function of the interface layer.

This is the first point in time where the concrete physical bus is relevant. The example signal shall be sent via the CAN bus; hence, the PDU will be handed over to the CAN interface. Defined by the configuration there exists a routing table entry for this PDU with a send function pointer to the CAN interface function *CanIf_Transmit*. The function *PduR_ComTransmit* itself is small and simple; it collects some information out of the configuration data structures and forwards all to the correct interface function. In this example, it is called the *CanIf_Transmit function.*

```
CanIf_Transmit(CanTxPduId, &pPduInfoPtr);
```

4.2.3.7 CAN Interface

The CAN interface is the last hardware-independent module in the AUTOSAR-layered architecture; it has the job to connect the hardware-dependent CAN driver. The CAN interface gets the PDUs from the above layers, adds the data the CAN driver needs and implements buffering functions (if the hardware is occupied) and acknowledgement functions for the send and receive signalling. Furthermore, the CAN interface controls the operating modes of the CAN controller (sleep mode and error modes) and signals changes of these modes to the higher layers (wake-up and bus-off).

The function *CanIf_Transmit*, which is called by the PDU router, takes the corresponding configuration for the given *PduId* and calls the CAN driver. The minimal configuration-related information is, for example, on which CAN channel the message shall be sent and which CAN ID shall be used. After assembling these data the driver function is called. It is possible that the CAN interface is able to call several drivers—e.g. if an external CAN module is used. In general, the on-chip CAN modules will be supported by one single CAN driver; hence, the use case with several CAN drivers is rare. The CAN driver function gets as parameter the unique hardware transmit handle (*Hth*) and a pointer to the data structure *PduInfo*, where the CAN ID, DLC and a pointer to the payload data are stored.

```
Can_Write(CanIf_CTxPdus[txPduIndex].hth, &canPduInfo);
```

4.2.3.8 CAN Driver

The CAN driver is the last piece of the communication path of an AUTOSAR compliant software platform. The driver initializes the register of the CAN module with correct values and sets the right bits for actions or data transfer. The access to these control and data registers will be provided by standardized functions to the higher software layers. Another important functionality is the implementation of all interrupt service routines Receive (Rx), Transmit (Tx), error, etc.). The driver receives the data to be transmitted by the CAN interface, copies it into a hardware-mailbox of the CAN controller and triggers the sending process. In the opposite direction, the CAN driver signals the reception of a new CAN message via callback mechanism, e.g. triggered by the receive interrupt.

Today's CAN controller provides complex, intelligent mechanisms implemented in hardware, for example, FIFO buffering or filtering mechanisms. Universal drivers, easy to port and maintainable on various hardware platforms, use these hardware functions not very often. For example, FIFO buffering implemented in software is not efficient; respectively, the valuable resource *CPU-time* of the microcontroller is dissipated in this case. With AUTOSAR, the hardware will become more comparable: With standardized interfaces, it is possible to do a benchmark test easier on the driver layer than on the direct hardware layer. With this comparability on driver layer, it becomes more important to have a highly optimized driver, which uses the available CAN controller hardware mechanisms perfect to save other resources. This is also valid for other protocol drivers, like the FlexRay driver, and of course for the peripheral drivers at all.

4.2.3.9 Conclusion

The AUTOSAR standard enables the possibility of microcontroller manufacturers to provide highly optimized driver for their hardware. These drivers can be used by the software vendors within their basic software platforms. With an overall accepted AUTOSAR standard, the huge investment into the drivers and higher software layers is safe—there is a chance to get a return on the investment. The CAN driver is now a piece of a much bigger picture and standalone nothing special, but together with the overall AUTOSAR concept, the model of a layered architecture makes sense. A legacy software was able to access direct the lowest software layer—the drivers. It prepares the data, implements all needed mechanisms and calls, e.g. *Can_Write* direct. This direct access was implemented in the past ECU development very often. Some developers already use quasi-standards for the driver layer to avoid double development of the same functionality. Within the higher software layers some modules were standardized, e.g. the OS was OSEK/VDX compliant. In future, it is needed to get a high total reuse rate for each single piece of software. A standardized platform like the described AUTOSAR platform gets more and more important for efficient software development. The use of this platform pays off as soon as there are first applications which can be completely reused for a second vehicle generation—not to be developed again due to a "simple" hardware change (due to cheaper hardware availability). Another contribution for cost reduction will be the reduction of integration effort for each single project: Due to standardized methods and increasing AUTOSAR knowledge in the overall development community, the integration will be faster. This cost reduction will only take effect if the AUTOSAR standard is used by many and every Tier 1 can use the same basic software platform for various OEMs. The success or failure of this standardization of automotive software architecture is in the hand of the AUTOSAR core members, especially the OEM. Only if all OEMs use the AUTOSAR standard "as it is", the overall AUTOSAR project will be successful.

4.3 Automotive Diagnostic Implementations on CAN

ISO 15765 Road vehicles—Diagnostic communication over Controller Area Network (DoCAN) was initially developed in the late 1990s. The purpose of the document set is to standardize the diagnostic communication-related OSI layers for the purpose of diagnostic and normal message communication.

Table 4.11 Enhanced and legislated OBD diagnostic specifications applicable to the OSI layers

Applicability	OSI layers	Vehicle manufacturer enhanced diagnostics	Legislated OBD (On-Board Diagnostics)		Legislated WWH-OBD (World Wide Harmonized On-Board Diagnostics)		
Seven layer according to ISO 7498-1 and ISO/IEC 10731	Application (layer 7)	ISO 14229-1 ISO 14229-3	ISO 15031-5		ISO 27145-3 ISO°14229-1		SAE J1939, -71, -73
	Presentation (layer 6)	Vehicle Manufacturer specific or ISO 22901 ODX	ISO 15031-2, -5, -6, SAE°J1930-DA, SAE°J1979-DA, SAE°J2012-DA		ISO 27145-2, SAE °1930-DA, SAE°J1979-DA, SAE°J2012-DA, SAE°J1939 Top Level		SAE J1939 Top Level
	Session (layer 5)	ISO 14229-2					N/A
	Transport (layer 4)	ISO 15765-2	ISO° 15765-2	ISO 15765-4	ISO 15765-4 / ISO°15765-2	ISO° 27145-4	SAE J1939-21
	Network (layer 3)						SAE J1939-31
	Data link (layer 2)	ISO 11898-1	ISO° 11898-1, -2		ISO 15765-4 / ISO°11898-1, -2		SAE J1939-21
	Physical (layer 1)	ISO°11898-2, -3, -4, -5					SAE J1939-15

ISO 15765 consists of the following parts, under the general title Road vehicles—DoCAN:

- Part 1: General information and use case definition.
- Part 2: Transport protocol and network layer services.
- Part 3: UDS on CAN implementation (UDSonCAN; will be replaced by ISO 14229-3 in 2010).
- Part 4: Requirements for emission-related systems.

Table 4.11 shows the relevant ISO and SAE standards as they are applicable to the OSI layers and how they relate to the following categories (columns in Table 4.1):

- Vehicle manufacturer-enhanced diagnostics.
- Legislated On-Board Diagnostics (OBD).
- Legislated World-Wide Harmonized On-Board Diagnostics (WWH-OBD): shows the legislator-approved ISO and Society of Automotive Engineers (SAE) implementations.

The application layer (OSI 7) services for the:

- Vehicle manufacturer-enhanced diagnostics are defined in ISO 14229-3 UDSonCAN (former ISO 15765-3). This part of ISO 14229-1 is an implementation of Part 1 onto CAN.
- Legislated OBD are defined in ISO 15031-5 Road vehicles—Communication between vehicle and external equipment for emission-related diagnostics—Part 5: emission-related diagnostic services. This standard is referenced by the European Commission (EC) directives for EURO 4, 5, 6 (passenger cars and

light-duty vehicles (LDV)) and directives for EURO IV, V, VI (heavy-duty vehicles (HDVs)). The technical equivalent standard of ISO 15031-5 is SAE 1979 E/E diagnostic test modes, which is referenced by the California Air Resources Board (ARB) and Environmental Protection Agency (EPA) for the federal states.
- Legislated WWH-OBD is defined in ISO 27145 Road vehicles—Implementation of emission-related WWH-OBD communication requirements. This upcoming standard is still under development (2010) and will long term replace the ISO 15031-5/SAE J1979 standards. The legislators in Europe and the USA have also approved several parts of SAE J1939 Recommended Practice for a Serial Control and Communications Vehicle Network:
  - SAE 1939-73, Application Layer—Diagnostics (includes DM—diagnostic modes and data items: parameter group numbers (PGNs) and suspect parameter numbers (SPNs).
  - SAE 1939-71, Vehicle Application Layer (includes normal in-vehicle communication messages and data items: SPNs).

The presentation layer (OSI 6) services for the:

- Vehicle manufacturer-enhanced diagnostics are:
  - either specific to the manufacturer or
  - the manufacturer defines the diagnostic data according to ISO 22901, Road vehicles—Open Diagnostic data eXchange (ODX).
- Legislated OBD are defined in:
  - ISO 15031-2, Road vehicles—Communication between vehicle and external equipment for emission-related diagnostics—Part 2: Guidance on terms, definitions, abbreviations and acronyms with reference to the Digital Annex of SAE°J1930-DA, electrical/electronic systems diagnostic terms, definitions, abbreviations and acronyms (includes all standardized terms and abbreviations).
  - ISO 15031-5, Road vehicles—Communication between vehicle and external equipment for emission-related diagnostics—Part 5: Emission-related diagnostic services with reference to the Digital Annex of SAE°J1979-DA E/E diagnostic test modes (includes all standardized data items: parameter identifiers (PIDs), test identifiers (TIDs), monitor identifiers (MIDs) and infoType identifiers (ITIDs).
  - ISO 15031-6, Road vehicles—Communication between vehicle and external equipment for emission-related diagnostics—Part 6: with reference to the Digital Annex of SAE°J2012-DA, diagnostic trouble codes (includes all standardized DTCs).
- Legislated WWH-OBD are defined in:
  - ISO 27145-2 Road vehicles—Implementation of WWH-OBD communication requirements—Part 2: Common data dictionary (CDD). This standard references the same SAE Digital Annexes as ISO 15031 (emission-related OBD):
    - SAE°J1930-DA, electrical/electronic systems diagnostic terms, definitions, abbreviations and acronyms (includes all standardized terms and abbreviations).

- SAE°J1979-DA, E/E diagnostic test modes (includes all standardized data items: PIDs, TIDs, MIDs, ITIDs).
- SAE°J2012-DA, diagnostic trouble codes (includes all standardized DTCs).
- SAE°J1939, top level (includes all standardized SPNs and PGNs).

The session layer (OSI 5) services are defined in ISO 14229-2 Road vehicles—Unified diagnostic services (UDSs)—Part 2: Session layer services. This standard provides a protocol-independent standardized service primitive interface between application layer services and transport protocol/network layer services. This standard is not limited to the standards as listed in Table 4.11.

The transport protocol layer (OSI 4) and network layer (OSI 3) services are defined in:

- ISO 15765-2 Road vehicles—DoCAN—Part 2: Transport protocol and network layer services. This part of the standard is applicable to all three categories. The legislated categories (OBD and WWH-OBD) have additional requirements defined in ISO 15765-4, which are related to CAN identifier definition, transport protocol timing, etc.
- SAE 1939-21, transport protocol is only applicable to WWH-OBD.
- SAE 1939-31, network layer is only applicable to WWH-OBD.

The data link layer (OSI 2) requirements are defined in:

- ISO 11898-1, Road vehicles—Controller area network (CAN)—Part 1: Data link layer and physical signalling. This standard is applicable to all CAN networks independent of the upper OSI layer implementations.
- SAE 1939-21, Data link layer is only applicable to WWH-OBD.

The physical layer (OSI 1) requirements are defined in various parts of ISO 11898.

- Vehicle manufacturer-enhanced diagnostics can be implemented on any of the following physical layer standards:
  - ISO°11898-2, Road vehicles—Controller area network (CAN)—Part 2: High-speed medium access unit (majority of automotive implementations).
  - ISO°11898-3, Road vehicles—Controller area network (CAN)—Part 3: Low-speed, fault-tolerant, medium-dependent interface.
  - ISO°11898-4, Road vehicles—Controller area network (CAN)—Part 4: Time-triggered communication.
- Legislated OBD is only allowed on ISO°11898-2, Road vehicles—Controller area network (CAN)—Part 2: High-speed medium access unit.
- Legislated WWH-OBD is allowed on:
  - ISO°11898-2, Road vehicles—Controller area network (CAN)—Part 2: High-speed medium access unit.
  - SAE J1939-15, Reduced Physical Layer, 250K bits/sec, Un-Shielded Twisted Pair (UTP).

Figure 4.19 illustrates the most applicable application implementations utilizing the DoCAN protocol.

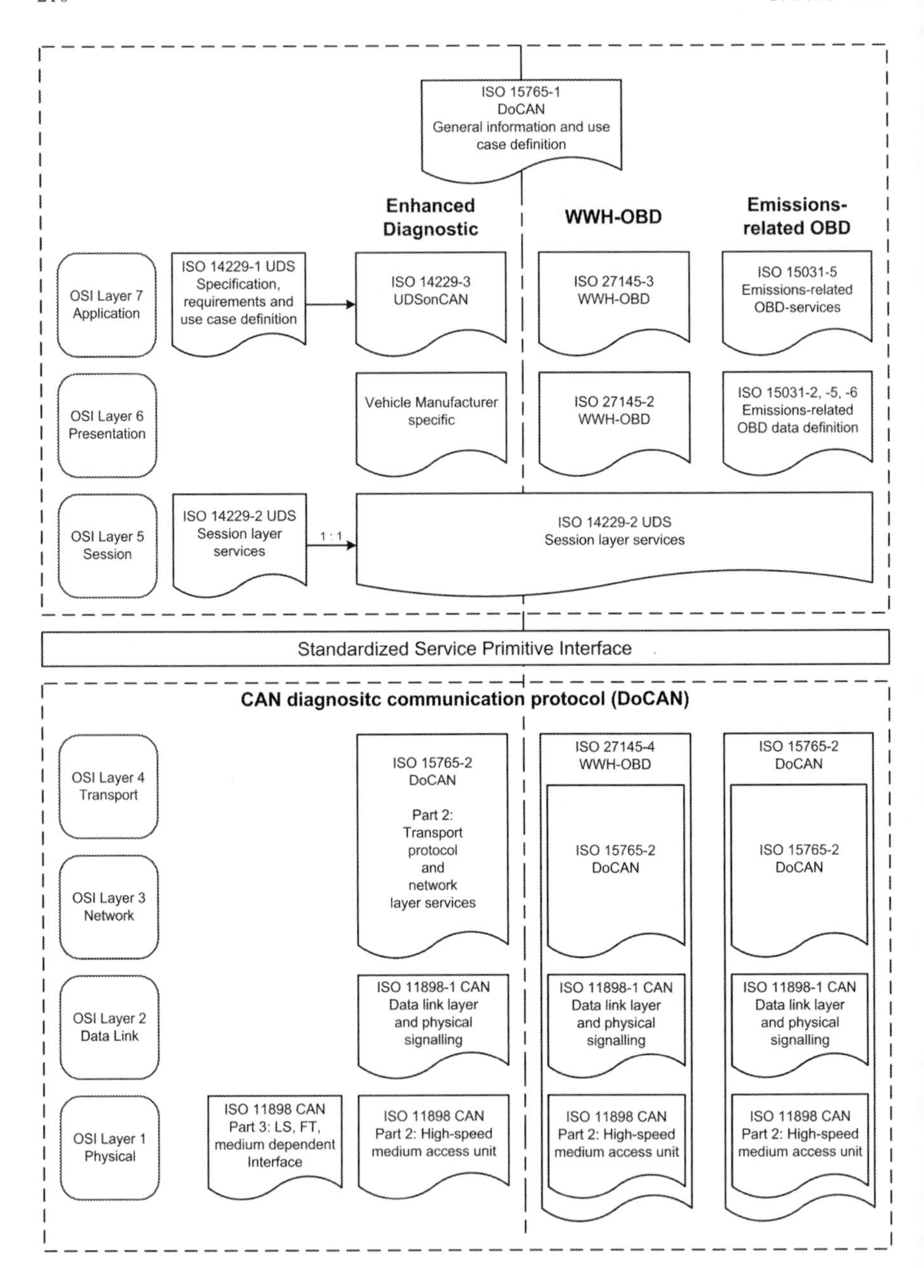

Fig. 4.19 Diagnostic communication over CAN document reference according to OSI model

4.3.1 OBDonCAN—ISO 15031 emissions-related OBD on ISO 15765-4 DoCAN

With the introduction of ISO 15765 DoCAN as a legislator-approved emission-related system OBD data link the ISO 15031-5/SAE J1979 emission-related diagnostic services/test modes have been slightly modified to benefit from the ISO 15765-2 DoCAN transport protocol and network layer services standard. ECU response messages, which used to be split into multiple response message for the non-CAN protocols (SAE J1850, ISO 9141-2 and ISO 14230-4), are now assembled into one response message, consisting of multiple CAN frames. The payload data of the PDU are identical between all ISO 15031-5/SAE J1979-defined diagnostic messages (DMs).

The requirements and features of the OBDonCAN protocol can be summarized as follows:

- Tester can request up to sic PIDs in a single functionally addressed request message. Non-CAN protocols allow only for one PID in the request message.
- Either baud rate is allowed:
  - 250 kBit/s
  - 500 kBit/s
- A maximum of eight emission-related OBD ECUs is allowed to respond on ISO 15031-5 requests.
- Tester is only allowed to send functionally addressed request messages (11 bit CAN ID=0×7DF). Flow control from the tester and response messages from the ECU use the physical request and response CAN IDs (Table 4.12).
- Vehicle manufacturer, which decides to use 29-bit CAN IDs on their CAN bus, must use the 29-bit CAN IDs as specified in ISO 15765-4 (Table 4.13).
- The ISO 15765-2 DoCAN transport protocol and network layer services standard supports up to 4.095 bytes in a message. Reception of messages up to six (6) or seven (7) data bytes is performed via reception of a unique N_PDU (Figs. 4.20 and 4.21).
- The ISO 15765-2 DoCAN multiple-frame message transfer. The flow control mechanism allows the receiver to inform the sender about the receiver's capabilities. Since different nodes may have different capabilities, the flow control sent by the receiver informs the sender about its capabilities. The sender shall conform to the receiver's capabilities (Fig. 4.22).
- The ISO 15765-2 DoCAN—Transport protocol and network layer services timeout and performance requirements are more stringent in ISO 15765-4 (Table 4.14).
- The external test equipment shall use the following network layer parameter values for its FlowControl frames, sent in response to the reception of a FirstFrame (Table 4.15 and 4.16).

Table 4.12 11-bit legislated OBD/WWH-OBD CAN identifiers

CAN identifier	Description
0×7DF	CAN identifier for functionally addressed request messages sent by external test equipment
0×7E0	Physical request CAN identifier from external test equipment to ECU #1
0×7E8	Physical response CAN identifier from ECU #1 to external test equipment
0×7E1	Physical request CAN identifier from external test equipment to ECU #2
0×7E9	Physical response CAN identifier from ECU #2 to external test equipment
0×7E2	Physical request CAN identifier from external test equipment to ECU #3
0×7EA	Physical response CAN identifier from ECU #3 to external test equipment
0×7E3	Physical request CAN identifier from external test equipment to ECU #4
0×7EB	Physical response CAN identifier ECU #4 to the external test equipment
0×7E4	Physical request CAN identifier from external test equipment to ECU #5
0×7EC	Physical response CAN identifier from ECU #5 to external test equipment
0×7E5	Physical request CAN identifier from external test equipment to ECU #6
0×7ED	Physical response CAN identifier from ECU #6 to external test equipment
0×7E6	Physical request CAN identifier from external test equipment to ECU #7
0×7EE	Physical response CAN identifier from ECU #7 to external test equipment
0×7E7	Physical request CAN identifier from external test equipment to ECU #8
0×7EF	Physical response CAN identifier from ECU #8 to external test equipment

While not required for current implementations, it is strongly recommended (and may be required by applicable legislation) that for future implementations the following 11-bit CAN identifier assignments be used:

0×7E0/0×7E8 for engine control module (ECM)

0×7E1/0×7E9 for transmission control module (TCM)

Table 4.13 Summary of 29-bit CAN identifier format—normal fixed addressing

CAN-Id ID bit position	28	24	23	16	15	8	7	0
Functional CAN IDId	0×18		0xDB		TA		SA	
Physical CAN IDId	0×18		0xDA		TA		SA	

Note: The CAN identifier values given in this table use the default value for the priority information in accordance with ISO 15765-2

4.3.2 UDSonCAN—ISO 14229-3

As mentioned earlier, UDSonCAN is originally based on ISO 15765-3. The revision of ISO 15765-3 currently under work at ISO will be published as ISO 14229-3. The reason for the renumbering of the well-established standard in the automotive industry is based on the fact that ISO 14229-1 UDS shall be implemented on many different protocols and data links (CAN, FlexRay, Internet Protocol, K-Line, etc.) according to the OSI model. Fig. 4.23 illustrates the implementation concept of UDS.

ISO 14229-1 UDS specifies the DMs in a protocol-independent manner. This document is currently under revision by the ISO TC22/SC2/WG1/TF5 diagnostic requirements task force. All lessons learnt and implementation feedback from system suppliers, vehicle manufacturers and tool suppliers is under implementation.

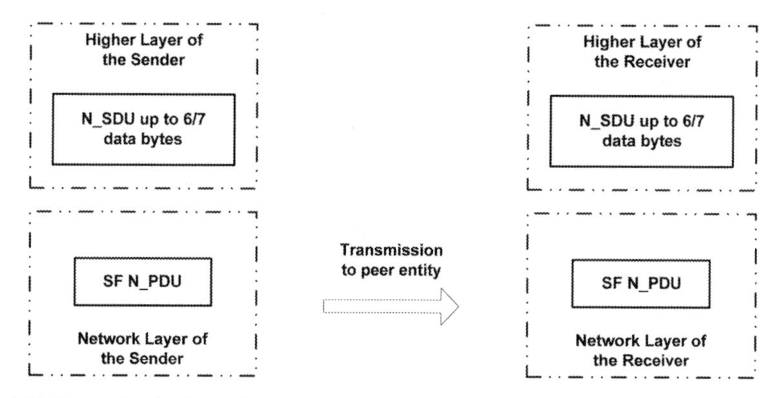

Fig. 4.20 Example of a single-frame (SF) transmission

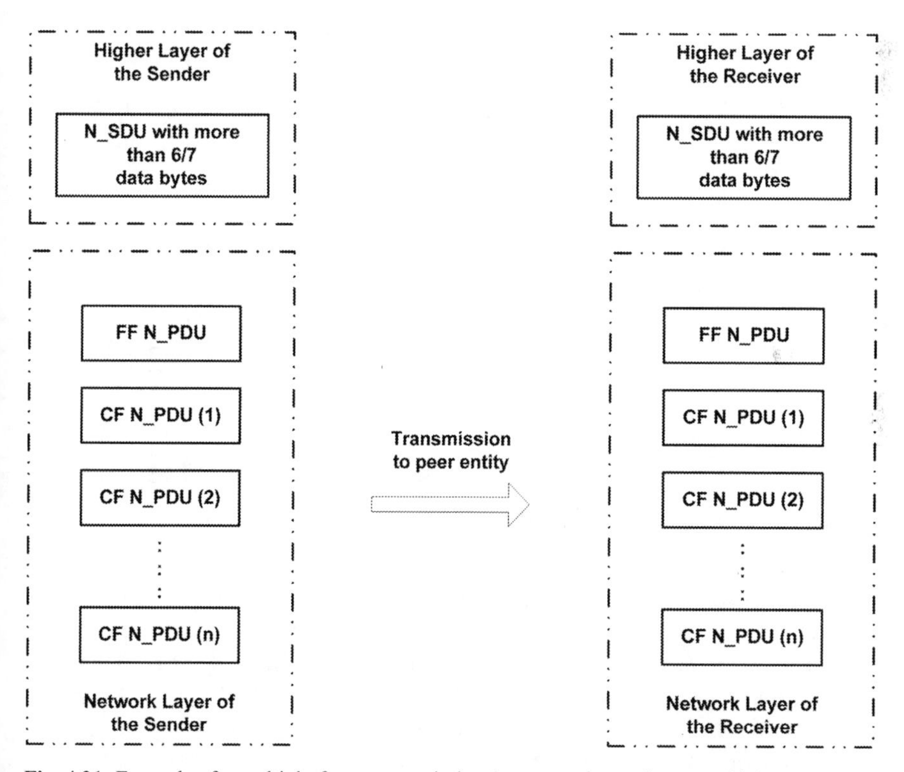

Fig. 4.21 Example of a multiple-frame transmission (segmentation and reassembly)

ISO 14229-2 UDS specifies common session layer services to provide independence between UDS (Part 1) and all network/transport protocols (ISO 15765-2 CAN, ISO/WD 10681-2 FlexRay, ISO/CD 13400-2 DoIP, ISO 14230-2 K-Line, LIN, MOST, etc.).

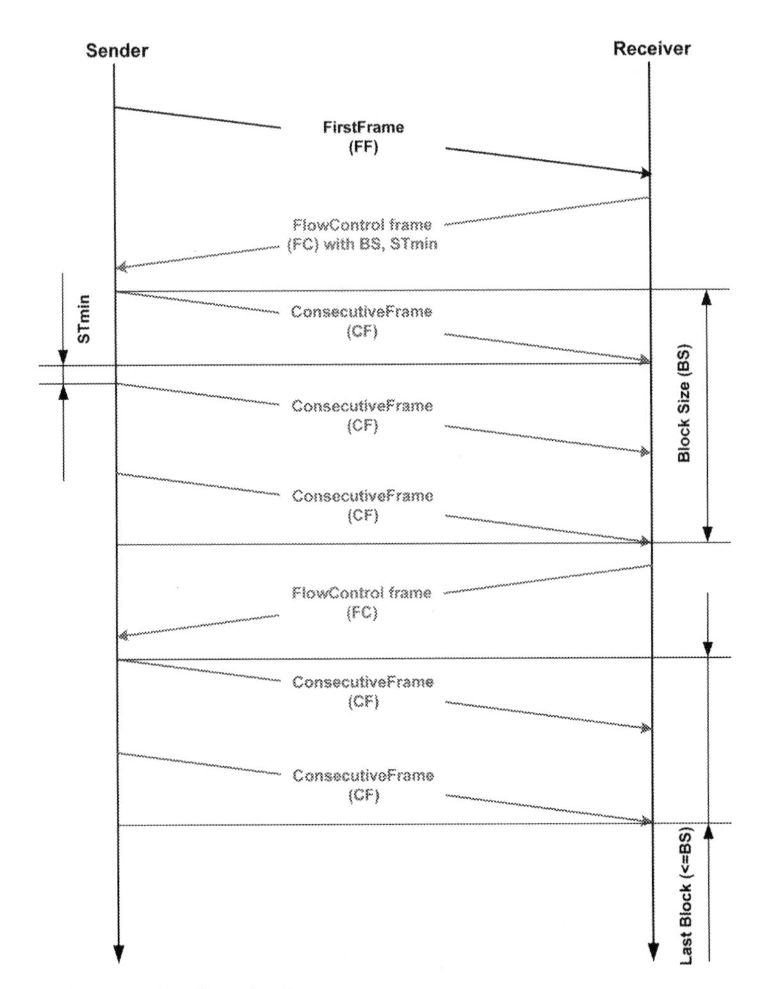

Fig. 4.22 Flow control (FC) mechanism

A common service primitive interface is specified between OSI layer 4 (Transport) and layer 5 (Session) via the so-called service request/confirmation/indication primitives. This interface allows seamless implementation of ISO°14229-1 UDSs with any communication protocol titled "DoXYZ/CoXYZ" like ISO°15765 DoCAN—diagnostic communication over controller area network, ISO°13400 DoIP—diagnostic communication over Internet protocol, ISO°10681 CoFlexRay—communication over FlexRay, ISO°14230 DoK-Line—diagnostic communication over K-line.

ISO 14229-3 UDSonCAN is based on a common document template for UDS implementations on a specific protocol like ISO 15765 DoCAN. This specific part of ISO 14229 specifies implementation requirements related to the following:

- The diagnostic services to be used for diagnostic communication over CAN.
- The server memory programming for all in-vehicle servers connected to a CAN network with an external test equipment.

Table 4.14 Network layer timeout and performance requirement values

Parameter	Timeout value	Performance requirement value
N_As/N_Ar	25 ms	–
N_Bs	75 ms	–
N_Br	–	(N_Br+N_Ar)<25 ms
N_Cs	–	(N_Cs+N_As)<50 ms
N_Cr	150 ms	–

Table 4.15 External test equipment network layer parameter values

Parameter	Name	Value	Description
N_WFTmax	WaitFrame transmission	0	No FlowControl wait frames are allowed for legislated OBD/WWH-OBD. The FlowControl frame sent by the external test equipment following the FirstFrame of an ECU response message shall contain the FlowStatus FS set to 0 (ClearToSend), which forces the ECU to start immediately after the reception of the FlowControl frame with the transmission of the ConsecutiveFrame(s)
BS	BlockSize	0	A single FlowControl frame shall be transmitted by the external test equipment for the duration of a segmented message transfer. This unique FlowControl frame shall follow the FirstFrame of an ECU response message
STmin	SeparationTime	0	This value allows the ECU to send ConsecutiveFrames, following the FlowControl frame sent by the external test equipment, as fast as possible

If a reduced implementation of the ISO°15765-2 network layer is done in a legislated OBD/WWH-OBD ECU, covering only the above-listed FlowControl frame parameter values (BS, STmin), then any FlowControl frame received during legislated OBD/WWH-OBD communication and using different FlowControl frame parameter values as defined in this table shall be ignored by the receiving legislated OBD/WWH-OBD ECU (treated as an unknown network layer PDU)

It does not contain information related to any requirement for the in-vehicle CAN bus architecture.

Figure 4.24 illustrates all referenced documents according to the OSI layers.

4.3.2.1 UDSonCAN Services Overview

The purpose of Table 4.17 is to reference all UDSs as they are applicable for an implementation of UDSonCAN. The table contains the sum of all applicable services. Certain applications using this part of ISO 14229 to implement UDSonCAN may restrict the number of useable services and may categorize them in certain application areas/diagnostic sessions (default session, programming session, etc.).

Table 4.16 Emission-related diagnostic service definition OBDonCAN (ISO 15031-5 on ISO 15765-4)

Service Id	Description
0 × 01	Request current powertrain diagnostic data The purpose of this service is to allow access to current emission-related data values, including analog inputs and outputs, digital inputs and outputs and system status information. The request for information includes a parameter-identification (PID) value that indicates to the on-board system the specific information requested. PID specifications, scaling information and display formats are included in SAE J1979-DA
0 × 02	Request powertrain freeze frame data The purpose of this service is to allow access to emission-related data values in a freeze frame. This allows expansion to meet manufacturer-specific requirements not necessarily related to the required freeze frame, and not necessarily containing the same data values as the required freeze frame. The request message includes a parameter identification (PID) value that indicates to the on-board system the specific information requested. PID specifications, scaling information and display formats for the freeze frame are included in SAE J1979-DA
0 × 03	Request emission-related diagnostic trouble codes The purpose of this service is to enable the external test equipment to obtain "confirmed" emission-related DTCs Send a Service 0 × 03 request for all emission-related DTCs. Each ECU that has DTCs shall respond with one (1) message containing all emission-related DTCs. If an ECU does not have emission-related DTCs, then it shall respond with a message indicating no DTCs are stored by setting the parameter of DTC to 0 × 00
0 × 04	Clear/reset emission-related diagnostic information The purpose of this service is to provide a means for the external test equipment to command ECUs to clear all emission-related diagnostic information. This includes the following: • MIL and number of diagnostic trouble codes (can be read with Service 0 × 01, PID 0 × 01) • Clear the inspection/maintenance (I/M) readiness bits (can be read with Service 0 × 01, PID 0 × 01) • Confirmed diagnostic trouble codes (can be read with Service 0 × 03) • Pending diagnostic trouble codes (can be read with Service 0 × 07) • Diagnostic trouble code for freeze frame data (can be read with Service 0 × 02, PID 0 × 02) • Freeze frame data (can be read with Service 0 × 02) • Status of system monitoring tests (can be read with Service 0 × 01, PID 0 × 41 • On-board monitoring test results (can be read with Service 0 × 06) • Distance travelled while MIL is activated (can be read with Service 0 × 01, PID 0 × 21) • Number of warm-ups since DTCs cleared (can be read with Service 0 × 01, PID 0 × 30) • Distance travelled since DTCs cleared (can be read with Service 0 × 01, PID 0 × 31) • Engine run time while MIL is activated (can be read with Service 0 × 01, PID 0 × 4D) • Engine run time since DTCs cleared (can be read with Service 0 × 01, PID 0 × 4E) • Reset misfire counts of standardized test ID 0 × 0B to zero (can be read with Service 0 × 06)

Table 4.16 (continued)

Service Id	Description
0 × 05	Request oxygen sensor monitoring test results Service 0 × 05 is not supported for ISO 15765-4. The functionality of Service 0 × 05 is implemented in Service 0 × 06
0 × 06	Request on-board monitoring test results for specific monitored systems The purpose of this service is to allow access to the results for on-board diagnostic monitoring tests of specific components/systems that are continuously monitored (e.g. misfire monitoring for gasoline vehicles) and non-continuously monitored (e.g. catalyst system)
0 × 07	Request emission-related diagnostic trouble codes detected during current or last completed driving cycle The purpose of this service is to enable the external test equipment to obtain "pending" diagnostic trouble codes detected during current or last completed driving cycle for emission-related components/systems. Service 0 × 07 is required for all DTCs and is independent of Service 0 × 03. The intended use of this data is to assist the service technician after a vehicle repair, and after clearing diagnostic information, by reporting test results after a single driving cycle. If the test failed during the driving cycle, the DTC associated with that test shall be reported. Test results reported by this service do not necessarily indicate a faulty component/system. If test results indicate a failure after additional driving, then the MIL will be illuminated and a DTC will be set and reported with Service 0 × 03, indicating a faulty component/system. This service can always be used to request the results of the latest test, independent of the setting of a DTC
0 × 08	Request control of on-board system, test or component The purpose of this service is to enable the external test equipment to control the operation of an on-board system, test or component The data bytes will be specified, if necessary, for each test ID in SAE J1979-DA, and will be unique for each test ID Possible uses for these data bytes in the request message are as follows: • Turn on-board system/test/component ON • Turn on-board system/test/component OFF • Cycle on-board system/test/component for 'n' seconds Possible uses for these data bytes in the response message are as follows: • Report system status • Report test results
0 × 09	Request vehicle information The purpose of this service is to enable the external test equipment to request vehicle-specific vehicle information such as vehicle identification number (VIN) and calibration IDs. Some of this information may be required by regulations and some may be desirable to be reported in a standard format if supported by the vehicle manufacturer. InfoTypes are defined in SAE J1979-DA
0 × 0A	Request emission-related diagnostic trouble codes with permanent status The purpose of this service is to enable the external test equipment to obtain all DTCs with "permanent DTC" status. These are DTCs that are "confirmed" and are retained in the non-volatile memory of the server until the appropriate monitor for each DTC has determined that the malfunction is no longer present and is not commanding the MIL on Service 0 × 0A is required for all emission-related DTCs. The intended use of this data is to prevent vehicles from passing an in-use inspection simply by disconnecting the battery or clearing DTCs with a scan tool prior to the inspection. The presence of permanent DTCs at an inspection without the MIL illuminated is an indication that a proper repair was not verified by the on-board monitoring system

Table 4.16 (continued)

Service Id	Description
	Permanent DTCs shall be stored in non-volatile memory (NVRAM) and may not be erased by any diagnostic services (generic or enhanced) or by disconnecting power to the ECU A confirmed DTC shall be stored as a permanent DTC no later than the end of the ignition cycle and subsequently at all times that the confirmed DTC is commanding the MIL on (e.g. for currently failing systems but not during the 40 warm-up cycle self-healing process) Permanent DTCs may be erased if: • The OBD system itself determines that the malfunction that caused the permanent fault code to be stored is no longer present and is not commanding the MIL on, e.g. three consecutive complete driving cycles with no malfunction, or as specified by the OBD regulations • After clearing fault information in the ECU (i.e. through the use of a diagnostic service or battery disconnect): • For monitors subject to minimum in-use ratio requirement, the diagnostic monitor for the malfunction that caused the permanent DTC to be stored has fully executed (i.e. has executed the minimum number of checks necessary for MIL illumination) and determined the malfunction is no longer present, e.g. one complete driving cycle with no malfunction or as specified by the OBD regulations • For monitors not subject to minimum in-use ratio requirement, the diagnostic monitor for the malfunction that caused the permanent DTC to be stored has fully executed (i.e. has executed the minimum number of checks necessary for MIL illumination) and determined the malfunction is no longer present, e.g. one complete driving cycle with no malfunction or as specified by the OBD regulations and the vehicle has completed a standard driving cycle used to increment the in-use general denominator • Permanent fault codes may be erased when the ECU containing the permanent DTCs is reprogrammed if the readiness status for all monitored components and systems is set to "not complete" in conjunction with the reprogramming event

4.3.3 Development Trends

4.3.3.1 History

Today two CAN-based protocols exist which are legislator-approved protocols for emission-related OBD:

- ISO 15031-5 on ISO 15765-4 OBDonCAN
- SAE J1939

ISO 15031-5 is technically identical with SAE J1979 and defines specific services/test modes to exchange data with the emission-related system in the vehicle. The services support the request of current PID information, stored freeze frame data, readout of DTC: current, pending and permanent status, clearing of DTC-related data, readout of monitoring test results, request control of on-board system (test or component) and request of vehicle information.

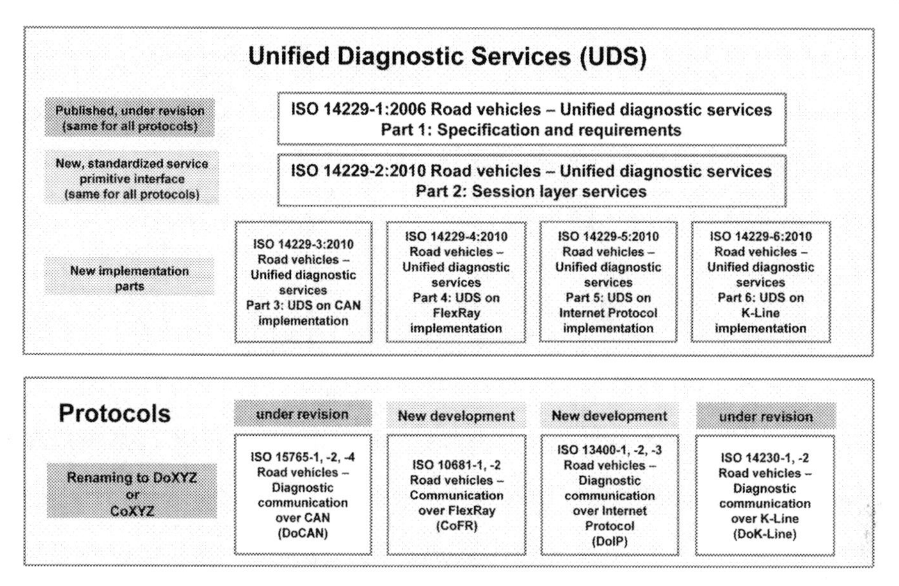

Fig. 4.23 Implementation concept of unified diagnostic services (UDSs)

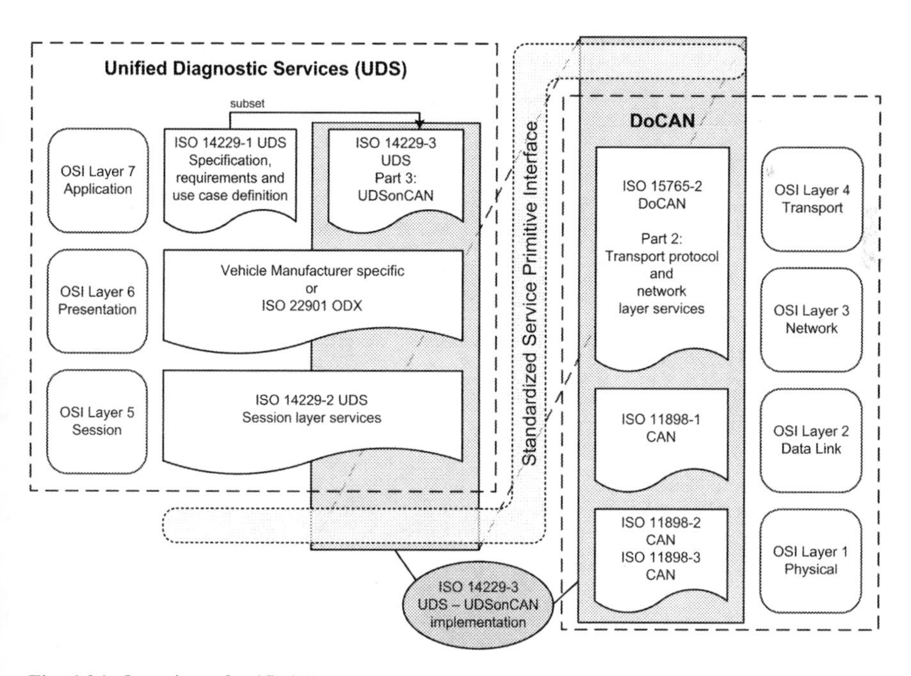

Fig. 4.24 Overview of unified diagnostic services to the OSI layers

Table 4.17 Overview of applicable ISO 14229-1 unified diagnostic services and data ranges

Diagnostic service name (ISO 14229-1)	SID value	SFID value	Sub-function name
Diagnostic and communication management functional unit			
DiagnosticSessionControl	0×10	0×01	defaultSession
		0×02	programmingSession
		0×03	extendedSession
		0×04	safetySystemDiagnosticSession
ECUReset	0×11	0×01	hardReset
		0×02	keyOffOnReset
		0×03	softReset
		0×04	enableRapidPowerShutDown
		0×05	disableRapidPowerShutDown
SecurityAccess	0×27	0×01	requestSeed
		0×02	sendKey
CommunicationControl	0×28	0×00	enableRxAndTx
		0×01	enableRxAndDisableTx
		0×02	disableRxAndEnableTx
		0×03	disableRxAndTx
TesterPresent	0×3E	0×00	zeroSubFunction
SecuredDataTransmission	0×84	N/A	N/A
ControlDTCSetting	0×85	0×01	on
		0×02	off
ResponseOnEvent	0×86	0×00	stopResponseOnEvent
		0×01	onDTCStatusChange
		0×02	onTimerInterrupt
		0×03	onChangeOfDataIdentifier
		0×04	reportActivatedEvents
		0×05	startResponseOnEvent
		0×06	clearResponseOnEvent
		0×07	onComparisonOfValues
LinkControl	0×87	0×01	verifyModeTransitionWithFixed-Parameter
		0×02	verifyModeTransitionWith-Specific-Parameter
		0×03	transitionMode
Data Transmission Functional Unit			
ReadDataByIdentifier	0×22	–	N/A
ReadMemoryByAddress	0×23	–	N/A
ReadScalingDataByIdentifier	0×24	–	N/A
ReadDataByPeriodicIdentifier	0×2A	–	N/A
DynamicallyDefineData-Identifier	0×2C	0×01	defineByIdentifier
		0×02	defineByMemoryAddress
		0×03	clearDynamicallyDefinedDataIdenti-fier
WriteDataByIdentifier	0×2E	–	N/A
WriteMemoryByAddress	0×3D	–	N/A
Stored data transmission functional unit			
ReadDTCInformation	0×19	0×01	reportNumberOfDTCByStatusMask

Table 4.17 (continued)

Diagnostic service name (ISO 14229-1)	SID value	SFID value	Sub-function name
		0 × 02	reportDTCByStatusMask
		0 × 03	reportDTCSnapshotIdentification
		0 × 04	reportDTCSnapshotRecordByDTC-Number
		0 × 05	reportDTCSnapshotRecordByRecord-Number
		0 × 06	reportDTCExtendedDataRecordBy-DTCNumber
		0 × 07	reportNumberOfDTCBySeverityMas-kRecord
		0 × 08	reportDTCBySeverityMaskRecord
		0 × 09	reportSeverityInformationOfDTC
		0 × 0A	reportSupportedDTC
		0 × 0B	reportFirstTestFailedDTC
		0 × 0C	reportFirstConfirmedDTC
		0 × 0D	reportMostRecentTestFailedDTC
		0 × 0E	reportMostRecentConfirmedDTC
		0 × 0F	reportMirrorMemoryDTCByStatus-Mask
		0 × 10	reportMirrorMemoryDTCExtend-edDataRecordByDTCNumber
		0 × 11	reportNumberOfMirrorMemoryD-TCByStatusMask-
		0 × 12	reportNumberOfEmissions-Relate-dOBDDTCByStatusMask
		0 × 13	reportEmissionsRelatedOBD-DTCByStatusMask
		0 × 14	reportDTCFaultDetectionCounter
		0 × 15	reportDTCWithPermanentStatus
		0 × 41	reportWWHOBDNumberOfDTCBy-MaskRecord
		0 × 42	reportWWHOBDDTCByMaskRecord
		0 × 55	reportWWHOBDDTCWithPerma-nentStatus
ClearDiagnosticInformation	0 × 14	–	N/A
Input/output control functional unit			
InputOutputControlByIdentifier	0 × 2F	–	N/A
Remote activation of routine functional unit			
RoutineControl	0 × 31	0 × 01	startRoutine
		0 × 02	stopRoutine
		0 × 03	requestRoutineResults
Upload/download functional unit			
RequestDownload	0 × 34	–	N/A
RequestUpload	0 × 35	–	
TransferData	0 × 36	–	
RequestTransferExit	0 × 37	–	

The ISO 15031-5 diagnostic services are very specific to emission-related OBD systems. The automotive industry has developed the so-called enhanced diagnostic protocols like ISO 14230 Keyword Protocol 2000 and SAE J2190 E/E Enhanced diagnostic test modes to be able to diagnose non-emission-related OBD systems as well as the functionality beyond the requirements included in the legislation.

At the time when the emission-related systems of the HDVs were referenced in the legislation (with EURO IV), the SAE J1939 protocol became an allowed data link in addition to ISO 15031-5. Both protocols support the CAN bus but are very different in their messages and functionality.

Since that time the HDV manufacturers have two choices to fulfil the emission-related OBD regulations. Two different diagnostic tester implementations are required to diagnose the HDV's emission-related OBD systems.

4.3.3.2 Requirements of the Legislators

The passenger car industry has harmonized the requirements for emission-related systems in the vehicle (connector, diagnostic services, trouble codes, communication protocol, etc.) during the past two decades.

The HDVs are using two alternative communication protocols:

- ISO 15765-4
- SAE J1939/73

Both will exist in parallel for some period of time.

The United Nations Economic Commission for Europe (UNECE) World Forum for Harmonization of Vehicle Regulations (WP.29) decided to develop a global technical regulation (GTR) concerning emission-related OBD systems for HDVs and engines (UNECE GTR No 5—Technical requirements for OBD systems for road vehicles). Consequently, a single OBD protocol is required to fulfil the communication requirements of this future regulation.

The emissions control systems on highway vehicles are not the only systems with OBD capability. The non-standardized diagnostics in all other systems in the vehicle cause negative implications on maintenance and inspection procedures. This was one of the driving factors of the WWH-OBD working group to design a modular structure of the GTR such that further OBD functionalities for, e.g. safety-related systems could be added any time in the future when appropriate.

The GTR consists of a base module (general requirements) and an emission-related system module.

4.3.3.3 ISO 27145 WWH-OBD

WWH-OBD is one of the objectives of the GTR No 5. A single protocol solution is highly desired by the legislators to be developed by the automotive industry.

ISO TC22/SC3/WG1 Data Communication, SAE and JSAE (Japanese SAE) set-up a joint task force to define the principles and concept for a future single protocol solution. This activity started in early 2003 with a document from the WWH-OBD informal group called:

GENERAL PERFORMANCE CRITERIA FOR HDV EMISSION-RELATED OBD

Communication protocols

WWH-OBD meeting 6/7 November 2002—DECISION 10

This document included requirements related to the following topics:

1. Needs for a common HDV OBD Communication Protocol:
   - Today, there exist two competing communication protocols for the application of OBD to HDVs—SAE J1939 and ISO 15765.
   - In the short term, it seems that both communication standards will exist in parallel but the primary aim must be to have one common protocol, which would be to the benefit of all sectors operating under the umbrella of "the automotive industry".
2. Needs for the legislator:
   - The scope of the standard must include both current chassis control and emission control systems and must provide for the seamless addition of further control systems (both simple and complex) as the market develops.
   - The standard must offer the capability to react to the wishes of the legislator in a quick and effective manner. This particularly encompasses the following likely future requests:
     a. the standard should be extendable to passenger cars and light commercial vehicles and
     b. the standard should offer the ability to retrieve the data necessary for in-use compliance testing, e.g. to identify vehicle operation in/out of a "Not to Exceed" (NTE) zone (if NTE remains a valid concept in the future), cumulative time/distance travelled in/out of an NTE zone, operation of auxiliary control devices, torque/load readings for engine testing or to enable the use of portable emission measurement systems (PEMS).
   - The standard must include the possibility for the application of wireless communication between the OBD system and a remote interrogation unit.
   - The standard must offer the ability to retrieve in-use OBD performance data, e.g. OBD monitoring frequency, vehicle operation frequency and time of operation.
   - Clear and precise specifications within the communication protocol, with minimal variations that will result in a minimum chance of difference in interpretation that could lead to vehicles being produced that are unable to communicate fully with a generic scan tool (note: some vehicles may not actually utilize hard-wired scan-tools in the future).
   - Availability of test equipment that can verify that communication protocol specifications are being adhered to on production vehicles.
3. Needs for inspection and maintenance (I/M) testing (roadworthiness testing or roadside spot-checks):

- Clear identification of the class of each malfunction, i.e.:
  - Hierarchical safety-related malfunctions.
  - Malfunctions that result in pollutant emissions exceeding a pre-set legislative threshold.
  - Malfunctions that do not result in pollutant emissions exceeding a pre-set legislative threshold, but which could result in pollutant emissions exceeding a pre-set legislative threshold at some time soon.
  - Malfunctions that do not need to be covered by legislation but are necessary for an efficient diagnostic and maintenance function.
- Ability to communicate 'readiness' data to confirm if the vehicle is ready to be inspected and has not had its fault memory cleared recently 1 (e.g. key starts/warm-up cycles/distance travelled since memory cleared, how many diagnostics have run and been completed, ability to see if a previously active fault has been cleared by a scan tool but not fixed).
- Ability to transmit roadworthiness-related fault information (e.g. malfunction indicator (MI) status and MI commanding 'on' fault codes, odometer readings, distance travelled with MI on, emission "severity/priority" of fault).
- Ability to transmit vehicle identification information 1 (e.g. vehicle identification number (VIN), software version, odometer reading, engine ID, transmission ID, vehicle weight rating/class information).
- Ability to help combat tampering 1 (e.g. unauthorized clearing of diagnostic information).
- Ability to help identify tampered or corrupted software at the time of inspection.
- Ability to help identify (potentially) tampered hardware at the time of inspection.
- Ability to identify and retrieve roadworthiness-related information from all electronic control modules (ECM, TCM, etc.) through a single process and by wireless connection.
- Compatibility with I/M equipment ([connector], hardware, software etc.).
- Additional specific inspection needs (e.g. mode $ 08-type commands for smoke opacity test etc.).
- Compatibility with potential future telematics-based vehicle systems, e.g. bluetooth, IEEE 802.11b (or later specification).

4. Needs for the technician (i.e. repairer, replacement part maker, tool maker, etc.) are as follows:
   - Data update rates (e.g. how fast can real-time sensor data be displayed for technicians, communication speed, ability to obtain multiple PID's with single requests etc.).
   - Access to established and extendable to non-established chassis control related fault codes and real-time data in a standardized manner.
   - Mode $ 06 test results (data available in a standardized, understandable format without need to refer to a service book).

- Freeze frame data (e.g. number of frames supported, data available in the frame, usefulness of data in the frame to technicians).
- Ability to clear memory and exercise monitors (e.g. post repair) to reset readiness codes for inspection and validation of repairs.
- Cost/compatibility/upgrade potential for new and existing service tools.

4.3.3.4 The WWH-OBD Task Force also Established Requirements Related to Diagnostics and Flash Programming of ECUs.

Objectives agreed by the task force:

1. To achieve the timetable proposed and have a DIS available by January 2007.
2. To enable the separation of vehicle-level technology and communication standards from the tool-level technology.
3. Single 'off-board' protocol with a single set of services to communicate at a minimum:
   - OBD legislated data
   - Enhanced diagnostic data
   - Reprogramming

4.3.3.5 Decisions on How to Proceed with the Development Work:

1. The solution must encompass:
   - OBD legislated diagnostics
   - Enhanced vehicle diagnostics
   - Reprogramming functionality
2. There is no desire to mandate something different for OBD legislated diagnostics than that which is used for other vehicle data access.
3. The benefits this brings are as follows:
   - Consistent use of services in development will improve the quality of the protocol implementation.
   - Consistent use of services will alleviate many of the currently identified communication and data-formatting problems.
   - Reduced development cost per vehicle module.
   - De-proliferation of protocols and service sets across the automotive industry will result in fewer implementation issues.
4. The combination of these functions will pave the way for the extension of OBD diagnostics across the whole vehicle and vehicle type (HDV, medium-duty vehicle (MDV) and LDV).
5. Existing automotive and IT standards will be recognized and analysed in determining the final solution:
   - ISO 14229-1 (UDSs), ISO 15765-x and SAE J1939-x.

   - Analysis of current IT standards, e.g. for transport and network protocols to current automotive standards.
6. Recognize the need for common services and data:
   - Analyse existing data definitions from SAE J1939–21/71/73 and ISO 15031–5/6.
   - Re-use OBD legislated definitions.
   - Explore the impact of hierarchical DTC's.
7. Our vision of a 'Single Solution' encompasses the need for CAN, wireless and wired Ethernet.
8. This vision provides a modular, structured methodology to achieve and support wireless communications for OBD.
9. Our view of a 'Single Solution' across all vehicles is as follows:
   - A single communication protocol for all off-board tester applications (e.g. OBD, enhanced and reprogramming).
   - A single set of services to retrieve and download information.
   - A single set of OBD legislated data.
   - A framework for the consistent 'look and feel' for the presentation of OBD legislated data to the user.

Based on above-mentioned objectives and requirements from all parties involved an implementation concept based on existing standards was developed and published 09/2006.

ISO/PAS 27145 Road vehicles—Implementation of WWH-OBD communication requirements consisting of four parts were developed:

- Part 1: General information and use case definition.
- Part 2: Common emission-related data dictionary.
- Part 3: Common message dictionary.
- Part 4: Connection between vehicle and test equipment.

Part 1 specifies three main use cases:

- Use case 1: Information about the emission-related OBD system state—The purpose of this information package is to provide the minimum data set specified as necessary by the WWH-OBD GTR to obtain the vehicle or engine state with respect to its emission performance as specified in the WWH-OBD GTR. A typical use of this information package may be a 'roadside check' performed by an enforcement authority.
- Use case 2: Information about active emission-related malfunctions—The purpose of this information package is to provide access to the expanded data set specified as necessary by the WWH-OBD GTR to determine vehicle readiness and characterize the malfunctions detected by the OBD system. A typical use of this information package may be a periodic inspection by enforcement authorities.
- Use case 3: Information related to diagnosis for the purpose of repair—The purpose of this information package is to provide access to all OBD data required by the WWH-OBD GTR and available from the OBD system. A typical use of this

information package may be the diagnostic servicing of the vehicle or system in a workshop environment.

Part 2: Common emission-related data dictionary defines all regulatory emission-related data elements of ISO/PAS 27145. A new part may be added in the future upon availability of new legislated WWH-OBD GTR modules. The data elements are used to provide the external test equipment with the diagnostic status of the emission-related system in the vehicle. All data elements are communicated with the UDSs as defined in ISO/PAS 27145-3 common message dictionary. Data elements are DTCs, PIDs, MIDs, TIDs/routine identifiers (RIDs) and ITIDs.

Part 2 defines three (3) different sets of data elements:

- A legacy (backward compatible) data set as defined in SAE J1939-71/-73 and ISO 15031-5/SAE J1979, ISO 15031-6/SAE J2012.
- A unified data set (new data definition according to ISO/PAS 27145-2).
- A manufacturer data set (defined by manufacturer).

Part 3: Common message dictionary definition of ISO/PAS 27145 specifies the implementation of a subset of UDSs as specified in ISO 14229-1. The diagnostic services are used to communicate all diagnostic data as defined in "ISO/PAS 27145-2 Common emissions-related data dictionary".

The subset of UDSs derives from the requirements stated in the WWH-OBD GTR. The common message set defined in this part is independent of the underlying transport, network, data link and physical layer. This document does not specify any requirements for the in-vehicle network architecture.

Part 3 includes a superset of a modified version of ISO 14229-1. Several significant modifications are included in this part in order to support the data set of SAE J1939, ISO 15031-5/SAE J1979 and ISO 15031-6/SAE J2012.

Part 4: Connection between vehicle and test equipment of ISO/PAS 27145 defines the requirements to successfully establish, maintain and terminate communication with a vehicle that implements the requirements of the WWH-OBD GTR. This requires plug and play communication capabilities of the vehicle as well as any test equipment that intends to establish communication with a vehicle. This document details all the OSI layer requirements to achieve this goal.

An ISO Publicly Available Specification (PAS) requires a worldwide ballot after 3 years of publication. The outcome of the ballot was to convert and establish ISO 27145 as an international standard with the addition of a Part 5 conformance test and Part 6 external test equipment.

Figure 4.25 illustrates all referenced documents according to the OSI layers.

Two protocols are supported:

- ISO 15765 DoCAN (diagnostic communication over CAN).
- ISO 13400 DoIP (diagnostic communication over Internet protocol).

ISO 27145-1 defines the general structure of the documents and the WWH-OBD applicable use cases as specified in the PAS.

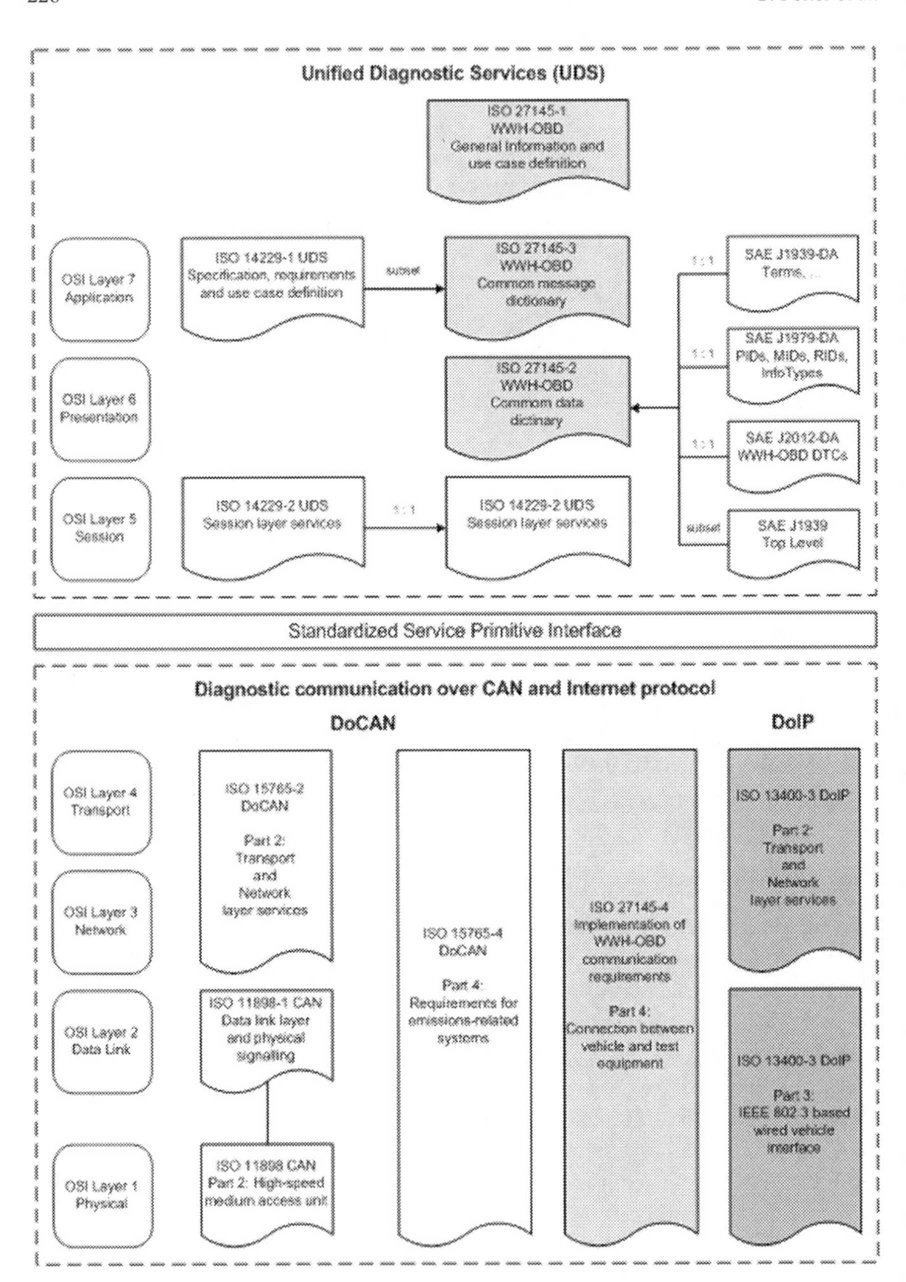

Fig. 4.25 Overview of all referenced UDS or rather WWH-OBD standards

ISO 27145-2 common data dictionary references the following documents containing emission-related data definitions:

- SAE°J1930-DA, electrical/electronic systems diagnostic terms, definitions, abbreviations and acronyms (includes all standardized terms and abbreviations).
- SAE°J1979-DA E/E diagnostic test modes (includes all standardized data items: PIDs, TIDs, MIDs, ITIDs).
- SAE°J2012-DA, diagnostic trouble codes (includes all standardized DTCs).
- SAE°J1939, top level (includes all standardized SPNs and PGNs).

ISO 27145-3 common message dictionary specifies the implementation of a subset of ISO 14229-1 UDS diagnostic services.

ISO 14229-2 UDS session layer services define the standardized service primitive interface between the OSI layer 5 (session) and OSI layer 4 (transport). Through this interface the implementation of the UDS diagnostic services is independent of the underlying communication protocols (CAN, IP). This is important when the vehicle manufacturer is required to transition from CAN to Internet protocol based on future legislation.

Table 4.18 provides an overview of the ISO 15031-5 OBD services and the mapping to ISO 27145-3 (ISO 14229-1 UDS) diagnostic services and associated subfunctions and data ranges.

4.4 SAE J1939

SAE J1939 is a set of standards for both in-vehicle normal ECU to ECU communication protocol and diagnostic communication protocol. The standards cover relevant OSI layers and specify physical link (cable), how the messages are built up, NMT, in-vehicle communication with data items, diagnostic communication with DMs and data, name claiming and conformance test specification. Parts of the set of standards are used in heavy-duty and medium-duty applications worldwide.

When SAE J1939 was introduced, CAN was not mentioned but it was soon included. First, the -71 layer was introduced for normal in-vehicle communication and later the J1939-73 diagnostic layer was introduced.

The protocol is used in HDVs including trailers, agricultural machines, off-road equipment, boats and stationary engines and it has been discussed for residential vehicles also. The main reason is that the protocol has been implemented in engine control modules (ECMs) for medium-duty/heavy-duty diesel engines as a standard.

The situation in the USA is that a truck may be a chassis from an OEM, with an engine from an engine manufacturer and a transmission from a transmission supplier. It is possible for the customer of the vehicle to equip the vehicle with systems from different suppliers. A fleet manager maybe have different truck brands (e.g. Navistar, Volvo and Freightliner) but wants to keep the same engine manufacturer (e.g. Cummings) for the complete fleet and the same for the different systems, e.g. transmission and brakes.

Table 4.18 Mapping emission-related system OBD with WWH-OBD GTR

OBD SID	ISO 15031-5/SAE J1979 service	WWH-OBD SID, SFID	ISO 27145-3/ISO 14229-1 UDS service name	ISO 27145-3/ISO 14229-1 sub-function name/data range (all data are referenced: SAE J1979-DA, J2012-DA)
0×01	Request Current powertrain diagnostic data	0×22, 2-byte DID (3 PIDs max.)	ReadDataByIdentifier	PIDs: 0xF400–0xF5FF (low byte=PID#)
0×02	Request powertrain freeze frame data	0×19, 0×04, 3 byte DTC#, Frame# 0×19, 0×06, 3 byte DTC#, ExtRecord#	ReadDTCInformation	reportDTCSnapshotRecordByDTC-Number reportDTCExtendedDa-taRecordByDTCNumber
0×03	Request emission-related diagnostic trouble codes	0×19, 0×42, FGID, DTCStatus-Mask, DTCSeverityMask	ReadDTCInformation	reportWWHOBDDTCByMaskRecord FGID=FunctionalGroup ID (e.g. emissions, safety, …)
0×04	Clear/reset emission-related diagnostic information	0×14, groupOfDTC	ClearDTCInformation	groupOfDTC=3 byte DTC 0xFFFFFF for all DTCs
0×05	Request oxygen sensor monitoring test results	0×22, 2-byte DID (MID)	ReadDataByIdentifier	MIDs: 0xF600–0xF7FF (low byte=TID# of monitor)
0×06	Request on-board monitoring test results for specific monitored systems	0×22, 2-byte DID (MID)	ReadDataByIdentifier	MIDs: 0xF600–0xF7FF (low byte=OBDMID#)
0×07	Request emission-related DTCs detected during current or last completed driving cycle	0×19, 0×42, FGID, DTCStatus-Mask, DTCSeverityMask	ReadDTCInformation	reportWWHOBDDTCByMaskRecord FGID=FunctionalGroup ID (e.g. emissions, safety, …)
0×08	Request control of on-board system, test or component	0×31, 0×01, 2-byte RID	RoutineControl	startRoutine, Routine ID (e.g. 0×01, 0×02, …) 2-byte RID (low byte=TID of service 0×08)
0×09	Request vehicle information	0×22, 2-byte DID (ITID)	ReadDataByIdentifier	ITIDs: 0xF800–0xF8FF (low byte=InfoType#)
0×0A	Request emission-related DTCs with permanent status	0×19, 0×55, FGID	ReadDTCInformation	reportWWHOBDDTCWithPermanent-Status FGID=FunctionalGroup ID (e.g. emissions, safety, …)

SAE J1939-71 makes the in-vehicle communication between the different ECUs possible, no or small (e.g. tuning of ECU addresses) adaptions are needed.

The ECUs are more or less "of the shelf".

The same situation for passenger car is that every OEM sets up his own network protocol and all ECUs must be adapted to co-exist in the vehicle.

4.4.1 Structure of SAE J1939

The structure of SAE J1939 is the same as the OSI layers. The following standards will be focused on in this document: SAE J1939-21, SAE J1939-71 and SAE J1939-73.

The lowest layers (SAE J1939-11 and SAE J1939-15) are almost similar to ISO 11898, i.e. -15 specifies an unshielded twisted cable and -11 specifies a shielded twisted cable and the CAN bus speed is currently 250 kbps and the CAN identifier length is 29 bits.

SAE J1939 is working on extending the CAN bus speed to 500 kbps.

SAE J1939-13 defines the connector, which is a nine-pin round Deutz connector with two pins for CAN and two pins for SAE J1708 which is the physical layer of SAE J1587. It is not allowed in the USA to use the SAE J1962 ("ISO" or D-shaped) connector together with SAE J1939 protocol. This specific requirement does not exist in Europe and Volvo truck and Volvo bus use SAE J1939 together with the D-shaped connector.

The usage of J1939 is for controlling vehicle or engine application. SAE J1939-71 is commonly used in heavy-duty applications for at least powertrain applications. Some OEMs use it also for complete vehicle applications.

There is an in facto agreement in the USA to use SAE J1939-73 for legislated OBD diagnostics, and some OEMs use the same protocol as an enhanced protocol for the complete vehicle and have implemented services for software download and everything which is needed for workshop fault tracing and repair.

Trucks in the rest of the world usually have adopted an ISO protocol, usually ISO 14230 ("Keyword Protocol (KWP) 2000 on CAN") or ISO 15765-3 (also known as ISO 14229-3 or DoCAN).

The ISO protocol does not interfere with J1939-71 which is used for vehicle control.

It is not allowed to implement both SAE J1939-73 and ISO 15765-4 for legislated OBD protocol. The reason is that independent scan tools will utilize an algorithm to detect as to which type of legislated OBD protocol is implemented in the vehicle. The algorithm is based on scanning through all allowed protocols and the tool stops when it has detected the first protocol.

It would be too advanced for the tool to try to use two completely different protocols and combine the data.

Table 4.19 OSI layers as a function of the SAE J1939

Application layer	SAE J1939-71	SAE J1939-73
Presentation layer	–	–
Session layer	–	–
Transport layer	SAE J1939-21	–
Network layer	SAE J1939-31	–
Data link layer	SAE J1939-21	–
Physical layer	SAE J1939-15	SAE J1939-11

The SAE J1939-73 is allowed for legislated OBD communication for US 10 emission legislation, Euro IV, Euro V and the upcoming Euro VI emission legislations.

Volvo is probably the only OEM in Europe which has implemented SAE J1939-73 as legislated OBD protocol for Euro IV and Euro V emission legislations. The company will transfer to ISO protocols in the future (Table 4.19).

4.4.2 SAE J1939-21 Data Link Layer

The SAE J1939-21 defines how the PDU is built up.

A SAE J1939 PDU consist of 3 bit priority (P), 1 reserved bit (R), 1 bit for data page (DP), 8 bit for PDU format (PF), 8 bit for PDU specific (PS) and 8 bit for source address (SA) plus up to 64 bit of data (8 byte) (Fig. 4.26).

- The priority bits set the priority during arbitration and 0 is the highest priority, 7 the lowest. A recommended priority is assigned to all PGNs listed in the standard, but the receiver should ignore the priority bits; this is due to the fact that the priority may be changed.
- Reserved bit is not the CAN reserved bit, but reserved for future expansion of the standard.
- Data page: All PGNs must be assigned to page 0 before page 1 is used.
- PDU format, PF, is used to determine the PGN.
- PDU specific, PS, can be either the destination address or a group extension. If the PS is below 240 then it is a destination address, otherwise it is a group extension.
  - Destination address: It specifies the ECU (or address) that should listen to the message. 255 is a global address for all ECUs.
  - Group extension: It provides 4,096 data groups per DP plus the 240 extra PDUs (PS<240). In total, there are (4096+240)*2 possible data groups.
- Source address is the ECU sending the message. The addresses are defined in SAE J1939-81.
- PGN is based on reserved bit, DP bit and then 16 more bits.
- Data field: up to 8 byte of information in a single frame. Non-used data bits should be set to non-available (padded to '1'), which would mean in practice that

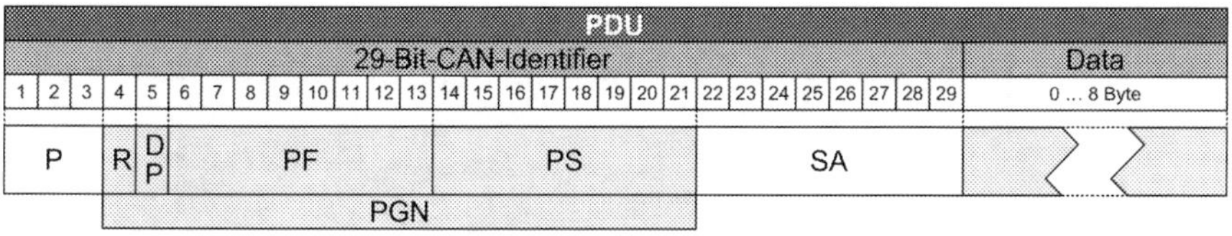

PDU
29-Bit-CAN-Identifier | Data
1 2 3 4 5 6 7 8 9 10 11 12 13 14 15 16 17 18 19 20 21 22 23 24 25 26 27 28 29 | 0 ... 8 Byte
P | R | DP | PF | PS | SA
PGN

Fig. 4.26 SAE J1939-21 PDU and the CN-identifier assignment

the CAN protocol will include bit stuffing and extend the number of bits in the message.

Some messages are longer than one single CAN frame (8 byte), e.g. VIN which consist of 17 ASCII characters. When a long message (up to 1785 bytes) should be transmitted, it is possible to send the message as a segmented message using the transport protocol function and there are two methods:

4.4.2.1 Transport Protocol (TP)

Method 1: Broadcast Announce Message, TP_BAM TP_BAM: A message with a global address, which means that all ECUs listen to the message. The message starts with a Connection Management (CM) message, PGN 00EC00 with a control byte indicating TP_BAM and then the PGNs with an inter-frame time of minimum 50 ms. This method should not be used if it is not specified in the applicable standard (i.e. SAE J1939-71, SAE J1939-73 or SAE J1939-03). The main reason is that all ECUs have to listen to something which may be a message between one ECU and a scan tool and therefore spend resources on a message which does not concern them.

One of the examples when a TP_BAM could be used is during scan tool initialization before the tool knows which ECUs are installed in the vehicle. It can at that time broadcast a message in order to identify all that support the service or data.

To identify if a vehicle is utilizing SAE J1939-73 as a legislated OBD protocol, the DM 5 (Readiness 1) is used (Fig. 4.27).

Method 2: Connection Management, TP_CM The other method is called TP_CM: A message is sent from point to point. The sending ECU sends a CM message indicating Request to Send (RTS). The receiver responds with a Clear To Send (CTS) with the number of packets (buffer size) it may accept and the sequence number of the expected packet.

The parameter group, together with the data is then transmitted in several data transfer messages (DT), wherein the first byte indicates the sequence number in each case. It is possible to pause the communication and to abort. This method is the preferred one when it comes to diagnostic communication to an off-board client (scan tool) (Fig. 4.28).

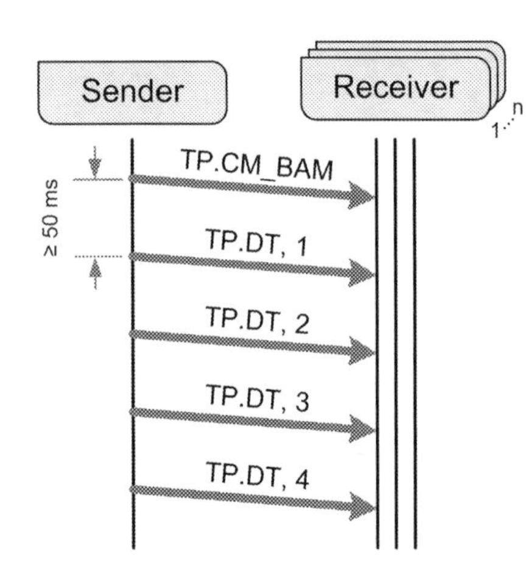

Fig. 4.27 J1939 transport protocol—broadcast announce message

Request: Normal in-vehicle control data are sent periodically on the bus but sometimes some data are needed and they are not usually sent on the bus. It is possible to request the data (or PGN) in those situations. One example is VIN which is used to identify the vehicle by a scan tool. The VIN is 17 characters and there is no need to send it periodically, so the scan tool needs to request the data from the ECU which has the information.

Data: The ECU either sends a negative acknowledge, NACK, if it does not have the data or respond to data. If one data element is not used in the PGN, then the bits for that suspect parameter number shall be set to 1.

The response time is 200 ms but the requesting equipment needs to wait up to 1.25 s before it times out. The main reason is that bridges (gateways or routers) can delay the message.

4.4.3 SAE J1939-31—Network Layer

The SAE J1939-31 network layer standard defines how a complete network should be designed. The standard describes gateway and router functionalities (Fig. 4.29).

Gateways can be within the vehicle to isolate different buses from each other (e.g. one private network for brake system, another network for the cab controller, a third for powertrain) or to act as bridges between, e.g. a tractor and a trailer.

4.4.4 SAE J1939-71 71—Vehicle Application Layer

SAE J1939 has relationship to SAE J1587 which is usually implemented on SAE J1708 bus.

J1939-71 defines signals for normal communication, point-to-point in-vehicle communication. The signals are defined as SPNs.

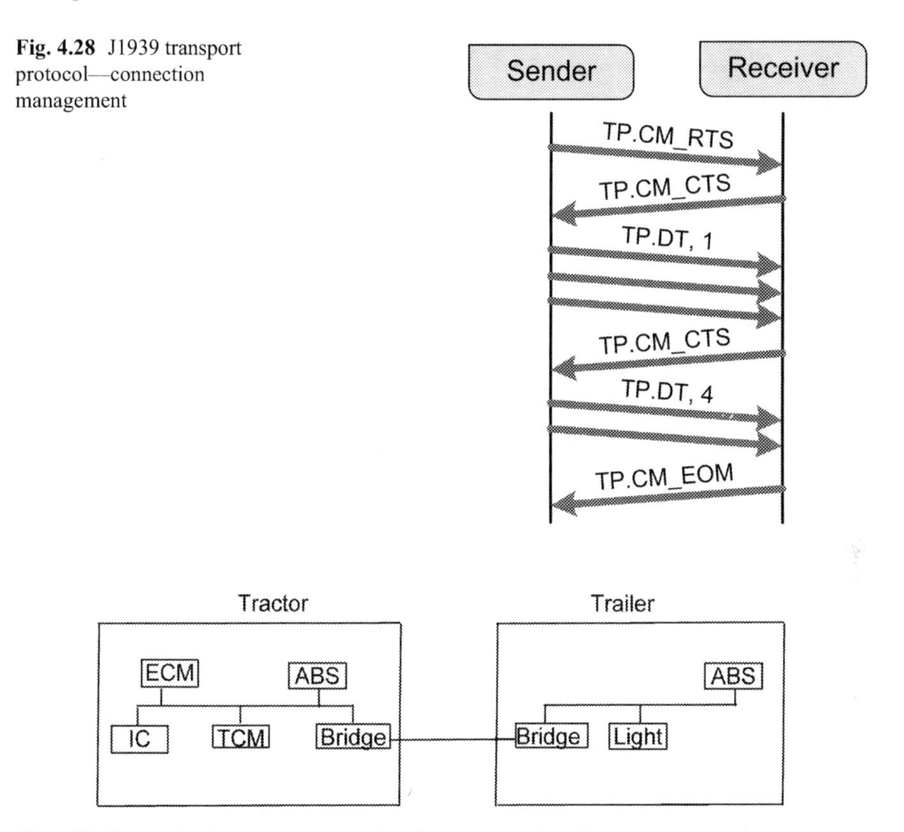

Fig. 4.28 J1939 transport protocol—connection management

Fig. 4.29 Connection between tractor and trailer via network bridges

The SPN is a fictive number which usually only exists in the standard and is not seen on the bus, except for some services, e.g. DM 24 which reports which SPNs are implemented as data stream, freeze frame parameters[5] or test values.[6]

The SPN is defined with length and scaling information and is connected to a PGN, which is more or less a part of the CAN identifier for the message.

Some SPNs, e.g. SPN 237 = VIN, are longer than a CAN frame and must be sent as a segmented message.

[5] Freeze frame is data which has have been frozen at the occurrence of the detection of a fault. The intention is that the information in the freeze frame may help the service technician to reproduce the conditions which existed when the malfunction was detected. Example of a freeze frame parameters are engine speed and ambient air temperature. The electronic control module will store these two parameters when it detects a malfunction.

[6] Test value is the value which is compared to the fault limits of a monitor to judge if the latest evaluation was pass or fail. It can be seen as an analogue value of a fault (diagnostic trouble code), e.g., if the test value is 0×8340 and the fault limit is 0×8000 then there should be a fault code stored in the electronic control module.

	PGN 61444 EEC1									
Databyte	1		2	3	4	5	6	7		8
SPN	899	4154	512	513	190		1483	1675		2432
Data	0	0	225	220	16000		33	4	15	225

Fig. 4.30 Example for number-dedicated SPNs

The SPN can be sent either on request (as for diagnostic information) or as a periodic transmission.

The request of an SPN is done by requesting the PGN and then decoding the SPN from the data field of the CAN frame (Fig. 4.30).

This specific PGN (EEC1) is transmitted as a function of engine speed, i.e. low engine speed = low transmission frequency, high engine speed = high transmission frequency. The SPN for engine speed is 190 and the value of SPN190 is 16,000.

In order to translate the information, which is received as the data field in a CAN frame, it is just to copy the content of data byte 4 and 5 and multiply it to 0.125 and the engine speed is given in rpm. The definition of EEC1 and the different SPNs can be found in SAE J1939-71.

The other SPNs in EEC1 are 899 (Engine Torque Mode), 4154 (Actual Engine Per cent Torque High Resolution), 512 (Driver's Demand Engine Per cent Torque), 513 (Actual Engine Per cent Torque), 1483 (Source Address of Controlling Device for Engine Control), 1675 (Engine Starter Mode) and 2432 (Engine Demand Per cent Torque). The value of 15 in data byte 7 indicates "no information".

4.4.4.1 SPN and Fault Information

Since the main purpose of SPN is for in-vehicle communication, i.e. from ECU A to ECU B, then a fault information concept has been developed. The valid data range for a 1-byte SPN is 0–250 and 0xFE is used to indicate that the source of the signal cannot be trusted due to a fault and 0xFF indicates that the signal is not updated. It is up to the receiving ECU to decide which default action it should take. It could be that the latest valid data are used or a default value, e.g. 20 degrees as ambient air temperature, is used in the application.

The method of using 0xFB–0xFF can be useful for normal in-vehicle communication, but it makes it impossible to store raw or un-defaulted data in a freeze frame or to report un-defaulted data stream data to a scan tool.

Un-defaulted data are defined as the data after linearization and converted to an engineering unit, e.g. the ADC measures voltage and uses a look-up table to convert the voltage to a temperature.

The defaulted data are when a malfunction is detected and the value of the parameter is changed to, e.g. 20 degrees, or, if sent on CAN, to 0xFE to indicate to

<table>
<tr><th colspan="32">Diagnostic Trouble Code (DTC)</th></tr>
<tr><td colspan="8">8 least significant bits of SPN</td><td colspan="8">Second byte of SPN</td><td colspan="8">3 most significant bits of SPN and the FMI</td><td colspan="8">4th byte of SPN</td></tr>
<tr><td colspan="19">SPN</td><td colspan="5">FMI</td><td>CM</td><td colspan="7">OC</td></tr>
<tr><td>32</td><td>31</td><td>30</td><td>29</td><td>28</td><td>27</td><td>26</td><td>25</td><td>24</td><td>23</td><td>22</td><td>21</td><td>20</td><td>19</td><td>18</td><td>17</td><td>16</td><td>15</td><td>14</td><td>13</td><td>12</td><td>11</td><td>10</td><td>9</td><td>8</td><td>7</td><td>6</td><td>5</td><td>4</td><td>3</td><td>2</td><td>1</td></tr>
<tr><td colspan="19">190</td><td colspan="5">2</td><td>1</td><td colspan="7">1</td></tr>
</table>

Fig. 4.31 Example for SPN 190 trouble code (engine speed)

another ECM that the signal is not valid. If the defaulted value is stored in a freeze frame, then the data will not help the service technician, i.e. he/she will not know at which temperature the malfunction was detected.

4.4.5 SAE J1939-73—Diagnostics

Diagnostic communications are based on SAE J1587 and include special PGNs called DMs. They can be considered as equal to diagnostic services in ISO 15031-5 or ISO 14229-1. Most of the DMs are sent on request. The main exception is DM1 which is used as communication message to the instrument cluster in order to illuminate warning lamps. If the ECU does not support the request DM, then it will respond with a NACK. The DTC contains the 19-bit SPN with a 5-bit failure mode indicator (FMI) (Fig. 4.31).

SPN is the suspect parameter number of the failed component. FMI is the failure mode indicator (comparable to failure type byte of ISO 14229-1). The CM is a conversion method, since there has been at least four different methods on how to represent the DTCs. At the moment there is just one method, #4 Intel format with 19 bit SPN + 5 bit FMI and CM0, which is allowed for legislated OBD communication. The other versions shall not be used. Finally, OC is an occurrence counter, which counts the number of times the fault has been gone from an InActive (previously Active) state to an Active state.

This method of using the signal as the part of the DTC is useful for simple diagnostic function, e.g. if an ambient air temperature sensor detects an electrical fault, then it will be clear for the service technician on which signal he should check or read. As said, this will work fine as long as there is a one-to-one relationship, i.e. the monitor is based on one signal/sensor. OBD II usually requires more advanced functions, e.g. the catalyst monitor for a gasoline engine maybe utilize the intake manifold airflow sensor and the rear oxygen sensor plus some other sensor values to check if the conversion ratio in the catalyst is OK. This is a multiple sensors to one DTC scenario and it is not possible to read a single sensor anymore and the SPN is more "fictive". The detection mechanism of a fault is usually called a monitor.

4.4.5.1 Fault Codes

The first DM for reading fault codes is DM1 to read active DTCs.

A DTC is active when it is detected after debounce filtering and the service is intended mostly to inform the instrument cluster to illuminate an appropriate lamp, red, yellow or white lamp. The DM1 service is sent periodically to the instrument cluster. It includes 4 bits, 1 bit per warning indicator, i.e. malfunction indicator lamp, red stop light, yellow warning and white information light. After the lamp information, the fault codes are sent.

DM2: Previously active DTCs. The service is sent on request and includes the warning light information plus all DTCs which are no longer detected as active faults.

The states for emission-related DTCs are more complicated after US10:

First, the DTC will show up as a pending DTC (DM6) in the first failed driving cycle.

If the monitor fails in the next consecutive driving cycle, then it will transit to an emission-related active DTC (DM12) and will not be reported in DM6. When the DTC heals, it will transit to previously active emission-related DTCs (DM23) and it will disappear completely after 40 fault-free warm-up cycles.

The DTC will also be reported as a permanent DTC (DM28) as long as the MI is commanded on.

N.B. It is allowed to report a DTC as both pending and emission-related active DTCs according to the standard, but any emission-related ECU which does will not fulfil California legislation (Table 4.20).

A note must be mentioned regarding DM24 and DM25.

DM24 reports which SuspectParameterNumbers are implemented as data stream parameters, as freeze frame parameters and as test values. It is not clear which SuspectParameterNumbers should be reported as data stream parameters: All according to legislation, all according to what is sent on the bus or those SuspectParameterNumbers that are broadcasted on request? The service is clearer on freeze frame parameters since the freeze frame is well defined.

Requesting latest test value from a non-continuous executing monitor is the same as Service $ 06 of ISO 15031-5, but first the DM25 is read and the SuspectParameterNumbers for test results are reported in the response.

The SuspectParameterNumber and the correct FailureModeIndicator (which is not reported by DM25) must be put into the DM7 request. It may also be possible to request FailureModeIndicator = 31 which means all FailureModes.

The response is sent by the ElectronicControlUnit as a DM30 response including the SuspectParameterNumber and the FailureModeIndicator, followed by a SLOT identifier which includes length and scaling information for the next parameter, the TestValue. The SLOT identifier should always report that the TestValue has a length of 2 bytes in DM30.

The TestValue is the value which the specific monitor compares with the fault limits to judge if the result from the monitor is a pass or fail. The TestLimits are also reported in DM30. The intention of DM7/DM30 is to show how close to the fault limit the monitor is, e.g. if the TestValue is 90 % of the test limit, then the monitor

Table 4.20 Relation of emission-based diagnostics standards

	ISO 15031-5	ISO 27145-3	SAE J1939-73 US10	SAE J1939-73 EURO VI
Request data stream	Service $ 01	Service $ 22	SPNs+DM21, DM5, DM 26, DM32, DM33, DM34	
Request freeze frame	Service $ 02	Service $ 19 04	DM25	
Request pending DTCs	Service $ 07	Service $ 19 42	DM6	DM41, DM44, DM47, DM50
Request confirmed DTCs	Service $ 03	Service $ 19 42	DM 12	DM42, DM45, DM48, DM51
Request permanent DTCs	Service $ 0A	Service $ 19 55	DM28	
Request previous DTCs		Service $ 19 42	DM23	DM43, DM46,
DM 49, DM52				
Clear DTCs	Service $ 04	Service $ 14	DM 11	DM11
Request test results	Service $ 06	Service $ 22	DM7, DM30	DM7, DM30
Command test	Service $ 08	Service $ 31	DM8	DM8
Request infoType	Service $ 09	Service $ 22	DM 19, DM20, +SPN	DM 19, DM20, +SPN

Note: ISO 15031-5 (SAE J1979) is allowed for legislated OBD communication for HDVs in the USA and for HDVs until Euro VI legislation in Europe. It uses Service $ 01–0A for retrieving the legislated communication. ISO 14229-1 (ISO 15765-3 DoCAN) is used for enhanced diagnostics in both passenger cars and some HDVs. The Euro VI column shows that there is a different set of DMs for fulfilling Euro VI legislation

could detect a failure during some conditions. The TestValues will usually differ each time the monitor is executed since they are affected by component deviation, component ageing and also driving and environmental conditions.

4.4.5.2 Scan Tool Initialization

The scan tool initialization procedure is easier than for a scan tool for ISO 15765-4 since there is currently only one allowed bus speed, 250 kbps and all Electronic-ControlUnits shall utilize 29-bit CAN identifiers.

First, the tool will check for PGN 61444 EEC1 which is always transmitted if there is an engine ElectronicControlUnit on the network. Then it will send a DM5 Diagnostic Readiness 1 request as a BAM request. All ECUs that utilize SAE J1939-73 will respond with the information of, e.g. the number of active DTCs, the number of previously active DTCs, the monitor groups that have executed since last clear DTC and the OBD compliance (similar to PID $ 1C of ISO 15031-5).

It is possible that vehicles prior 2010 in the USA do not utilize SAE J1939-73 for diagnostic communication, and in Europe it is quite common to use SAE J1939-71 together with ISO 15765-4/ISO 15031-5 for OBD communication. It is therefore not enough to check the ParameterGroupNumber 61444 EEC1 to distinguish if the vehicle utilizes SAE J1939-73.

The scan tool will build a list of the ElectronicControlUnits that has responded to this request and the OBD compliance will be an input of the services that are implemented. Table 1 of SAE J1939-73 states which DMs and which information are needed to comply with different OBD legislations. It is not clear how to handle smart sensors or smart actuators since these simple devices have a limited implementation of the diagnostic protocol. They have maybe only implemented DM1 active DTCs, DM11 clear DTCs, DM 19 CALibrationIDentification/CalibrationVerificationNumber and DM5 for scan tool initialization, but the vehicle needs to comply to FinalRegulationOrder 1971 which is for US10 legislation in California.

4.4.6 SAE J1939-81 81—NMT

SAE J1939-81 includes NMT, i.e. how to handle if two ECUs are connected to the same network with the same ECU address. Normally two ECUs shall never share the same SA, but in some type of vehicles, this may happen, e.g. a tractor with two trailers with both trailers having a brake system.

It uses a NAME field and a method to claim addresses. If there are multiple ECUs with same address, then the ECU with the lowest NAME will win and the others have to claim other addresses. The claiming can be seen as an ECU address arbitration. This works as the best in theory but there are problems in real life; the ECUs detect bus-off before they could claim the address, they misunderstand the response since maybe both ECUs respond to the same Name claim response and get the same response, i.e. they cannot know if the data are intended for the own ECU or the other ECU and both will react as for communication faults.

4.4.7 SAE J1939-84

The standard is a conformance test, but not equal to SAE J1699-3 for LDVs. The J1699-3 is a conformance test for a complete vehicle and J1939-84 is a very thin test specification for a single ECU. The implementation of the standard can also be downloaded as a Dynamic Link Library (DLL) on www.sourceforge.org. The standard is used to verify that the application in the ECU fulfil the SAE J1939-73 standard.

4.5 CanKingdom

CanKingdom (CK), first published in 1990, is considered the ancestor of the CAN-based higher layer protocols (HLP). In many respects, it is quite different from later CAN HLPs:

- It is not really a HLP; it is a meta-HLP, i.e. a protocol for constructing a protocol.
- It is not a communication protocol, but a system control protocol.
- It is designed to maximize the composability of a system.
- It is designed for achieving high performance at low cost throughout the life cycle of a system.
- It is intended for hard real-time and safety critical systems.
- It is designed to allow a mix of time-triggered, event-triggered and sequential schedules.
- It is designed to minimize development time of systems and modules by separating the system design and module design tasks as far as possible.
- It is designed to allow a system to be individually optimized at any time during its lifetime.
- It is designed to allow changes between modes during run time, e.g. from normal mode to a limp home mode.
- It is not compliant with the OSI model.
- CK uses a unique vocabulary for essential terms to make them unambiguous.

CK is based on an unorthodox approach to Distributed Embedded Control Systems. Most CAN systems are based on the seven-layer OSI model, where the different applications in a network are separated by an independent communication layer. However, this approach can lead to severe timing and synchronization problems because the timing of CAN messages is not controlled. A great advantage of CK is that it allows the system designer to take full control of message transmissions by scheduling them in time or sequence, as well as having unschedulable messages (as alarms) to be transmitted when needed. In the approach for CK the whole system is viewed as a combination of devices, such as joysticks, actuators, sensors, etc., all controlled by one imaginary application. This imaginary application is broken down into a number of real sub-applications, with each sub-application residing in a module of its own and integrated into the respective device. Sub-applications have two parts: A local part takes care of everything needed for the device it resides in and the other part interacts with other devices. In this way, we have two clear layers: an imaginary system layer and a module layer. The imaginary system layer is brought to reality by a third "glue" layer that integrates the sub-applications with each other. This glue layer provides a common application programming interface (API) that is partly based on a serial communication. Thus, the communication is not a separate layer as in the OSI model; it is an inherent part of the system, gluing sub-applications together to run as one common real-time application. The restrictions imposed by the serial bus communication make the interfaces between the respective sub-applications very clean and predictable in time and sequence.

CAN is the serial communication of choice and the basis of the glue layer in CAN Kingdom. Each module has primitives of the glue layer that are completed by information sent during a configuration phase. A specific system module containing all necessary information needed to harmonize each module to the system requirements controls this phase. In this way, generic modules can be easily combined into complex, high-performance systems. In a simple system, the system module can be disconnected after the setup procedure. However, in more advanced systems, the system module is an integrated part of the system, taking a monitoring and supervising role during run time, allowing for hot swaps of modules, changing of modes, etc. CK can be implemented in a modular way. A full implementation requires roughly 5.5 K flash and 100 K RAM.

4.5.1 Background

In the early 1980s, the Swedish company Rovac developed an advanced factory automation system based on distributed embedded controllers. The whole factory was seen as one big application that included robots, material supply, mould positioning, heat control, etc. This application was broken down into as small pieces as possible and each piece was executed in a separate micro-controller, physically integrated into the device it controlled or monitored. As an example, a robot was constructed of a set of actuators connected by structural parts, tubes and swivels. In this way, special robots could be easily designed using standard parts. The micro-controllers were grouped according to their functions and the members of each group were connected to each other and to a central computer by a serial communication bus. The central computer coordinated and supervised the groups to act in concert in a safe manner. The concept, designated 'Trainet', turned out to be very flexible and efficient, but the bit rate of the communication, 9,600 bps, limited the update frequency to 25 Hz, i.e. any feedback control loop had to be executed locally.

When CAN became available in 1988, high-speed bus communication was available at a reasonable cost. As the concept of CAN is to minimize the need for run-time bandwidth by defining as much as possible off-line, it was a perfect match with the Rovac ideas. The combination of Trainet features and CAN resulted in CK, first published in 1990 by the company Kvaser. It was developed further and version 3 was made available at CiA 1992 and reached a broader audience. CK v. 3 formed the basis for the US DoD CDA 101 project for a common CAN-based protocol for different types of airborne and seaborne targets. During the development period of CDA 101, from 1996 until 2001, CK was further improved to meet the high requirements for any aspect of an embedded CAN, such as adding improved support for hard real-time and synchronization, composability, membership verification, safety, troubleshooting, etc.

Being a meta-HLP, CK is used as a base for proprietary HLPs, e.g. the Mercury "SmartCraft" for pleasure boats and the Sauer–Danfoss "Plus+1" for off-highway machines.

4.5.2 The Concept Behind CK

A cornerstone of the concept is the notion of a system: An electro-mechanical system is constructed of a number of modules, each with an ECU that is connected to a serial communication of some kind. Each module has a specific role in the system, e.g. a steering wheel or joystick, a gearbox, a motor, an actuator, a sensor, etc. They are all connected as nodes in a system network. A system designer has to combine a number of modules and make them perform in concert.

In this way, we have three concise layers:

- System layer
- Glue layer
- Module layer

The system is also seen as one big application, broken down into sub-applications that reside in respective nodes and are coordinated by the glue layer. The glue layer can then be seen as an API built upon a specific serial communication protocol. The qualities of the glue layer are therefore highly dependent on the qualities of the chosen communication protocol.

This concept has many advantages:

- The glue layer makes a clean and simple interface between interacting modules as it is only control and data messages that are transmitted and received according to need.
- Module designers can concentrate on the performance of the module and need not know much about the system.
- The system designer has only to see to it that each module receives and transmits messages as needed to perform in concert with the other nodes and does not need to know much about each ECU.
- The performance of each module can be easily checked individually as a stream of messages can be used to simulate the rest of the system. If the module responds to commands and is able to transmit messages according the specification, it is OK.
- Modules and systems can be developed in parallel, saving time and money.

This concept does not fit the OSI model. The OSI model is based on a concept where modules just need to exchange information and do not require any form of coordination between communication sessions.

CAN is very well suited as the basis for a glue layer:

- It conveys 11–93 bits at a time from one node to all other connected nodes in a safe, predictable way.
- Any bit rate between 10 kbps and 1 Mbps is supported.
- It can be used for any scheduled and/or unscheduled transmissions.
- Time scheduled, sequence scheduled and unscheduled messages can be transmitted simultaneously with a guaranteed latency time.

4.5.3 Overview

CK is a glue layer based on CAN. It contains a set of rules—all in all 18—that separate the module layer and system layer as much as possible. Only three of these rules are mandatory (Table 4.21).

Table 4.21 Rules of the CanKingdom glue layer

Rule	Description	
1	Start/stop modes. To force a module to stop and go into silent mode	Mandatory
2	Initiate. To establish an exclusive communication between a module and the configuration tool	Mandatory
3	Assign CAN IDs to receive and transmit data in a module	Mandatory
4	Assign groups. Make a module a member of a group to receive group commands	–
5	Remove groups. To expel a module from a group	–
6	Trigger setting. To make a module trigger a task on a message or an event	–
7	Assigning modules to product or producer-specific groups	–
8	Assigning a physical address to a module identified by its serial number	–
9	Change the physical address of a module	–
10	Bit timing register setting	–
11	Inhibit time. To prevent a module from retransmitting a message until a certain time has elapsed	–
12	Circular time base setup. To create a global clock	–
13	Repetition rate and open window setup. To set up a time-triggered communication	–
14	Giving common system wide identifications to messages or groups of messages	–
15	Create CAN messages from local parameters	–
16	Create CAN messages where the data field is extended into the ID field	–
17	Creating bit filter masks	–
18	Creating advanced message filters	–

A full implementation of CK requires roughly 5.5 K flash memory and 100 K RAM.

4.5.4 CK Vocabulary

In order to make the CK rules and functions unambiguous, it uses a unique vocabulary and specific CK terms are spelt with a capital letter. The description is based on a simile of a kingdom where the King in his Capital sets the rules for the Kingdom. The Kingdom has Cities, each of them ruled by a Mayor. The Kingdom is designed by a Kingdom Founder, i.e. the system designer and Cities by City Founders, i.e. module designers. Any information exchange between Cities in the Kingdom is made via a Postal System. The Capital and each City has a Post Office with a Postmaster (a CAN controller) (Fig. 4.32).

The only way to communicate within a Kingdom is to use Letters (CAN messages). Each Letter has an Envelope (CAN ID) and a Page (CAN data field). A Page is built up of 0–8 Lines (bytes in the CAN data field) and a Line can be constructed of 0–8 Dots (bits in a byte in the CAN data field). Pages are organized in Docu-

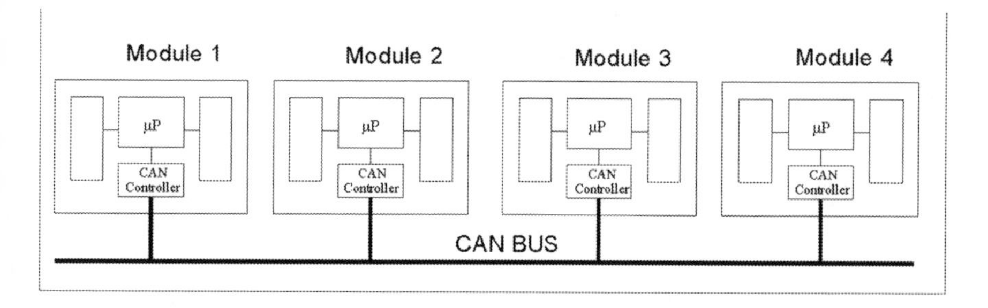

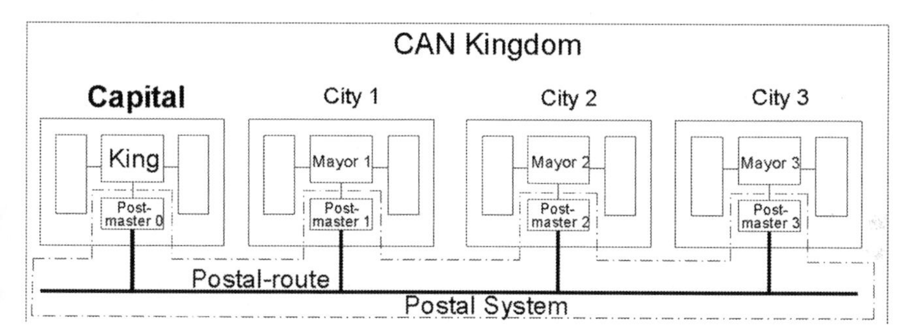

Fig. 4.32 Structure of CanKingdom

ments. The Document is the key for the Cities to encode or decode the Pages. A Document can contain one un-enumerated Page or more enumerated Pages. Cities have matching Documents for coordinated tasks; one is set up for transmission and the other ones for reception. The King uses a King's Document to configure each City. In this process, he assigns Envelopes to matching Documents. Then no CAN ID (except the one for the King's Document) is predefined and a system designer is free to give any message its proper priority. A Document can contain not only data but also tasks, e.g. a Letter with a blank Page or even a Letter for another City can be used for triggering the execution of tasks in one or more Cities. A programmer may see the transmission entities as threads (Fig. 4.33):

4.5.5 King's Document

The King's Document contains many Pages, one for each rule. The Kingdom Founder has to implement all King's Pages supported by any City he/she will use in his Kingdom in order to set up each City in a proper way. The King's Document contains then at least three Pages (the mandatory rules) but also any Pages corresponding to additional rules implemented by selected Cities. All King's Pages use the Envelope 0 (CAN ID 0 Std) as default. (This number can be changed if neces-

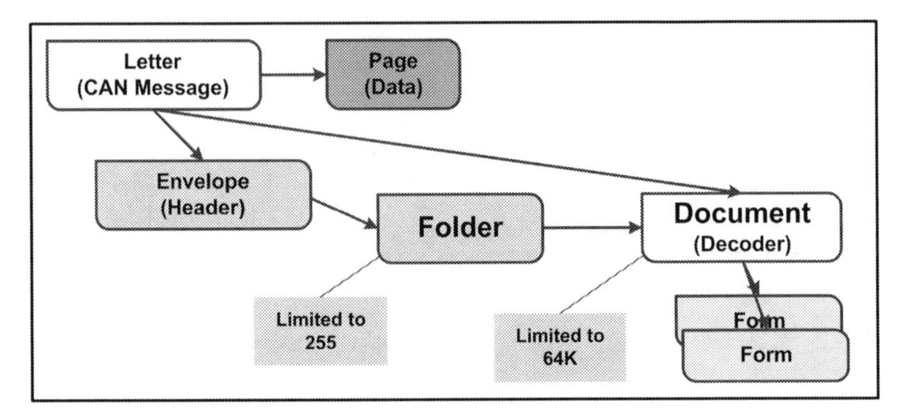

Fig. 4.33 Organizational structure of the CanKingdom messages

sary, but then each module has then to be updated before it is integrated into the system.) The first Line (first byte in the CAN data field) is a Group or City address. Any module in a system needs to have a physical address, a number between 1 and 255, before it is connected to the system. All Cities in the Kingdom belong to a Group with the address 0. This address can be used for broadcasting purposes. It is also possible to assign a City to additional Groups. The Group address is given by the King and can be any number between 1 and 255 not used as a City address. This feature can, for example, be used to freeze a part of the Kingdom in an emergency situation. The second Line is the King's Page number. Numbers 0–31 are reserved for the core CK rules, 32–127 for future additions needed to enable the integration of other HLPs. The numbers 128–255 can be used for City-specific needs. King's Page 0 is shown in Table 4.22.

4.5.5.1 Action Mode

It relates to actual City Mode and Communication Mode. The reset behaviour may be different in different Communication Modes regarding stored bit-timing register settings and stored parameter values. Action Modes have to be defined by the City Founder. It is recommended that the City supports at least the following Action Modes:

- Run: The City is functional and operating.
- Freeze: The City takes a safe state, still responding to King's Letters.
- Reset: The City performs a restart including the Start-up procedure in Sect. 9. The 200 ms at 125 Kbit/s sequence may be omitted.

Table 4.22 Form of the King's Page 0

Document name: King's Document
Document: DL_0.0
Form: PL_2.0

Page description.

Page number: 0
Number of Lines: 8
Data description: The King's Page 0. Terminates the setup phase. Orders a Mayor to set his City into a specific working mode, e.g., in a Run or Freeze mode.

Line description.

Line	Bits	Field	Meaning
Line 0:	City or Group address		
Line 1:	00000000	(Page 0)	
Line 2:	rrrrrrAA	Action Mode	
		AA = 00	Keep current mode
		AA = 01	Run
		AA = 10	Freeze
		AA = 11	Reset
		r = 0	Reserved
Line 3:	rSSLLRCC	Communication Mode	
		CC = 00	Keep current CC mode
		CC = 01	Silent
		CC = 10	Listen Only
		CC = 11	Communicate
		R = 1/0	Reset the Communication Mode Yes/No (go through the Startup sequence using the current settings)
		LL = 01/10	Skip listen for good message during the Startup sequence Yes/No
		LL = 00	Keep current LL setting
		SS = 01/10	Skip wait 200 ms Yes/No
		SS = 00	Keep current SS setting
		r = 0	Reserved
Line 4:	MMMMMMMM	City Mode	
		M = 0	Keep current Mode.
		M ≠ 0	Modes according to the City specification.
Line 5:	rrrrrrrr	r = 0	Reserved
Line 6:	rrrrrrrr	r = 0	Reserved
Line 7:	rrrrrrrr	r = 0	Reserved

4.5.5.2 Communication Modes

- Silent: The Postmaster will be silent but still receives Letters and notifies the Mayor when Letters are accepted. The Postmaster will transmit neither acknowledgement bit nor error or overload frames.
- Listen Only: The Postmaster will be fully active but the Mayor will send Letters only upon the King's request.
- Communicate: Normal communication.

4.5.5.3 City Mode

City-specific modes, e.g. configuration mode, service mode, working mode, etc. For each City Mode the City Founder has to define how the City will work on Action Mode and Communication Mode commands from the King.

4.5.6 Mayor's Document

Each Mayor has a Document with some Pages and an Envelope of his own by which he can respond to the King. The King will broadcast a Base Number and the Mayor will use the Envelope that equals the sum of the City Address and the Base Number. The Mayor's Pages 0 and 1 are mandatory (Table 4.23 and 4.24):

A City is fully identified by the Mayor's Page 1 and 2. Any King's Page can be mirrored by a Mayor's Page and the current setup of a City can be checked by asking for those Pages.

4.5.7 City Organization

The City Founder (the module designer) always knows what information his City must receive and what it can transmit in different situations but frequently he does not know how his City will be used in a specific Kingdom. It is only known by the Kingdom Founder (the system designer). A convenient way for the City Founder to escape the problem of how to receive and transmit information is to leave it to the Kingdom Founder. This is done by organizing the City information into Lists where selectable information blocks are referenced by records. Lists are enumerated from 0 to 253 and each list can hold up to 256 records. The King can use some King's Pages to order the Mayor to construct new Pages by referring to records in the Lists and place them in new Documents. In this way, modules can be optimized to the system requirements and the use of bandwidth optimized as only required data are transmitted. No profiles like the ones in CANopen and DeviceNet have to be defined.

A full-blown City may have all of the following Lists:

Table 4.23 Form for the Mayor's Page 0

Document name: Mayor's Document
Form: PL_2.0

Page description.

Page number: 0

Number of Lines: 8

Data description: City Identification, EAN-13 Code
Diagnostics Line

Line description.

Line	Value		Description
Line 0:	00000000		Mayor's Document, Chapter 0
Line 1:	00000000		Page 0
Line 2:	xxxxxxxx	LSB	
Line 3:	xxxxxxxx		Product Identification Code
Line 4:	xxxxxxxx		(EAN-13, Check-code omitted)
Line 5:	xxxxxxxx		Unsigned 40-bit integer.
Line 6:	xxxxxxxx	MSB	
Line 7:	Rccccfes	s	self-test failed = 1, passed = 0
		e	runtime error detected Yes = 1, No = 0
		f	fatal error, main task cannot be performed Yes = 1, No = 0
		c	error code (City defined) No errors or undefined cccc = 0000
		R=1/0	request King's identification Yes/No The King responds with his Mayor\s Document, Chapter 0, Page 1 **and** 2.

Note:
"fatal error", "runtime error", "main task" and the semantics of self-test are City specific and have to be defined and documented by the City Founder.

1. Document List: The King selects Documents for transmission and reception by placing Documents into Folders. Available Documents are listed in Document Lists. These Lists are of two different types: Receive or Transmit.
2. Page List: A City can offer the King the opportunity to construct Documents from predefined Page Forms. These Forms are then listed in one or more Page Lists.
3. Line List: A City can offer the King the opportunity to construct Pages by predefined Line Forms. These Forms are then listed in one or more Line Lists.

Table 4.24 Form for the Mayor's Page 1

Document name: Form: PL 2.1	Mayor's Document		
Page description.			
Page number: 1			
Number of Lines:	8		
Data description:	City Identification, Serial Number.		
Line description.			
Line 0:	00000000		Mayor's Document, Chapter 0
Line 1:	00000001		Page 1
Line 2:	xxxxxxxx	LSB	
Line 3:	xxxxxxxx		
Line 4:	xxxxxxxx		Serial Number
Line 5:	xxxxxxxx		Unsigned 40-bit integer.
Line 6:	xxxxxxxx	MSB	
Line 7:	rrrrrrrr	r = 0	reserved

4. Dot List: A City can offer the King the opportunity to construct Lines by predefined Dot Forms. These Forms are then listed in one or more Line Lists.
5. Item List: In a Compressed Page, data can be placed not only in the CAN data field but also in the CAN ID field. A City supporting Compressed Pages or Letters has an Item List where available constants, parameters, variables, etc. and information about them can be found. By using references to List and Record numbers, the King can instruct the Mayor where to place or read specific data anywhere he likes in a CAN message. Data can then be extended into the CAN identifier or integrated as a part of the identifier field.

4.5.8 The Folder

A Folder is the link between a Document and the Postal System. One or more Envelopes are assigned to a Folder that contains one Document. A Folder Label contains all necessary Postal information for the exchanging of Letters between Cities.

A City can have up to 256 different Folders for incoming or outgoing Documents. The Mayor puts the Documents for the information that he will send or receive into these Folders. The King will then assign Envelopes to Folders containing a Document of interest for the Kingdom. A Folder can be fixed, i.e. the Document in the Folder is predefined or dynamic, i.e. the King can order the Mayor to put a Document into a given Folder. The advantages of letting the King to decide which Documents will be put into which Folders are that matching Documents in different Cities will have a common identification throughout the Kingdom. The disadvantage is that this requires some software and a City Founder may find it too expensive and choose to have the Documents placed in fixed Folders to save memory.

4.5.9 Folder Label

A Folder always has a label, the Folder Label. It contains the following information:

- Folder Number.
- Document List Number.
- Document Number.
- Transmit/Receive mark.
- The CAN Control Field according to the CAN specification.
- Remote Envelope(s).
- An Envelope can be set as "remote" according to the CAN specification by the RTR bit. How RTR set to 1 is interpreted is dependent on the application corresponding to the Document in the Folder and has to be defined in the City documentation.
- Enable/disable the Folder.
- By disabling the Folder, any CAN communication by the application corresponding to in this Folder is interrupted.
- Envelope(s) assigned to this Folder.
- Envelope(s) enable/disable.
- The use of an Envelope can be switched on or off with an enable/disable tag.

How a Page is identified in a Transmit Form List and put into a transmit Document is depicted in Fig. 4.34. Receiving Cities use the same Page Form in their corresponding Receive Document.

As shown, a City Founder (module designer) does not have to care about how his module will exchange information with other modules in a system. He has defined what information his module needs and the timing constrictions. He has also specified what kind of information the module can make available to other modules. A Kingdom Founder can later on adapt the module to the needs of his system by transmitting King's Pages at a start-up procedure and even dynamically tune it during run time. Thus, the system designer can adapt and control the system using the King in the Capital. Some diagnostics can only be made at the system level by monitoring the traffic and getting internal information from modules. Any internal

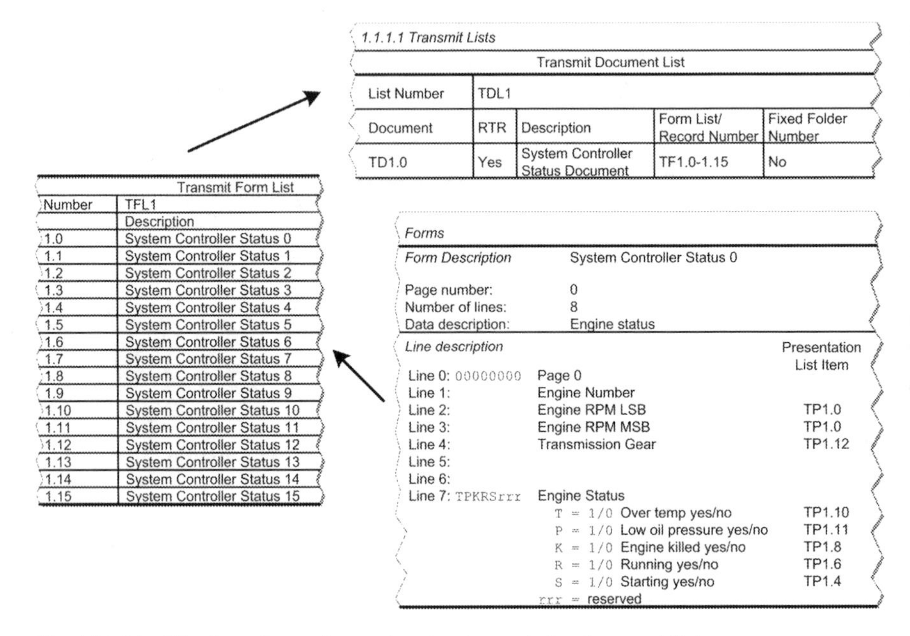

Fig. 4.34 Identification of a sending system in a transmission document list

status and parameter value in a module can be made available to the King in the Capital. Consequently, any decision on what to do in specific situations can be left to the system designer to decide among a set of alternatives.

A small CK system is depicted in Fig. 4.35. City 1 measures oil temperature and City 2 measures water temperature. Each City only measures temperature and makes the measured values available to the system. They do not have to know what they are measuring and when to transmit. City 3 receives both temperature values and has to distinguish between the two. This is easily done by assigning an Envelope to the Temperature Transmit Document in City 1 and the same Envelope to the oil temperature Receive Document in City 3. The temperature in City 2 is connected to the water temperature in City 3. With CK, no profiles are necessary as the King can set up a City to match the needs of the system. The module designer has great freedom to integrate a variety of options to meet the requirements of different HLPs and profiles. The module can then be adapted to a specific HLP and profile using proper King's Pages at an end of line programming. A system designer can also integrate modules made for other HLPs into his system by adapting other modules to the runtime behaviour of the integrated ones. Any HLP-specific start-up procedure, e.g. the "Duplicate MAC ID check" in DeviceNet or "Address Claim" in J1939 can be simulated by the King to make the integrated module happy.

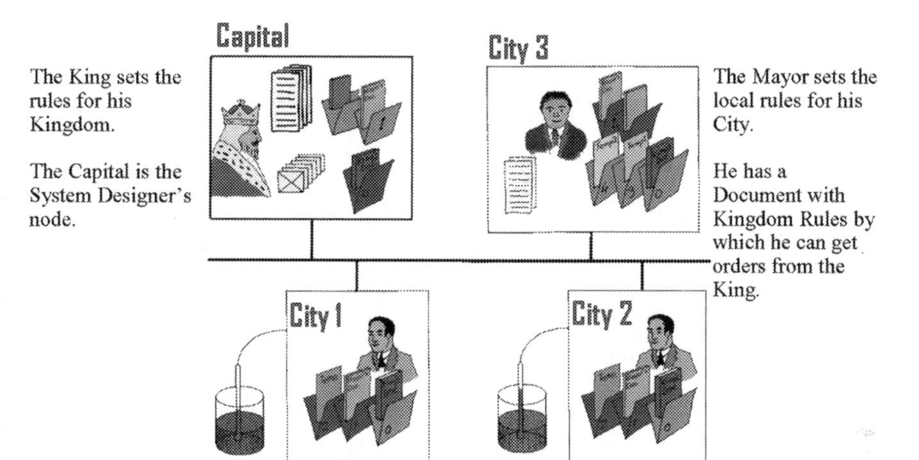

Fig. 4.35 CanKingdom basics

4.5.10 Composability and Membership

The composability in CK relies on the King, i.e. the system needs to have a Capital (a system node) and the start-up procedure. During the start-up procedure, all connected Cities are identified. Any new City can be properly configured. For a safe cold start, the following are necessary:

- Each module conducts a self-test.
- Each node connects to the CAN network in silent mode, i.e. without transmitting ACK bits, and listens for a specific message at 125 kbps for 2 s.
- Switch to stored bit rate for the current system.
- When a valid message is received, the module switches to listen-only mode, i.e. it participates in the CAN error checking and acknowledgement, but does not transmit and waits for the King's Page 1 with the Base Number.
- The King sends King's Page 1, either with the Group Address 0 or with individual addresses, to make every module respond with its European Article Number (EAN) and serial number (Fig. 4.36).

In this way, the King has complete control of the system. The King can check that all anticipated modules are connected and working correctly before the system is turned into runtime mode. During runtime mode, the King can supervise the system and check for different types of errors as missed schedules, unauthorized bus traffic, mismatching values, etc. The King can also act as a gateway to external tools or systems and provide full documentation of the settings of each node and their respective states.

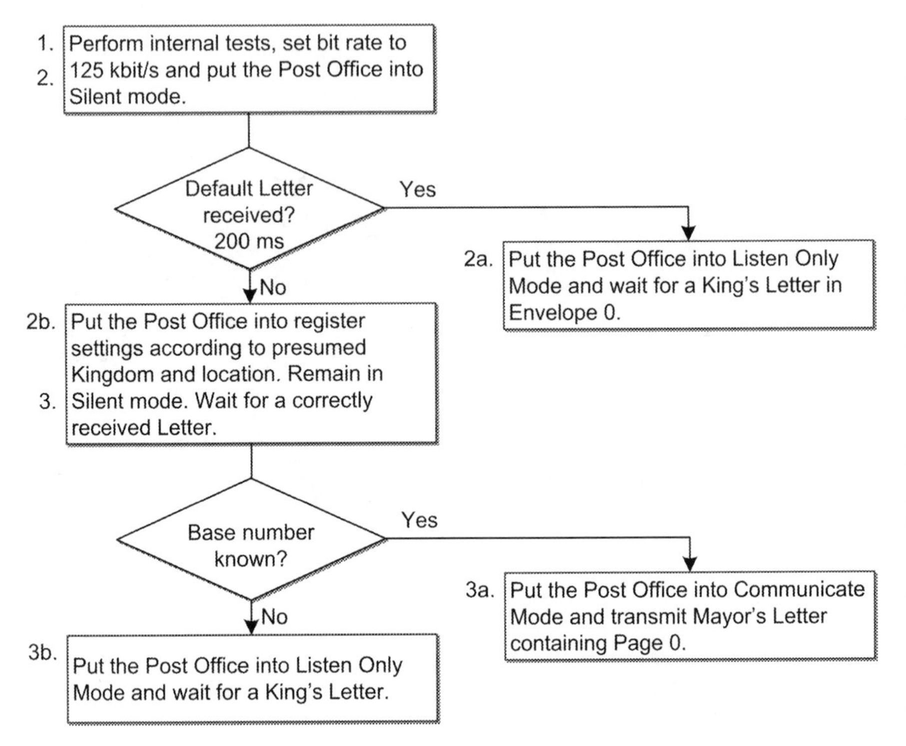

Fig. 4.36 CanKingdom setting

If any module is set to a false bit rate, it will not disturb the network as it stays in silent mode. An external setup tool (or the King) will always be able to connect at 125 kbit/s and correct the bit-timing register settings.

Chapter 5
Applications

Guenter Reichart, Gabriel Leen, Nathalie Courmont, Ralph Knüppel, Christian Schmid and Markus Brockmann

5.1 Electronic system architectures of Automobiles Application of CAN Bus

5.1.1 Bus Systems in Automobiles

Bus systems in automobiles allow the communication, which means the exchange of data between the electronic control units (ECUs), smart sensors and actuators. Depending on the respective requirements, different bus systems are used. Typical requirements consist of required data rate, allowed message length, number of connectable nodes (ECUs), required topologies, requirements on deterministic transmission capability and in further reliability, availability or safety-oriented requirements. Further requirements address aspects of physical characteristics, like tolerance against voltage deviations, temperature stability, impacts on wiring harness (electromagnetic compatibility (EMC), copper wire, plastic or glass fibre,

N. Courmont (✉)
Airbus France S.A.S., 316 Route de Bayonne, 31060 Toulouse Cedex 03, France
e-mail: Nathalie.Courmont@airbus.com

G. Reichart
BMW AG, Petuelring 130, 80788, Munich, Germany

G. Leen
University of Limerick PEI, Limerick, Ireland
e-mail: gabriel.leen@ul.ie

R. Knüppel · C. Schmid
Airbus Deutschland GmbH, Hünefeldstr. 1-5, 28199 Bremen, Germany
e-mail: ralph.knueppel@knueppel-online.de

C. Schmid
e-mail: Christian.Schmid@airbus.com

M. Brockmann
WILO AG, Nortkirchenstrasse 100, 44263 Dortmund, Germany

W. Lawrenz (ed.), *CAN System Engineering,* DOI 10.1007/978-1-4471-5613-0_5,
© Springer-Verlag London 2013

unshielded twisted pair (UTP) cabling or shielded twisted pair (STP) cabling) and last but not least, cost aspects.

In automotive engineering, bus systems are differentiated according to the so-called Society of Automotive Engineers (SAE) classes:

5.1.1.1 Class A

Bus systems for simple applications with low data rates of up to 10 kbit/s, e.g., sensor data or simple control commands. The main application domains are relatively simple functions without safety relevance in the body domain. The transmitted messages are mainly short and event triggered with a low data rate. The application area is relatively cost sensitive and demands therefore a rather cheap interconnection technology.

5.1.1.2 Class B

Bus systems for applications with data rates of 10 kbit/s and up to 125 kbit/s (e.g., many more complex body functions).

5.1.1.3 Class C

Bus systems for applications, which require real-time behaviour with data rates of 125 kbit/s and up to 1 Mbit/s (engine domain and chassis domain). In these applications, domains at high data rates with defined low latencies of data transmission are required.

5.1.1.4 Class D

Bus systems for the data transmission of long data streams with high bandwidth. These requirements prevail mainly in the area of infotainment and entertainment, e.g., for the transmission of audio/video streams.

International Organization for Standardization (ISO) differentiates bus systems only in two steps:

- Low-speed communication (bit rates < 125 kbit/s) and
- High-speed communication (bit rates > 125 kbit/s).

All these classifications are not really satisfying to adequately describe all the relevant requirements. A classification which is mainly focused on bandwidth is not sufficient to describe the requirements of the different application domains. Due to the development towards higher bandwidth and towards wireless data transmission, this traditional classification concept has to be reconsidered anyway.

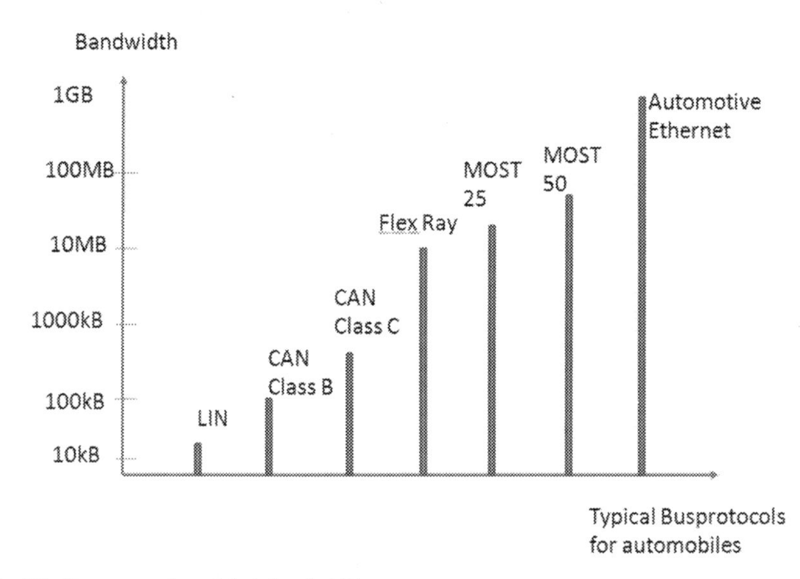

Fig. 5.1 Bus protocols and their bandwidth

The controller area network (CAN) bus protocol is currently applied in two different variants which correspond to class B and class C of the SAE logic.

Figure 5.1 shows typical bus protocols for automotive applications which are either already in use or in development. They are ranked according to their bandwidth.

Simple functions can be covered by the Local Interconnect Network (LIN) bus which allows for a data transmission of up to 20 kbit/s. The low-speed CAN bus can operate with a data rate of 5 kbit/s and up to125 kbit/s and in a network of up to 32 nodes. The strengths of the low-speed CAN are fault tolerance and the possibility to transmit over a single wire connection.

Due to the ever-increasing data rates and the ongoing trends towards higher functional integration along with a decreasing cost advantage of the low-speed CAN versus the high-speed CAN (Class C), low-speed CAN will soon go out of use. There is a trend visible that, in future vehicle architectures, the cost-effective LIN will be used for rather elementary functions. More demanding functions will be realized using the high-speed CAN and/or FlexRay, especially for time-critical or safety-critical applications. Media Oriented Systems Transport (MOST) will be used for multimedia applications in the infotainment and entertainment domain. In a broader perspective, Ethernet will play a significant role for system interconnection and can replace some of the traditional bus protocols. First applications for vehicle flash and diagnostic access are already in the market.

The electronic system architectures which we find today in modern cars will not change suddenly. Different bus protocols will be used even in the coming years since a radical change of the architecture would create huge costs and high quality risks. Even if the goal of the system architect remains to establish a more homogeneous network and therefore a reduction of the number of different protocols, the only solution can be to establish a clear and feasible migration plan.

5.1.2 The Application of CAN in Today's Vehicle Networks

The CAN protocol is, in today's vehicle networks, primarily used for the following three domains:

- Body electronics and active systems of passive safety,
- Chassis domain and driver assistance and
- Engine domain.

The main applications in the body domain deal with the control of windows, doors and flaps, mirror adjustment, control of lights, seat adjustments, climate control and comfort access. For cost reasons, the low-speed CAN plays a significant role but is in competition with the LIN bus. The safety electronics require a fast and safe data transmission; thus, the interconnection of the ECUs is, in most cases, realized by the high-speed CAN.

Chassis control systems as well as driver assistance functions put rather demanding requirements on the safety of the data communication and on the timing. Even if the high-speed CAN does not allow for a deterministic data transmission, high bandwidth can provide a sufficiently low latency in many applications. This implies, however, that only 50 % of the maximal data-transfer capacity can be exploited. Experience has shown that, beyond this level, non-deterministic latencies begin to rise. The CAN protocol contains a number of supervisory functions and error recognition concepts:

- Cyclic redundancy check (check of test sums),
- Frame check (check of frame length and structure),
- ACK error (proof of acknowledgment),
- Bit stuffing (error check on bit level by stuff bits) and
- Level monitoring (monitoring of the bus level by the connected ECUs).
- These features and the multi-master concept for the bus access are the foundations why the CAN bus has become a very reliable interconnection technology which has extended, beyond its original scope, the automobile into field bus applications in automation technology.

5.1.3 CAN and AUTOSAR

For the software architecture of ECUs, the international AUTOSAR standard has become increasingly widespread. The acronym AUTOSAR means AUTomotive Open System ARchitecture (see also Sect. 6.2). Within the software, one can speak of architecture if the application level as well as the system basis level is realized in a defined, structured manner. One speaks about an open architecture if the interfaces are standardized and disclosed. Usually, a certain independence from technologies

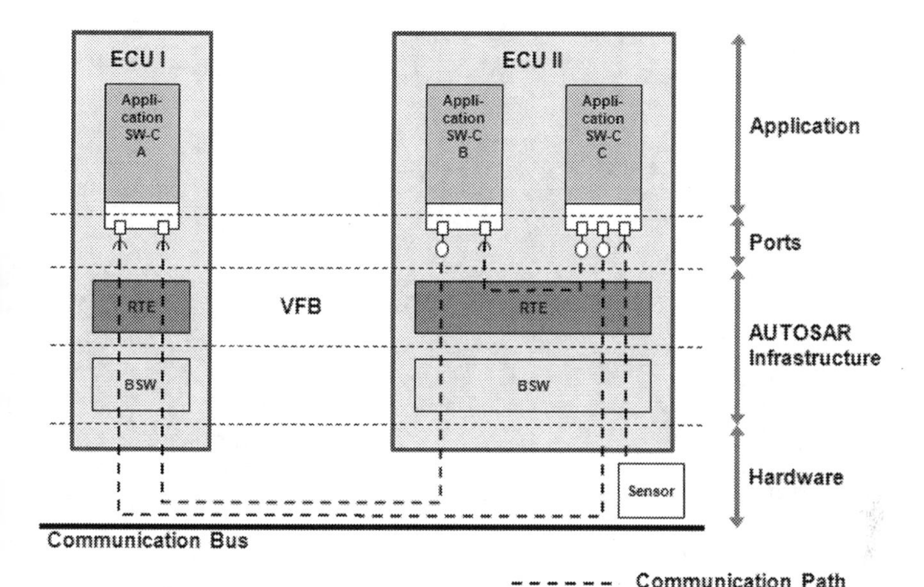

Fig. 5.2 Communication within and between ECUs according to the AUTomotive Open System ARchitecture (AUTOSAR) Standard. (According to Simon Fürst, AUTOSAR Guided Tour 2010)

by the introduction of abstraction layers is also a prerequisite. Based on these characteristics, AUTOSAR allows a transferability of software modules within or between ECUs. Moreover it can support, in the longer run, the exchange of software modules between different original equipment manufacturers (OEMs) if they comply with the AUTOSAR standard.

The system basic functions comprise, e.g., system services (operating system (OS), memory, network, diagnostic and ECU management), the microcontroller abstraction, device driver, driver for communication and communication services, communication hardware abstraction, etc.

The application layer docks on the so-called *Run Time Environment* (RTE) by means of standardized interfaces. The RTE is frequently called a *Virtual Function Bus*, a middleware layer, which allows the communication of the software modules within and between the ECUs (Fig. 5.2).

Ports implement the interface according to the communication paradigm (here: client–server-based).

They are the points of interaction of software components. The communication works through the RTE. The communication layer of the system basis software is encapsulated and not visible at the application layer.

To support the existing variety of bus technologies, one has to establish tailored standard solutions for the system basis functions. These solutions are packed in so-called stacks (e.g., CAN stack and FlexRay stack) and are offered by a number of first and second tiers. Up to now, it was not possible to standardize the system basis functions to such a level that one standard solution would meet the requirements of

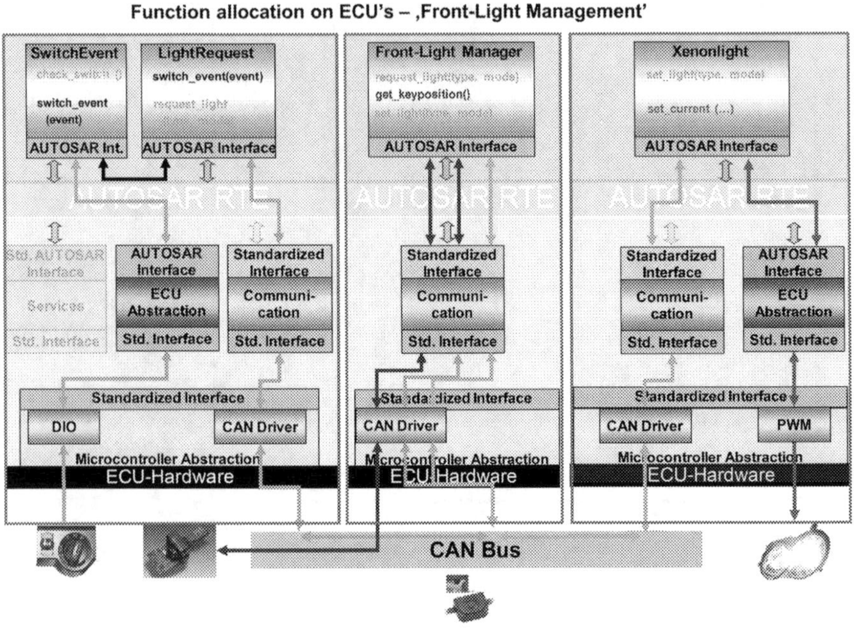

Function allocation on ECU's – ,Front-Light Management'

Fig. 5.3 Function allocation on electronic control units (ECUs)

all OEMs. For that reason, those parts of the system basis functions which are still company specific are allocated in the so-called *Complex Device Drivers*.

Figure 5.3 shows a solution for the control of headlights.

The interconnection of ECUs with AUTOSAR software architecture can be realized without any problem.

5.2 Time-Triggered Controller Area Network (TTCAN)—Applications

The following application example is a research educational prototype steer-by-wire and brake-by-wire system which is built on a basic software implementation of the time-triggered controller area network (TTCAN) protocol.

5.2.1 Software Implementation of TTCAN X-by-Wire

A current trend in the automotive industry is to replace certain mechanical components in vehicles with ultra-dependable fault-tolerant electronic systems, referred to as X-by-wire systems. Mechanical components such as drive belts, water pumps, hydraulic brakes and steering columns can be replaced with electronic systems.

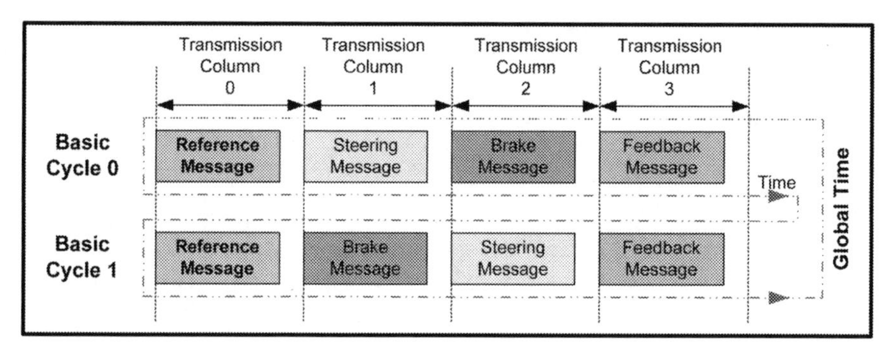

Fig. 5.4 Message cycle matrix

This initiative should result in a lighter, safer, more fuel-efficient and less expensive X-by-wire vehicle which exhibits additional functionality. In such vehicles, there are fewer environmentally unfriendly fluids to contend with, and the systems are self-diagnosing, reconfigurable and easily adapted across vehicle platforms. X-by-wire systems allow the tightest possible integration of distributed functionality within the vehicle, in contrast to the discrete, often disjoint, operation of conventional mechanical systems. The introduction of an X-by-wire vehicle infrastructure facilitates the implementation of many active safety improvements, based on advanced electronic systems; examples include autonomous cruise control, collision avoidance, automated parking assist and autonomous driving. The European SPARC Project is an example of an X-by-wire accident-avoiding vehicle with a Safety Decision Control System (SDCS). A necessary prerequisite to such highly integrated X-by-wire systems is a fault-tolerant communication infrastructure. The following section describes a prototype experimental brake-by-wire and steer-by-wire system based on TTCAN.

5.2.2 TTCAN Network Implementation

At the time this X-by-wire prototype was developed, there were no TTCAN protocol engines available in silicon. As a result, a system based on the Infineon C515C microcontroller and an application layer based on the TTCAN protocol with level 1 synchronization was implemented in software. Figure 5.4 illustrates the TTCAN message matrix used. The cycle matrix consists of two basic cycles. Each basic cycle commences with a reference message, which is followed by either a steering wheel position message or a brake pedal angle message, and terminates with a feedback message. The reference message is used to synchronise the network by resetting the cycle time in each network node. The reference message also contains the current basic cycle count, which is used to help to ensure that all nodes observe the correct schedule pattern. The TTCAN local clock is implemented using the microcontroller's on-chip timers and the TTCAN triggers are implemented using real-time interrupts generated by the overflow of these timers.

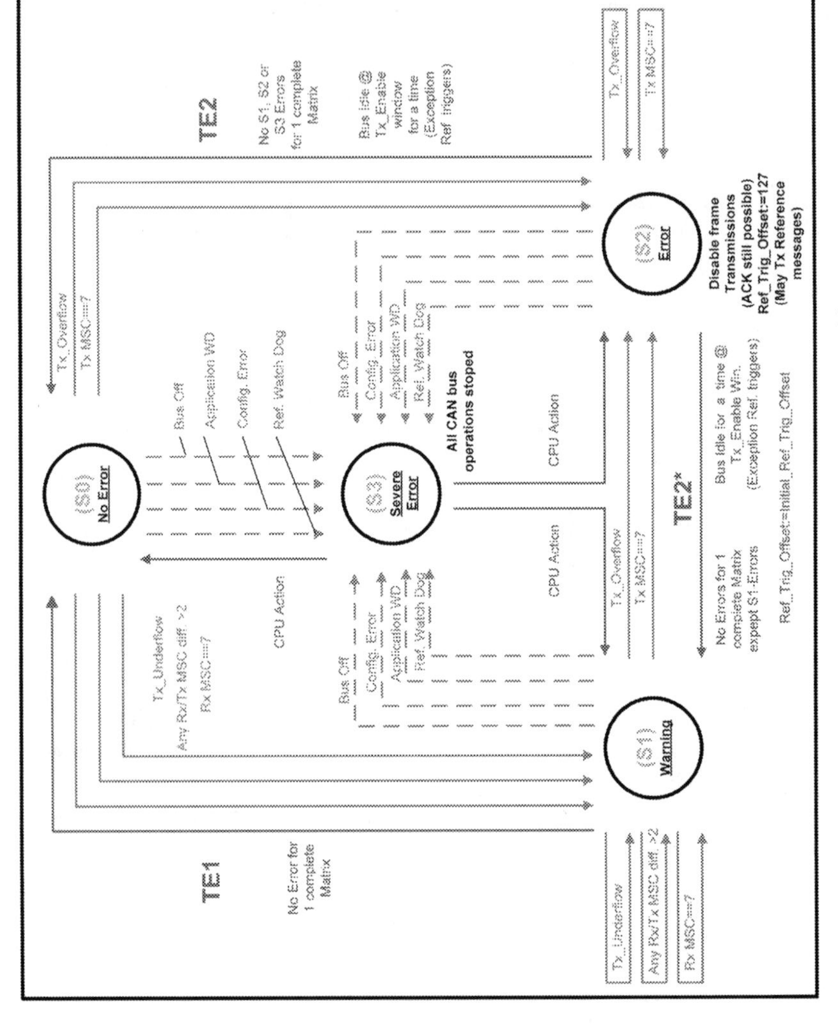

Fig. 5.5 Time-triggered controller area network (TTCAN) error state machine

Each network node monitors the transmission and reception of relevant messages; this and additional information are used as input to an error state machine. For example, if an expected message is not transmitted or received, a corresponding message status counter (MSC) is incremented. If any MSC reaches a predefined limit, an error is flagged and an appropriate action is taken. In this case, the node is reconfigured and attempts to rejoin the network; see Fig. 5.5 for an overview of the TTCAN error state machine.

Fig. 5.6 Steer-by-wire system

5.2.3 Steering Implementation

In the prototype, the vehicle steering column was removed and replaced with a position sensor as seen in Fig. 5.6. A rotational absolute encoder is used to measure the angular position of the steering wheel. The sensor measures 128 positions per revolution and, therefore, has a resolution of 2.81 degrees. The sensor is read by the user interface microcontroller, and its position sent over the TTCAN network in the steering wheel position message. More accurate sensors with greater resolution were considered along with the possibility of gearing this sensor for greater resolution; however, for this concept demonstration model, such accuracy was not considered necessary. A real steer-by-wire system would probably require a sensor with a resolution in the order of 0.5 degrees, or better.

At the rack and pinion end (actuator end), a servo-controlled 12-V direct current (DC) motor is used to change the road wheel steering angle. A rotational position sensor connected to the rack and pinion drive provides feedback. Magnetic limit switches at the extremities of motion are used to detect the left and right maximum steering lock positions. Figure 5.6 shows the steering system used. The actuator ECU receives the requested wheel angle via the TTCAN network and rotates the wheels to the required position. In a production implementation, sensors, actuators and communication channels would most likely be replicated to provide redundant backup systems.

5.2.4 Brake Implementation

The system brake pedal uses a linear potentiometer to measure the extent to which the brake pedal is pressed as seen in Fig. 5.7. The voltage drop across the potentiometer is read by an 8-bit analog-to-digital converter (ADC) and its value transmitted over the network in the brake position TTCAN message.

The floating calliper brake unit is modified and incorporates a linear stepper motor to adjust the position of the brake pad, thus applying the braking force. In practice, a servo motor would be more appropriate. The actuator ECU receives the brake pedal position via the TTCAN network and adjusts the position of the brake pad accordingly.

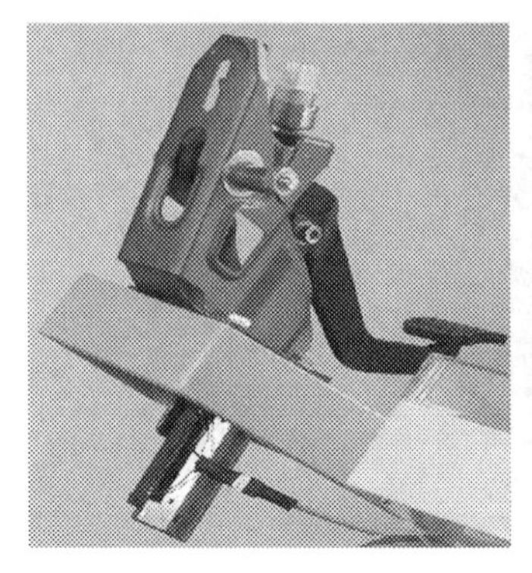

Fig. 5.7 Brake-by-wire system

5.2.5 Feedback Message

No physical force feedback was implemented in this prototype. Nevertheless, a feedback message is sent from the rack and pinion to the steering wheel, containing information relating to the steering angle. This message was included to demonstrate that a feedback message could be easily incorporated into TTCAN's network traffic.

5.2.6 Final System

The final prototype is illustrated in Fig. 5.8. The steer-by-wire and brake-by-wire systems were implemented on a single TTCAN network. The network was operated at 250 Kbaud with a Network Time Unit (NTU) of 1.2 μs. The time windows are 2 ms long. The reference and feedback messages take 0.236 ms each to be transmitted, and the steering and brake messages take 0.276 ms each to be transmitted. This results in a total network bandwidth usage, excluding error frames, of 12.8 %. It should be noted that in this configuration the network is only running at one-quarter of its maximum speed of 1 Mbaud.

5.3 CAN in Aircraft World

5.3.1 Why CAN?

A large amount of information crosses through an aircraft. Many systems coexist ranging from high critical avionics systems (displays, flaps command, engine fire detection, etc.) to cabin systems such as ventilation, water and galleys (kitchens).

Fig. 5.8 Time-triggered controller area network (TTCAN)-based X-by-wire prototype

Networking is already an old story: aircrafts have been using ARINC 429 since 30 years. Why then a change towards a "non-avionics" network as CAN?

5.3.1.1 From ARINC 429 to CAN

First of all, a few words on the Aeronautical Radio Incorporated (ARINC) label:

The Airlines Electronic Engineering Committee (AEEC) is an international standards organization, comprising major airline operators and other airspace users. The AEEC establishes consensus-based, voluntary form, fit, function and interface standards that are published by ARINC and are known as ARINC Standards. ARINC Standards specify the air transport avionics equipment and systems.

ARINC 429 is very well defined and largely used and known communication system. The first specification was delivered in 1977.

The physical layer is robust to the aeroplane environment and is characterized by:

- Rreturn-to-zero (RZ) bipolar modulation and tri-state modulation consisting of "HI," "NULL" and "LO" states,
- Nnominal voltages values as described in Table 5.1 and
- Cables and nodes with 75 Ω impedance.

Nevertheless, main drawbacks have limited the application and increased wires:

- It has a low bit rate, with high-speed operation at 100 kbits/s and low-speed operation around 12 kbits/s.
- Labels (equivalent to CAN identifiers) are too strictly defined.

Table 5.1 ARNIC 429 emitter voltage values

	HI (V)	NULL (V)	LO (V)
Line A to line B	+10	0	−10
Line A to ground	+5	0	−5
Line B to ground	−5	0	+5

Table 5.2 ARINC 429 word structure

32	31	30	29 11	10	9	8 1
P	SSM		Data	SDI		Label

The label is used to determine the data type of the remainder of the word and, therefore, the method of data translation to use
P parity bit, *SSM* sign/status matrix, *SDI* source/destination identifier

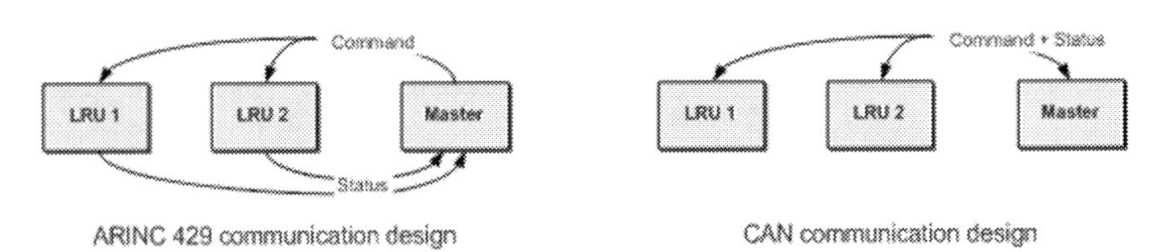

Fig. 5.9 ARINC 429 and controller area network (CAN) communication design

- Components are handled by the aeronautics industry.
- The ARINC 429 word is 32 bits with 20 bits maximum for data field as shown in Table 5.2.

Moreover, the communication happens through one transmitter/multiple receivers. It is highly reliable but increases number of wires (Fig. 5.9).

System designers and aircraft manufacturers therefore decided to apply an already worldwide established important standard for their increasing communication demands - that is why CAN is chosen.

The main advantage seen with CAN is that it is the automotive standard. It is not that airframers "copy" automotive ideas but component obsolescence is a very critical factor. The long life of aircrafts (30 years) but the small amount of units (around 1 per day gets out from Boeing and Airbus assembly lines) pushes us to follow a size market that gives quantities.

The "open" standard, the large number of tools offered and the price have contributed to CAN's success in aircrafts. CAN also offers good error detection and high electromagnetic immunity.

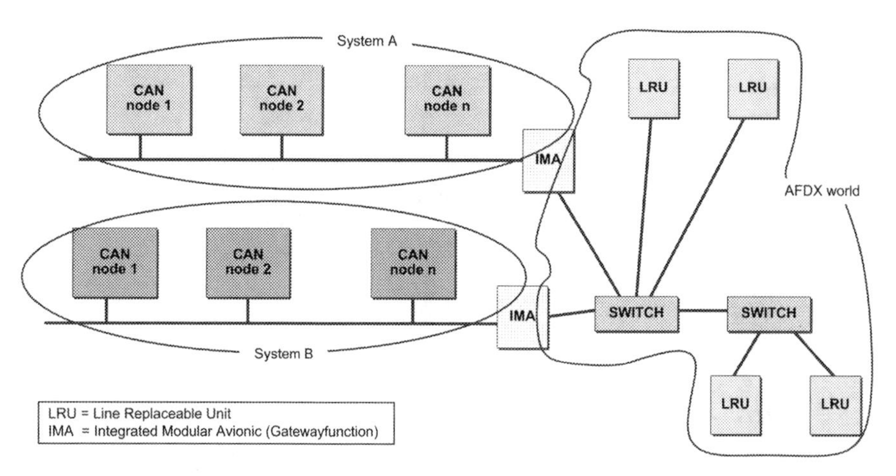

Fig. 5.10 A380 CAN (controller area network) in the global avionics network architecture

5.3.1.2 History and Future ...

Two different worlds coexist in aviation: general aviation (GA) with small aircrafts and helicopters and airframers (Airbus and Boeing).

Airframers CAN started with cabin systems (ventilation, smoke detection and water/waste) on A318 and A340, developed by a unique supplier with low bit rate (83 kbps). It was so appealing that the use was largely extended on A380 within avionics high critical systems (power distribution, control panels, engine fire detection, door monitoring, etc.) leading to more than 500 CAN nodes and 75 busses per A380. A380 was also the starting point of the backbone avionics communication with AFDX (avionics full duplex Ethernet; switched Internet ARINC 664). The redundancies are not shown in Fig. 5.10.

The redundancies are not shown on this drawing.

General Aviation Uses CAN for backbone communication and major avionics buses. Therefore, it has to fulfil all the requirements of a flight safety network. National Aeronautics and Space Administration (NASA) has also used CAN for research program.

A specific application layer was developed: *CANaerospace*, the initial version created in 1998. The story does not stop there as the Boeing B787 Dreamliner also hosts a large number of CAN systems, and Airbus as well as Boeing have chosen CAN as a basis for subsystems communication. A CAN standard ARINC 825 is ready for all applications on board aircraft.

Why is an ARINC standard needed, one could ask. The following sections give the answer: aircrafts are specific….

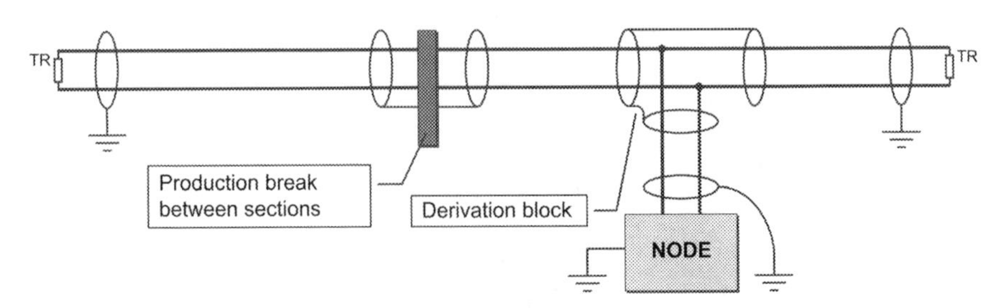

Fig. 5.11 Controller area network (CAN) wires layout

5.3.2 *Aircraft-Specific Physical Layer Constraints*

Compromise between wire length, number of equipment and baud rate is critical and aircraft systems reach the CAN physical layer limits.

5.3.2.1 EMC and Lightning Stress

Exposition zones are defined depending on the system installation area (pressurised and non-pressurised) and the EMI and lightning stress levels. Maximum bulk current injection (BCI) levels are 75 mA for A380, which are not that far from automotive constraints. Installation without shielding is possible. On A400M military aircraft, the BCI maximum level is 300 mA and installation with efficient and maintained shielding is necessary. Future aircrafts will apply GLAss-fiber REinforced aluminum (GLARE) for weight reduction purpose. This will increase the design constraints even more.

Lightning protection, designed with transorbs, is added in each equipment, which increases the node internal capacitance up to 300 pF.

5.3.2.2 Installation

The total length of A380 cables is 500 km. Routes are defined for cable and side segregation—M1 and M2 with no specific requirements and S1 and S2 for specific signals constraints: for example, few millivolts of audio signals.

The installation of a CAN network is a large part of the CAN adaptation to aircraft world. The constraint of maximum 30 cm stub length can not be met sue to cable routeing weight reduction constraints. This is especially true for cabin installation. Some systems are more than 150 m long, which is out of automotive use.

Some CAN wires go from the cockpit to the aircraft tail (Fig. 5.11) and go through different sections, which are produced in different European sites. The wire connection between sections is called a production break. At this point, the wire is untwisted and impedance is then modified.

Fig. 5.12 ARINC 600 and EN3646 connector type

Signal quality estimation by simulation and mock-up are run in order to anticipate potential risks for system design.

At node termination, the connector type is a regular avionics one: ARINC 600 or circular EN3646 type as shown Fig. 5.12.

5.3.3 DAL, Safety, Certification

DAL (stands for Design Assurance Level from A to E. DO178B) classifies effects of a functional failure on aircraft safety. Safety analysis is run for all systems and loss of equipment/component (wire, etc.) is classified. Redundancies are built up to reach the system safety requirement. The loss effects classification is as follows:

- E—no safety effect,
- D—minor,
- C—major,
- B—hazardous and
- A—catastrophic.

A loss of a CAN network is no more than major. In specific high critical networks, we have up to three CAN wires and equipment redundancies.

5.3.3.1 Certification

The US Federal Aviation Administration (FAA) and the European Aviation Safety Agency (EASA) are two independent administrations that allow an aircraft to transport passengers. Numerous CRIs (Certification Review Items) will assume that CAN is related to CRI-F40 for A380 and the objectives are:

1. To ensure that the bus perform its intended function under the most demanding conditions and
2. To evaluate effect of abnormal behaviour and ensure the safety consequence.

The following issues are documented for all systems:

- Safety study,
- Data integrity,
- Performance,
- Design assurance,
- EMC,
- System configuration management and
- Continued airworthiness.

For more information, contact the FAA or EASA. Another CRI is related to CAN with the CRI-F09 for critical components; the CAN controller enters this category.

5.3.4 Example: Smoke Detectors Interfaced by a Safety-Critical Aircraft-Based CAN-Bus Network

5.3.4.1 Abstract

Classic architectures of aircraft systems contain equipment using interfaces with digital, analogue or discrete signals. The electrical network to interface the equipment varies between the applications. Some equipment require a dedicated power supply and provides information on an analogue current loop, while others use proprietary digital busses or discrete input/outputs (I/Os) for information exchange.

Initially, CAN was developed for use in the automotive industry, but is nowadays being used in an increasing number of applications. One of these areas is aviation, where in the past 5 years CAN has grown from being an exotic newcomer to an established and widely accepted solution. Within the Fire Protection System on an Airbus, smoke detectors are installed in various areas overall in the pressurised zones of the aircraft like lavatories, equipment bays and cargo compartments. As the CAN bus defines only layers 1 and 2 of the Open Systems Interconnection (OSI) communication model, additional higher layer features are necessary to achieve the level of operational assurance required for a safety-critical application, namely fire protection on an aircraft.

This example is particularly focused on the development of a safety-critical CAN bus network with strict configuration control of smoke detectors in the scope of an aircraft application. In 2003, international airworthiness authorities approved the application in the frame of the Airbus A318 Type certification.

5.3.4.2 Introduction

The objective of the new smoke detection system was to replace the proprietary current modulated supply and communication loop with an open, non-proprietary bus standard. The overall system reliability and performance were aimed to match or

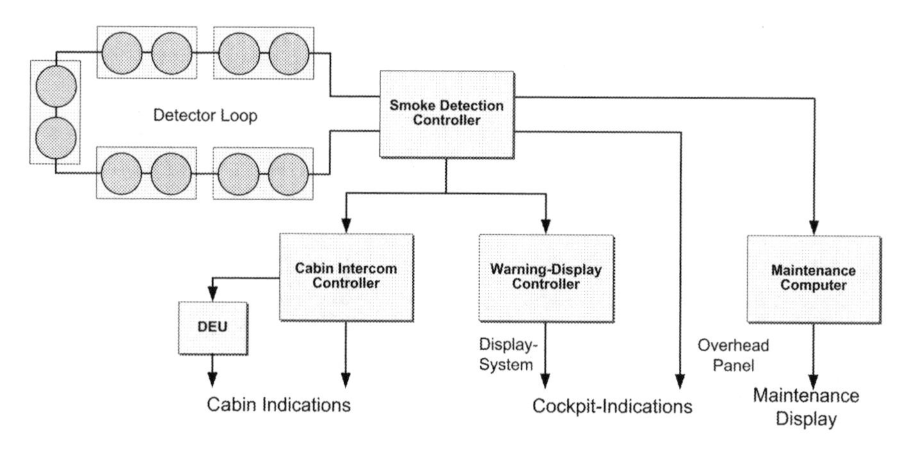

Fig. 5.13 Smoke detection system using proprietary detector supply and communication loop

surpass the existing architecture while keeping development and purchasing costs at a comparative level.

The latter was feasible by reusing the existing smoke detector core and fitting it with an altered communication and power interface (see Figs. 5.13 and 5.14).

The communication medium had to meet a number of requirements for eligibility in a safety-critical application:

- Advanced data integrity and error detection features,
- Deterministic behaviour,
- Operability in challenging EMC environments and
- High degree of flexibility in choice of network size and topology.

Considering the 30-year design life of a modern passenger aircraft, the long-term availability of electronic components was scrutinized in order to minimize the risk of equipment redesign resulting from component obsolescence throughout the life cycle of the aircraft.

The CAN bus was deemed the most suitable communication medium capable of fulfilling the above requirements.

5.3.4.3 Protocol

The CAN protocol, as defined in ISO 11898 [ISO11898], covers layers 1 and 2 of the OSI communication model. The remaining layers, up to layer 7, have to be managed by additional services up to the application. Various standardised higher layer protocols such as CANopen are available and widely used in industrial applications. Instead of selecting a generic high layer protocol, a specific to-type application layer protocol was developed and documented in a System Interface Document [SCHMID] in order to ease compliance with RTCA/DO-178 [DO178B] guidelines.

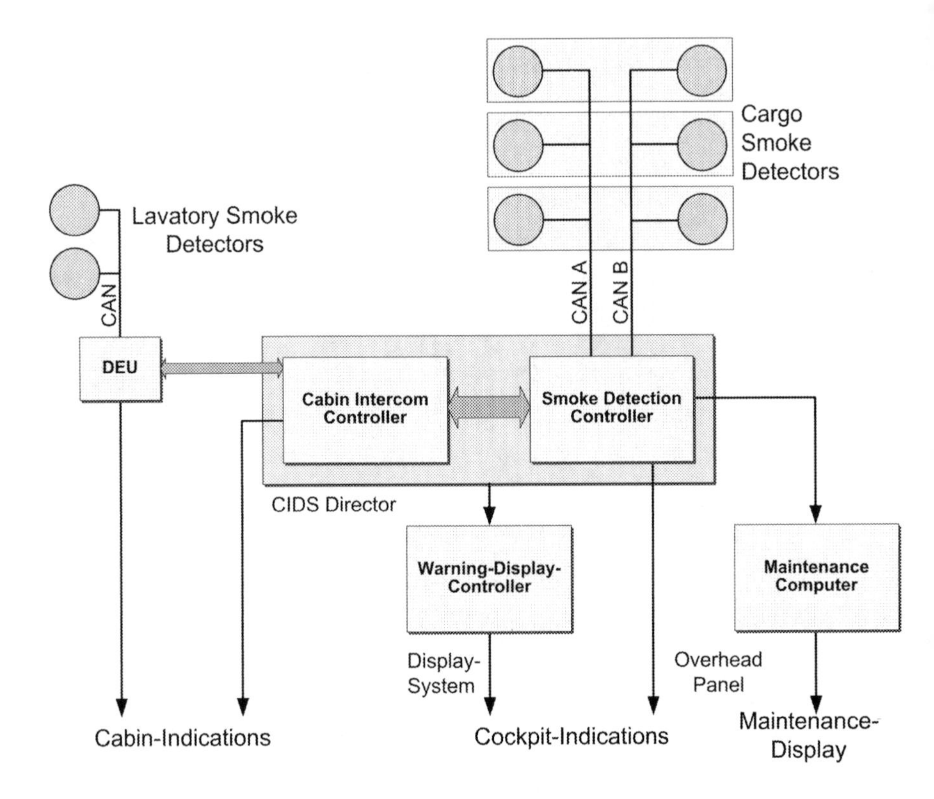

Fig. 5.14 Smoke detection system using an open standard controller area network (CAN) bus-to-interface detector

Analysis of the communication needs to result in the following protocol requirements:

- Every individual smoke detector on the network must be uniquely identifiable,
- Messages generated by a smoke detector must contain information about its identity and
- The detector must support a master–slave communication model.

CAN Identifier The 29-bit extended identifier is utilized and partitioned into the subfields as shown in Fig. 5.15.

Message Type The purpose of the Message Type is to categorize messages according to their overall relative priority and indicate whether the Module ID contains a transmitter or receiver address. Two classes of Message Type, Process Data Object (PDO) and Service Data Object (SDO) are instantiated either as Transmit or as Receive objects, T_PDO and R_PDO as well as T_SDO and R_SDO, respectively. A Transmit Data Object (T_xDO) denotes that Module ID contains the network address of the transmitter, whereas a Receive Data Object (R_xDO) contains the network address of the intended receiver in the Module ID field.

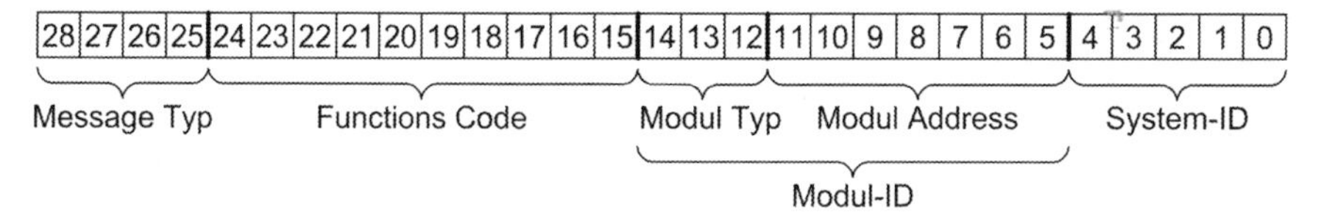

Fig. 5.15 Controller area network (CAN) identifier

Function Code (bits 24...15) Every application function is designated a unique Function Code within its respective Message Type. In addition to describing the next level of arbitration priority, the Function Code is used to transport logical data without the use of the actual CAN data fields. In this case, the Data Length Code (DLC) is 0, enabling efficient use of data bandwidth, particularly for R_PDOs and R_SDOs which contain mostly status requests directed at smoke detectors and do not carry any further information than the request itself.

Module ID (bits 14...5) The Module ID field contains the unique network identification of the CAN node. This may also be a broadcast identification when a message is directed at several nodes simultaneously. Two subfields Module Type and Module Address split the Module ID into equipment classes and their individual addresses. The entire Module Address space may be reused for every Module Type on the network.

System ID (bits 4...0) The System ID is used to tag the CAN identifier with a unique system identification code. All smoke detectors and other fire protection components are assigned a fixed value.

Data Frames A Data Frame is generated by a transmitter to transfer application data to one, or in the case of a broadcast, several receivers. Within the Data Frame, the Data Field consisting of 1–8 bytes carries the application data. A Data Frame may contain an empty Data Field (DLC=0). In this case, data are carried through the Function Code alone.

A smoke detector's 8-byte status Data Field is as defined in Table 5.3 with the meaning of the data bits in Table 5.4.

5.3.4.4 Network Management

It is of utmost importance that the system configuration and availability of resources (smoke detectors) are known to the network master. Lack of configuration control through the network master device would jeopardize safety and disqualify the system. From a safety assessment point of view, the worst-case condition is an undetected configuration error leading to an incorrect compartment designation in case of fire; an alarm reported in the aircraft's forward cargo compartment while the real fire occurrence is in the aft cargo compartment and vice versa. Such a case is

Table 5.3 Smoke detector data field

Data byte	MSB	LSB	Description	Format
1	7	5	Spare/not used	-
	4	4	Detector Warning	Discrete
	3	3	Prefault threshold exceeded	Discrete
	2	2	Detector standby	Discrete
	1	1	Detector alarm	Discrete
	0	0	Detector failure	Discrete
2	7	0	Trouble shooting data	Binary
3	7	2	Spare/not used	-
	1	0	MSB contamination level	Binary
4	7	0	LSB contamination level	Binary
5	7	2	Spare/not used	-
	1	0	MSB smoke level	Binary
6	7	0	LSB smoke level	Binary
7	7	2	Spare/not used	-
	1	0	MSB temperature	Binary
8	7	0	LSB temperature	Binary

Table 5.4 Meaning of the status bits

Designation	bit	Meaning
Failure	0	The smoke detector is no longer able to detect smoke or to communicate this information in a reliable manner
Alarm	1	Smoke is detected and confirmed
Standby	2	The smoke detector is able to detect smoke and communicate this information in a reliable manner
Prefault	3	The smoke detector optical cell contamination level has exceeded the internal threshold for triggering a corresponding maintenance message
Warning	4	The CAN TX error counter has exceeded 96

classified as catastrophic. A catastrophic event is defined as an occurrence leading to total loss of the aircraft and occupants and must be ruled out with a defined level of probability of failure $< 1 \times 10^{-9}$. Therefore, various network management mechanisms are necessary to ensure proper system configuration during initialization and normal operation.

5.3.4.5 Power-Up Configuration Control

The normal expected configuration of smoke detectors is fixed in a lookup table within the network master's operational software. At power up or system initialization, the current configuration is compared with the expected through a mechanism called Configuration Check Request. During the Configuration Check Request process, the network master broadcasts the Configuration Check Request as an R_PDO

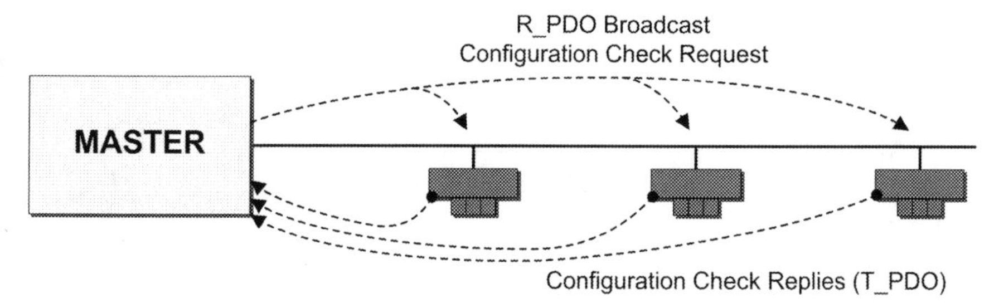

Fig. 5.16 Normal configuration check request/reply process

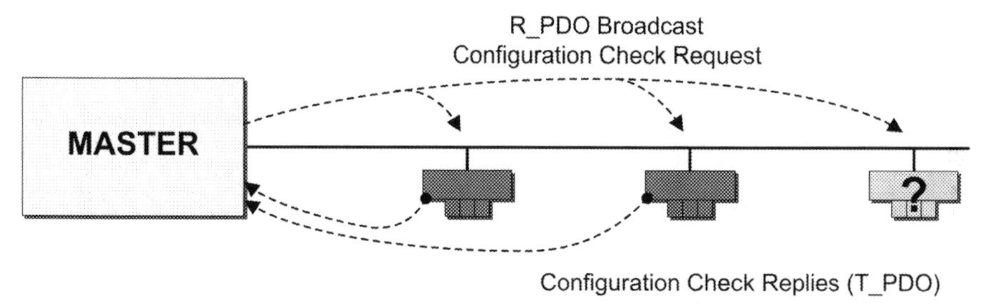

Fig. 5.17 Failure of expected smoke detector to reply

with the broadcast Module ID to all smoke detectors. These in turn reply with T_PDOs containing their individual Module Address, enabling the network master to make a comparison of the received replies with the expected replies and, thereby, detect the following failure cases (Figs. 5.16, 5.17 and 5.18):

- The network master is incorrectly configured for the intended installation,
- An expected smoke detector has not replied (missing smoke detector on network) and
- An unexpected smoke detector has replied (excessive smoke detector on network).

Thus, comparison of the configuration present on the network with the expected configuration is a prerequisite for determined network behaviour.

5.3.4.6 Normal Polling Operation

The CAN bus is operated in the master–slave mode (Fig. 5.19). The network master cyclically acquires the status of each smoke detector by an explicitly addressed request frame. Not to be confused with CAN remote-request frames, the request message is a regular data frame of type R_PDO containing the individual Module ID of the subject smoke detector and is replied to by a T_PDO data frame containing the Module ID of the replying transmitter.

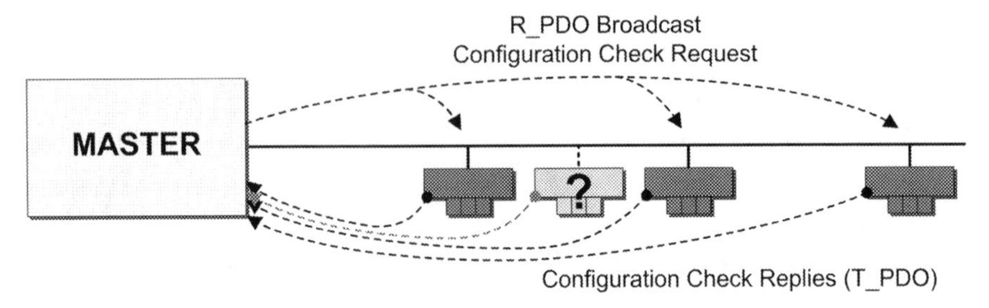

Fig. 5.18 Unexpected smoke detector reply

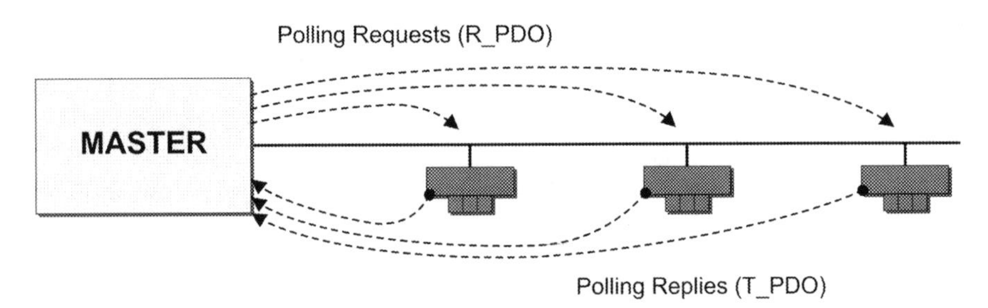

Fig. 5.19 Normal polling operation

Each polling request is monitored by a timeout in which the reply is expected. The polling cycle is repeated every 2 s.

5.3.4.7 Failure Detection/Reconfiguration

The response time of the smoke detector is 60 ms, including internal processing time and retry mechanisms inherent to CAN. A reply is considered timed out by the network master when not received prior to the following polling cycle, 2 s later. An outstanding reply increments a counter C in the network master. The reception of a normal polling reply while the counter is $1 \leq C < 5$ leads to a reset of the counter to 0 and the smoke detector is restored to normal operation. Once the counter reaches 5 outstanding replies (10 s), the smoke detector is declared inoperable and is no longer polled, thereby resulting in a reconfiguration of the system. System determinism is ensured through the request–reply time window and the polling cycle as shown in Fig. 5.20.

In summary, the polling process abides by the following rules:

- Only expected smoke detectors are polled,
- A smoke detector determined missing during power up is not polled,
- A smoke detector is no longer polled following five consecutive timeouts and
- A smoke detector is no longer polled when declared failed.

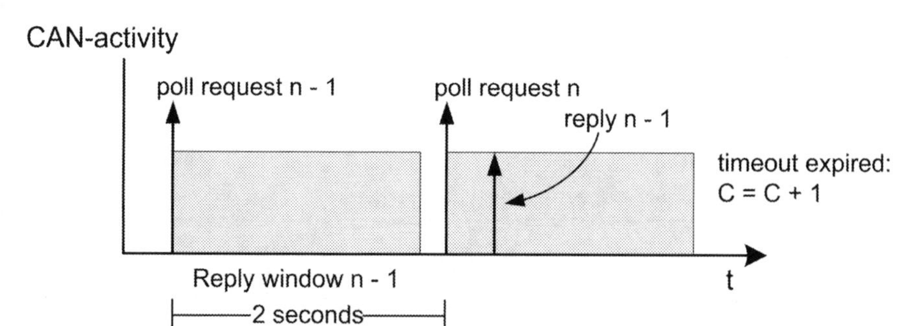

Fig. 5.20 Timeout expired

5.3.4.8 Smoke Detector Monitoring

In addition to the network-based configuration and time monitoring, the smoke detector is monitored for proper functional behaviour by the network master.

Normally, the smoke detector is in the standby condition (bit 2 on data byte 1 is TRUE). In case of alarm, the standby bit becomes false while the alarm condition (bit 1 on data byte 1) is TRUE. These conditions are by definition mutually exclusive and are therefore monitored for proper behaviour. If two consecutive polling replies are received with neither the standby nor the alarm bit set to TRUE, or both bits set to TRUE, the smoke detector is declared failed and is no longer polled. In Boolean terms:

$$Failed = (alarm * s\tan dby) + (\overline{alarm} * \overline{s\tan dby}). \quad (5.1)$$

5.3.4.9 Network Topology

The smoke detectors are connected with the network in a linear bus topology with stubs departing from a central bus line. Bus termination is accomplished through resistors implemented within the network master at one end of the network and the last smoke detector at the other end (Fig. 5.21). Each item of equipment is qualified to operate on a CAN bus of length 150 m, with 32 nodes connected through 2-m-long stubs to the main bus line.

Depending on the aircraft compartment being monitored, either a single- or dual-redundant bus line is incorporated depending on the reliability requirements and whether the compartment is accessible or not during flight. The dual-redundant architecture implies two smoke detectors at each location within a compartment. This is the case for the cargo compartments in the lower deck of the aircraft. Each lavatory, on the other hand, is fitted with a single smoke detector.

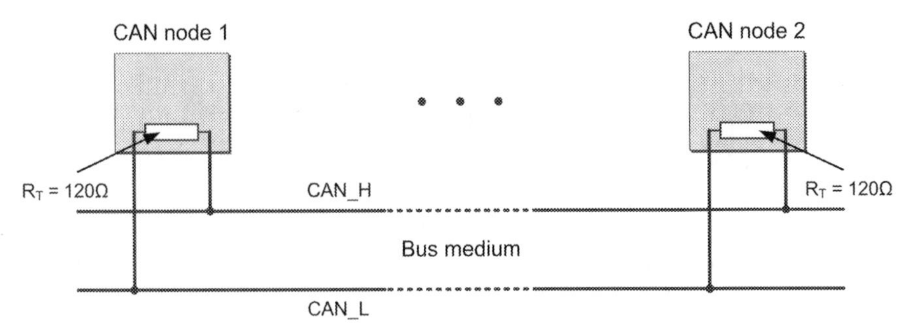

Fig. 5.21 Network topology

5.3.4.10 Development Process

The safety philosophy in aviation defines quantitative safety objectives and assigns acceptable probabilities. The overall probability for a failure with catastrophic consequences must be extremely improbable. This must be demonstrated to the airworthiness authorities for certification. The demonstration is endorsed through a complete, detailed and documented safety analysis, which is one of the integral parts of the software development process.

Guidelines for development of aviation software in the USA are defined in the DO-178B. Since its production by the RTCA, the DO-178B has become a de facto standard. The FAA's Advisory Circular AC20-115B established DO-178B as the accepted means of certifying all new aviation software.

DO178-B is primarily concerned with development processes. As a result, certification to DO178-B requires delivery of multiple supporting documents and records. The quantity of items needed for DO178-B certification, and the amount of information that they must contain, is determined by the level of certification being sought.

The higher the consequences of a potential failure of the software (catastrophic, hazardous-severe, major, minor, or no-effect) are, the higher is the DO178-B certification level. The levels are from A for the highest certification level through B, C and D to E.

This aviation-specific development process had to be followed on an equipment and on a system level.

5.3.4.11 Conclusion

Through clever system design and network management, a CAN bus-based safety-critical smoke-detection system with deterministic behaviour, capable of fulfilling the safety and reliability requirements, was developed and approved by airworthiness authorities. The robustness and reliability of CAN in this airborne application are being closely monitored, with some 1.45×10^7 accumulated flight hours (including multiple equipment factor) having been accumulated in the period between mid-2003 and February 2006.

5.4 The Geniax System Decentralised Heating Pumps

"Geniax" is a real technical revolution in the field of heating technology. It is based on several miniature pumps at the heating surfaces or in the heating circuits instead of using thermostatic regulating valves. The conventional "supply-oriented heating" with one central heating pump is replaced in this way by "demand-oriented heating"—pumping only takes place when heat is needed (see Fig. 5.22, which demonstrates the basic principle of Geniax).

Also new is a central control intelligence for the whole heating system. It maintains the heating system in a hydraulically optimal state and generally improves precision, speed and energy efficiency. Fields of application include new buildings and upgrades of older buildings. The system can be installed in both single- and multi-family houses as well as in commercial properties such as office buildings. The central advantage—besides improved hydraulics and comfort—is the significant reduction in heating energy consumption by an average of 20 %.

A further decisive component of the decentralised pump system—besides the miniature pumps and their pump electronics—is a central management unit with an interface to the heat generator: the Geniax server. It is responsible for the coordination of heating needs in the individual rooms and, using the present specifications from the room user interfaces, the management of all components in the entire heating system. The Geniax server's control signals to the pump electronics are used to variably control the pump speed and therefore also the mass flow of the pumps and the heating performance in a needs-based fashion. Beyond this, the server controls the displays of the room user interfaces, monitors all connected components, collects data for diagnostic purposes and controls the heat generator via the 0–10 V interface (see Fig. 5.23).

As already indicated above, the Geniax system is a master–slave system in which all slaves are dependent on the communication with the Geniax server. With this concept, Wilo SE also takes an unusual path for a traditional individual room temperature regulation system, where all the rooms to be regulated typically work on their own.

In the Geniax system, the server works with all the system's information so that individual pumps, if need be, can even be handled according to preference or can perform anticipatory work as a result of learning processes.

As is the case for any master–slave system, the Geniax system requires a suitable communication medium. Since a Geniax pump has a maximum power consumption of approximately 3.5 W, there is no need for a permanent battery supply for the components integrated in the system. This led to the decision to use a cabled system at the start of the newly developed system.

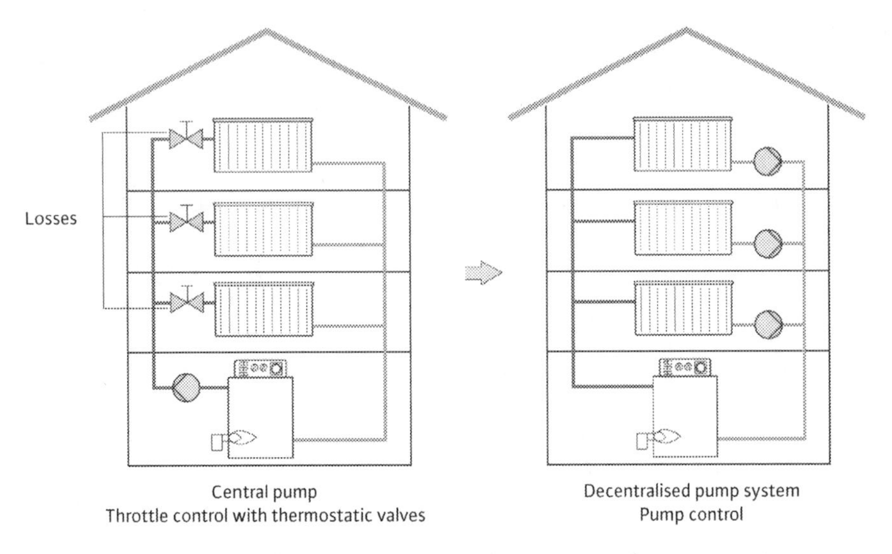

Fig. 5.22 Basic principle—throttle control compared to pump control

From the many established bus systems, a preselection was made and a benchmark was carried out on this selection.

The selection was to be made among the following bus systems: EIB, LON, BACnet and CAN.

The following reasons finally led to the decision in favour of CAN as the medium for the Geniax system.

- Comparatively low costs per communication node of about 1 €,
- EIB and LON are partially subject to license costs, leading to node costs of more than 10 €,
- Long-term availability of standardized transceivers from the automotive industry,
- The energy-saving functions of the CAN transceiver to be used for further energy saving at every individual node and
- The possibility to develop an exact proprietary protocol to meet the needs of the Geniax system.

Besides the relatively high costs for the hardware design of each communication participant, EIB, BACnet and LON already have a fixed protocol format which cannot be used directly with the Geniax system.

Not the least for this reason, a bus system was selected for the Geniax system which is adaptable when it comes to speed and protocol format and is also becoming better established in other areas of house and building automation.

After the decision was made to use the CAN bus, the transfer speed had to be determined. This would implicitly determine the maximum network expansion. By using low baud rates, a huge expansion of the topology of more than 2,000 m is possible, theoretically. This expansion exceeds the expected network expansion in

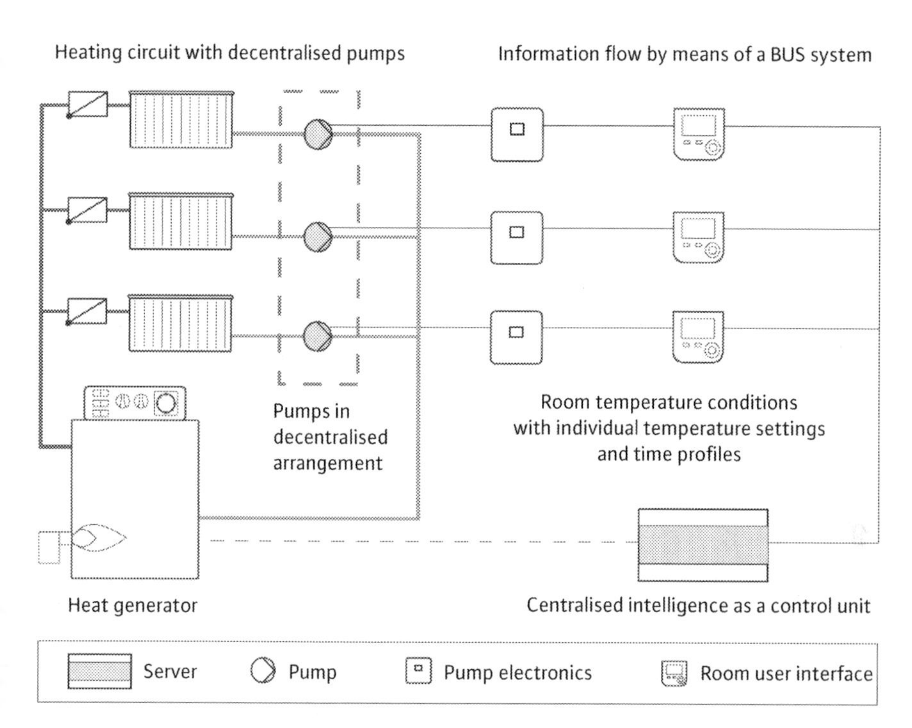

Fig. 5.23 Schematic representation of the main system components

a single- or multi-family house. Even for the installation in a functional building, such a greatly expanded system is not supportable due to reliability.

In addition, the greatly extended lines will inevitably lead to voltage drops and the associated earth offset, causing problems for reliable operation and communication between the participants linked by the bus. Beyond this, building construction factors place demands on the installation, which deviate from the linear bus topology, is specified by the CAN standard.

For this reason, the Geniax system is supplemented by the so-called bus coupler. With the help of the bus coupler, it is possible to segment a greatly extended system into logical subsegments. Every subsegment formed with a bus coupler is galvanically isolated from the upstream subsegment and has its own power supply. Due to the galvanic isolation of individual CAN segments, it is possible to implement nearly any topology without violating the CAN principles. This allows even a convoluted star topology to be done without complicated calculations of termination resistance.

This means that clever planning of bus couplers can lead to a system with higher overall availability of the existing heating surfaces than would be the case for traditional operation of a single pump in the basement.

Taking such a system into consideration, if, for example, the power supply in a subsegment fails or there is a short circuit in a particular subsegment, the subseg-

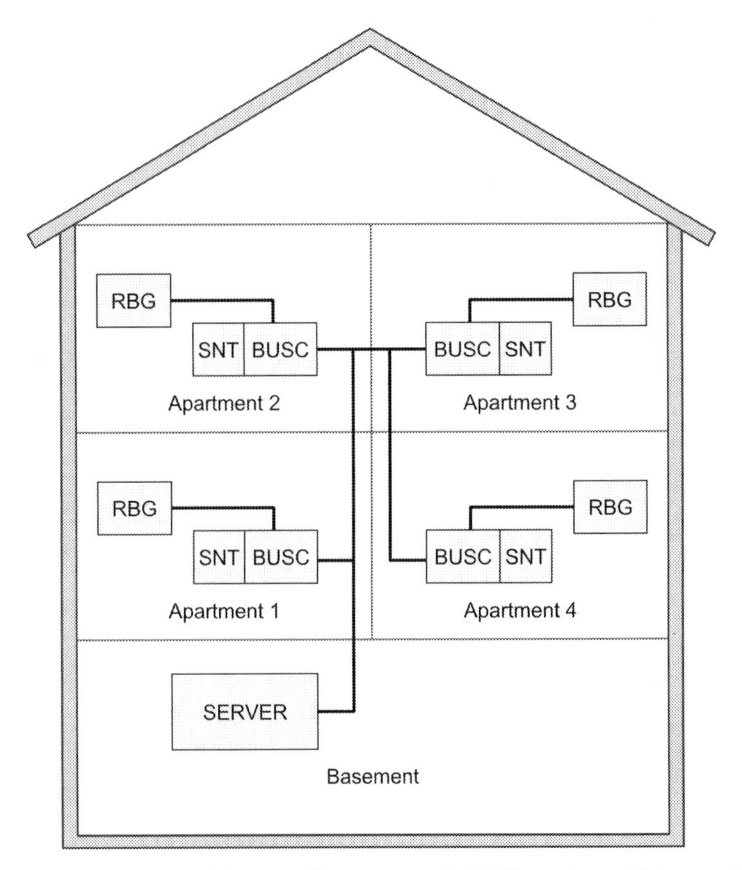

Fig. 5.24 System overview of the controller area network (CAN) topology with bus couplers

ments located upstream of this bus coupler will continue to work without a problem (see Fig. 5.24). Naturally, the Geniax server recognises that there is a problem in the system and passes this information on accordingly.

Besides the physical advantages of using the bus coupler, its protocol-level operation produces system-stabilising effects.

Since every message from every bus coupler must first be completely received before it is forwarded to upstream or downstream subsegments, CAN "Error Frames", for example, in one subsegment are automatically filtered out and, therefore, will not be propagated in the entire system.

To summarise, the following can be stated:

In order to guarantee CAN communication in extended networks with high expected load currents in particular, dividing up the network into smaller segments that are galvanically isolated from each other is a recommendable measure. It is important for the individual CAN segments that all communication nodes within a subsegment are connected to the same reference GND.

Chapter 6
Testing

Wolfhard Lawrenz, Federico Cañas, Maria Fischer, Stefan Krauß, Lothar Kukla and Nils Obermoeller

6.1 Conformance Test Methodology

Technical products are tested in different ways sometimes even already in the concept design phase, and finally while the production process is ongoing and at the end of the production itself. Various test methods are applied such as module tests, integration tests, system tests and qualification tests in order to assure that systems or system components finally provide the desired functionality and comply with their related requirements. The need for conformance tests especially is obvious recognizing the immense progress in technologies over the last decades.

In order to master complexity of current systems, modularization as well as composition of systems from standardized components, being supplied by different manufacturers, became key factors. Currently, it is almost impossible to provide all technical knowledge about a system as a whole just within one single company. The example of cars industries clearly demonstrates this fact: Almost none of the car manufacturers provide an in-house production depth of 40 %. Not only components but also more and more complete (sub-) systems are supplied by third parties. Even tier 1 suppliers integrate components supplied by tier 2 suppliers and so on.

M. Fischer (✉) L. Kukla · N. Obermoeller
C&S group GmbH, Am Exer 19b, 38302, Wolfenbüttel, Germany
e-mail: Maria.Fischer@cs-group.de

F. Cañas
Quellweg 27, 13629 Berlin, Germany
e-mail: federico.canas@gmail.com

W. Lawrenz
Waldweg 1, 38302, Wolfenbuettel, Germany
e-mail: W.Lawrenz@gmx.net

S. Krauß
Vector Informatik GmbH, Ingersheimer Strasse 24, 70499 Stuttgart, Germany
e-mail: stefan.krauss@vector-informatik.de

W. Lawrenz (ed.), *CAN System Engineering,* DOI 10.1007/978-1-4471-5613-0_6,
© Springer-Verlag London 2013

The application of standard components is reasonable because of economic reasons as their wider usage implies higher quantities and this, in conjunction with pressure from inter-companies' competition, results in reduced cost. In order to make standard components from different suppliers acceptable, they must be compatible and functionally homologous. That is, what standards shall guarantee? Unfortunately, it is not a self-fulfilling prophecy ensuring that different suppliers provide identical products, while applying the same standard—why?

The driving factor for the development of a product basically is a need for that product, being compliant to specific requirements and desired functions and properties. The specification of requirements is the first step towards a product. In a next step, a component specification is derived from the requirements and the possibly already-existing (partial) solutions, which then serves as the basis for all implementation processes. Requirements must be expressed in one way or another and from there unique definitions and description of the functionalities and their implementations must be derived.

The difficulties when specifying the requirements, properties and functions come from the limitation of resources in human imagination, human linguistic capabilities and description means. It is almost impossible to describe requirements in an arbitrary way. Furthermore, none of the product designers is able to express all of his or her ideas and describe them correspondingly in a formal language. The number of designers for a specific product mostly is very limited. As such, it is quite natural that these persons are not able to express the personal knowledge completely or they even are not willing to do so. Specifications are textual documents in most cases. This is mostly due to the fact that purely formal languages are not widely known or correspondingly needed tools are not available. Quite often, it seems to be easier to describe certain mechanisms by words than to express them in formulas and diagrams. On top of this, the application of a formal language such as Specification and Description Language (SDL) or Unified Modelling Language (UML) does not guarantee per se that such a description is understood in the same way by everybody.

When mapping ideas into text and diagrams as well as into formal languages as there are the system description languages and modelling languages, there is a gap between the author's thoughts and the reader's comprehension. Any of the requirements descriptions and any of the specifications contain certain ambiguities which may have more or less effect depending on the quality of the description and the qualification of the reader. This consequently results in different implementations, which should be avoided.

Currently standards are developed in cooperation with multiple companies. As such, the quality of the specification documents under development is checked by a consortium of persons. However, the implementation of the specified component mostly is done by single companies only. The interpretation of a specification of only a few persons consequently bares the risk that small deviations between specification of a component and its implementation may occur. This effect shall be called "interpretation deviation". It can be observed too that when regarding the definition of requirements and their specification, difficulties while implementing

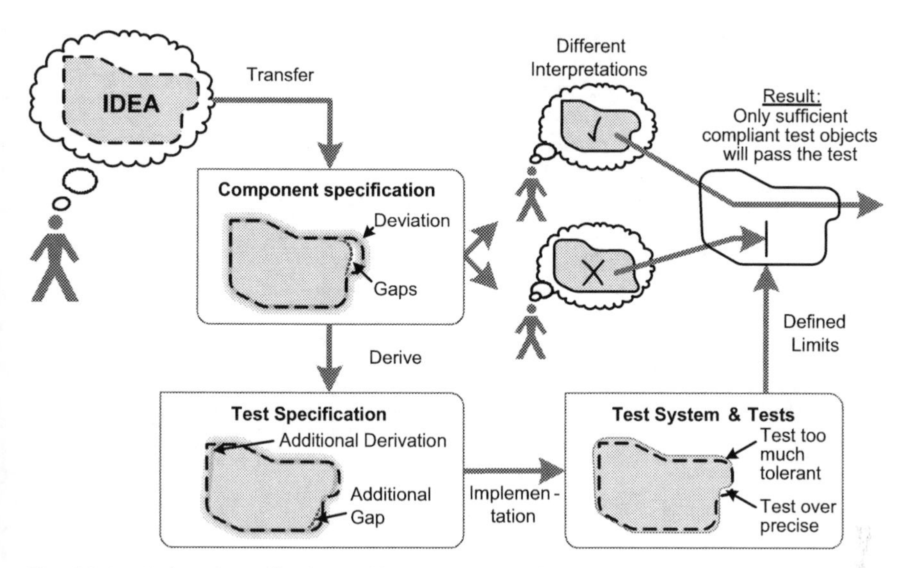

Fig. 6.1 Deviation of specification and implementation of components

certain requirements may become an important issue, referring to the previously mentioned step (see Fig. 6.1).

The implementation of component specifications may include (unfavourable) misinterpretations and "real" mistakes. If a mechanism was not understood in the same way as originally intended, it may be difficult to call it a mistake. A specification may lead to a logically correct conclusion, which, simply said, had not been intended by the designer. Nevertheless, the major goal must be kept in mind that products of different suppliers shall be sufficiently compliant and in accordance with the standard.

6.1.1 Important Terms

- Conformance

According to ISO 9000 (see [ISO9000], §§ 3.6.1, 3.8), conformance corresponds to "fulfilment of a requirement", while a mistake is "non-fulfilment of a requirement". Furthermore, this standard emphasizes testing as an appropriate means for objective proof of conformance. An implementation is proven to be conformant if it is compliant to all specified requirements.

- Conformance Tests

Conformance tests check the behaviour of an implementation against the correspondingly specified standard. These tests are functional tests, belonging to the group of dynamic test methods (see Fig. 6.2).

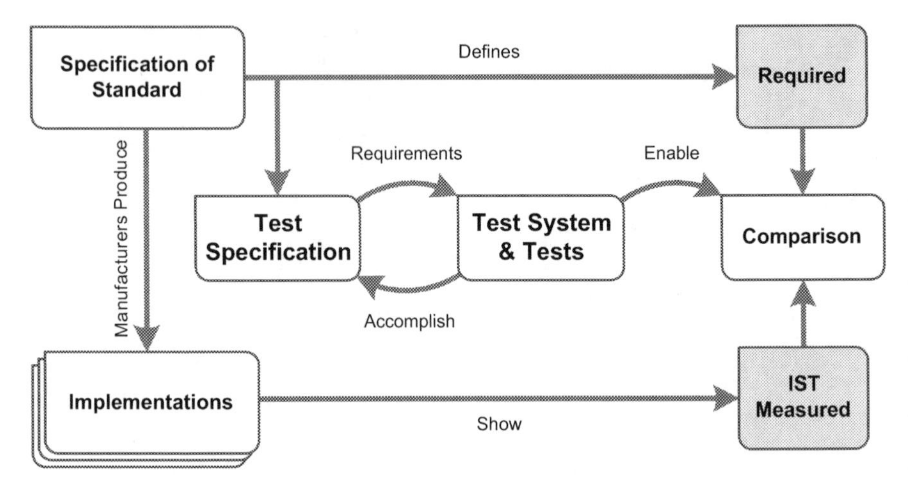

Fig. 6.2 Principle of conformance tests

- Conformance Test Specification

The conformance test specification lists test cases (TCs) as well as constraints for the tester implementation and the test execution. When generating the conformance test specification, the derivation and precise definition of the TCs are important as well as formal requirements such as the independence of the test description from any specific implementation, from test houses, as well as from any specific test environment, if not explicitly required.

6.1.2 Purpose, Benefits and Intention of Conformance Tests

Distributed systems consisting of components from different suppliers can only operate reliably if all components provide a compliant functionality in a correspondingly timely manner. Any deviation from that functionality and timing may lead to failures in a distributed system.

As such, a conformance test directly aims at the interpretation deviations as previously mentioned. However, not to be misunderstood, conformance tests too are derived from requirements, descriptions and specifications by humans. Therefore, conformance tests possibly may contain false interpretations. However, it is good to know that conformance tests typically are generated under the supervision of a group of users, cooperating companies or consortia of the related standard. Insofar, quality of conformance test specifications is comparable to the quality of the related component specifications. Particularly, long discussions take place on restrictive, demanding TCs, which, at the end, result in how the related parts of the component specifications are to be understood. As such, the precise meaning is clarified and put into a binding conformance test.

Quite often weaknesses of the component specification are discovered, when developing the related test scenarios. As such, even the process of TCs generation per se makes sense, as for that purpose especially checks of the individual functions of the implementation are done, which require a unique interpretation of the component specification. When performing a conformance test only unique verdicts are allowed, which discriminate clearly whether or not a test candidate performs as required by the component specification. Thus precisely defined limits are given, within which an implementation must perform, if it shall be compliant to the standard. The benefits of this kind of tests obviously help to discover and avoid interpretation vagueness. This vagueness inevitably would lead to problems in a system, in the event that components must be exchanged and/or implementations of different manufacturers are used in the same networked system.

Well-known companies integrating those standardized components require conformance tests as the basic precondition, if those components of a specific supplier are to be applied. This is not only the case in automotive industries, but also in most technical areas. Especially, conformance tests apply when using such components in communication-related areas such as telecommunication up to high-speed computer interfaces. Again this results from the application of components in networked systems, which are supplied by different manufacturers.

When applying tests in general and especially conformance tests, problems can be detected in an early phase and not just in the late phase of integration into the final end product. At that late phase, problems may be even difficult to discover; maybe they even can only be observed under very specific conditions. When performing conformance tests on a single component partially, very specific and detailed test scenarios are applied, while systems tests typically focus on known application cases and error cases. As such, a potentially non-standard-conform component may fail when used by the end user. This can lead to a recall action and thus cause tremendous cost for correcting that problem. Typically, before going into the integration phase, the supplier of a component himself applies various tests. However, these tests are based on the interpretation of that specific supplier. As such, these tests do not guarantee this component to be compliant to the standard (see Fig. 6.3).

For currently developed standards, the specification of related conformance tests is done in parallel. Quite often, this is done by consortia, which are put together especially for that purpose (in automotive industries, there are, e.g. FleyRay consortium or AUTOSAR consortium). This procedure makes sure that from the beginning, only components are applied which are conforming to the standard. In addition, this procedure avoids that standards must be altered in a later phase, to make them compliant to the already-available implementations. Obviously this procedure does not make sense.

- Some purposes of conformance tests
  - Proof of functions and timely behaviour of different implementations as required by the standard.
  - When developing conformance tests, the related component specification is analysed very deeply, discovering various ambiguities and problems as

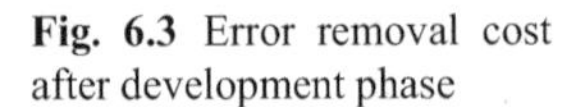

Fig. 6.3 Error removal cost after development phase

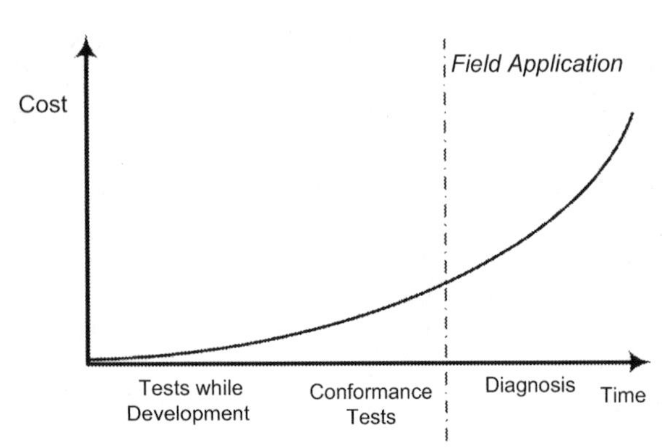

well as clarifying the original intention of its designer. Unfortunately, it is not always possible to find retroactively unique expressions for that specification. Nevertheless, the test specification describes clearly the expected behaviour and the limits within which conforming implementations must be operating.

- Conformance tests are developed in cooperation of many companies and persons. This may be possibly supervised by consortia. Clear limits between compliant and non-compliant behaviour are defined. This effectively protects against supplier-specific interpretation vagueness. Supplier tests, however, check the supplier's interpretation of a "correct" behaviour.
- Clear instructions concerning the test-system implementation and the test process ensure a standardized test execution, generating comparable results. Therefore, all implementations are tested under the same constraints.
- While applying identical evaluation criteria, a high degree of correspondence of different implementations is ensured. If the component specification describes interoperable components, then the compliant implementations very likely will be interoperable, too. However, the following paragraph explains in detail why anyhow conformance tests as well as any other test cannot guarantee interoperability.
- Implementations which do not correspond to the standard are reliably discovered. The correctness as a whole cannot be proven, but there is absence of failures in all areas which are covered by the test scenarios.

6.1.2.1 Relation Between Conformance and Interoperability

A conformance test provides evidence that the behaviour of an implementation of a component exclusively and comprehensively complies with the corresponding specification of that component. This fundamental issue is the reason for the above statement that no conformance test can ever guarantee interoperability of conforming components: Conformance tests absolutely only focus on the component specifications. As such, a component specification could define a component which is not interoperable with other components of its kind.

An undiscovered error in the component specification could result in components conforming to the specification but, nevertheless, not working appropriately. As such, the quality of implementations not only depends on the coverage and excellence of the conformance tests, but also depends significantly on the quality of the component specification itself. This specification is responsible for the definition of a component with an appropriate functionality corresponding to the system requirements. Failing this, the conforming implementations could not operate appropriately as this was a contradiction to the specification.

As for the end customer, finally interoperability is the utmost goal; they mostly require tests proving interoperability. However, this cannot be achieved by mere testing. Nevertheless interoperability tests do exist, mostly connecting various implementations under laboratory conditions resulting in the verdict “OK” or “NOT OK”, when applying specific test scenarios. The value of this verdict, however, is rather limited with respect to the applied test scenarios, as due to this concept, technically and practically it is not feasible to check any possible scenario. As such, any effort claiming to *guarantee* interoperability by testing is dubious.

Assuming a specification is properly done, in general, the risk of non-interoperable specified components is low. Consequently, the chance for interoperability is increasing with the degree of compliance between the implementation and its specification. The compliance is proven by conformance tests.

As mentioned above, the quality of the specification is supported by development of conformance tests while precisely analysing the component specification and pushing for its non-ambiguous interpretation. As such, a high-quality component specification and related conformance tests will result. The risk of non-interoperable components is estimated to be rather low, especially when findings from the development of the conformance tests are fed back into the subsequent revision of the component specification resulting in non-ambiguous statements. Unfortunately, the constructive feedback very often does not quite easily have its effect, because standards quite often rather quickly become “untouchable”.

In conclusion, conformance tests as well as any other tests cannot guarantee absolute interoperability, but they increase the chance to it.

6.1.3 Test Methods, Test Standards and Test Rules

Various test objects require different implementations of test systems, especially related to the coupling of the test object to the test system. Most of the test systems are based on the principle of accumulating findings from experience in order to come to better solutions. Many of the characteristics, excellence and advantages of existing test systems can be transferred to other test systems.

The standard ISO 9646 (see [ISO 9646]) specifies conditions and the methodology for the design and execution of conformance tests according to well-defined

Table 6.1 Typical, basic terms related to conformance tests

Abbrevations	Term	Meaning
IUT	Implementation under test	Test object—the implementation of a component, a protocol layer particularly, which is to be tested
SUT	System under test	In most cases, the IUT is embedded in a complementing system, which cannot be separated from the IUT when performing the tests on the IUT
PCO	Point of control and observation	Interface of the IUT for its stimulation and monitoring
UT	Upper tester	Entity accessing the IUT from next higher protocol layer in order to control and monitor the IUT
LT	Lower tester	Entity accessing the IUT from next lower protocol layer in order to control and monitor the IUT
TCPs	Test coordination procedures	TCPs serve for management and control of UT and LT
ASP	Abstract service primitive	Commands to UT and LT for stimulation and monitoring of specific services. An implementation-non-depending description between service client and service provider

procedures. This standard has been developed for conformance tests of layered protocol implementations which are oriented on the well-known seven layers model corresponding to ISO/International Electrotechnical Commission (IEC) 7498. Standardized test procedures and related constraints are required in order to achieve reproducible, comparable test results. For that purpose, ISO 9646 defines guidelines for the specification of abstract TC collections, and gives recommendations for the description language of tests as well as for the test implementation. Furthermore, it specifies requirements and guidelines for test laboratories and their conformance test customers and it lists further support such as how to handle alternative versions and optional characteristics of test objects and how to handle declarations of the customers on the abilities of its implementation, which are to be tested (See Table 6.1).

The ISO 9646 standard exclusively defines general rules and explicit guidelines but no specific test techniques nor test systems. Nevertheless, those definitions are applicable on almost any specific case and, therefore, to be considered correspondingly.

As a starting point, the ISO 9646 standard defines an abstract test architecture surrounding and embedding the implementation under test (IUT). Only the external behaviour of the IUT can be accessed by the tests; its internal realization is invisible—like a black box. The test object is assumed to be part of a layered architecture, the behaviour of which shall be tested on its interaction towards the borderlines to the higher and lower layers. The Upper Tester (UT) controls and monitors the interface to the next higher layer. The Lower Tester (LT) does the same on the next lower-layer interface. These interfaces are called the Points of Control and Observation (PCOs). UT and LT are both responsible for the translation of the Abstract Test Services (ASP) into real commands for the IUT. They are both controlled by the Test Coordination Procedure (TCP) and they both communicate back the data received from the PCOs to the TCP. Comparing the monitored data with the reference

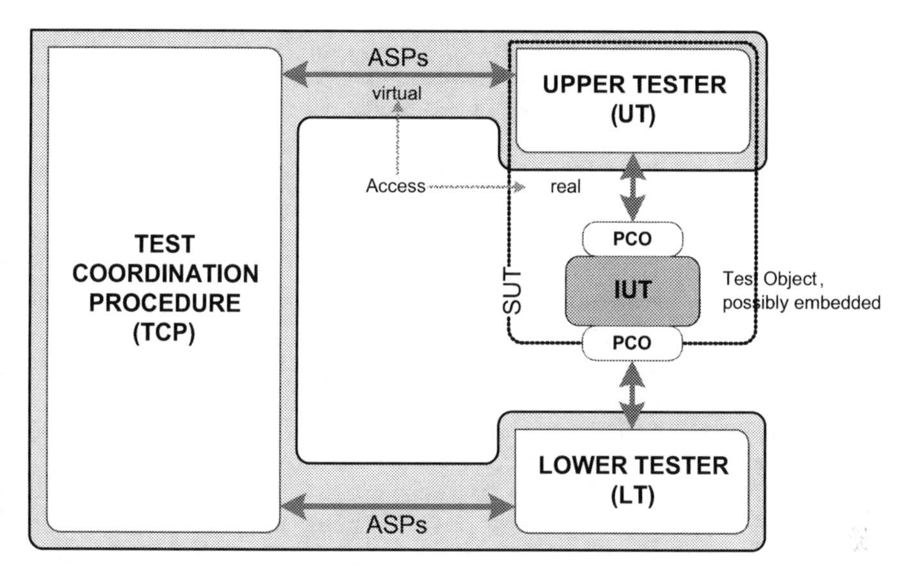

Fig. 6.4 Generic test-system architecture according to ISO 9646

data results in the test verdict in the TCP. All data are stored for the purpose of later evidence of correct test execution. There are some test-system implementations in which the UT operates within the system under test (SUT) virtually as an embedded system (see Fig. 6.4).

In addition to the generic tester architecture ISO 9646 specifies various conditions and requirements for conformance tests and the corresponding specification. These criteria are of major importance for the quality and significance of conformance tests. Some of the most important ones are listed below and their meaning is explained:

- The test description language must be human readable and preferably computer executable. For instance, *Test and Test Control Notation* (TTCN) is such a recommended language for which there are compilers and interpreters which allow execution of test descriptions on computers. The representation is abstract; it is not real time executable. When applying current Personal-Computer-technologies (PC-technologies) specific operating systems, a time resolution in the range of some milliseconds can be achieved, unfortunately, quite often, with a large jitter. Alternatively, human-readable script languages are applied for configuring real time capable measurement devices. Resolution and accuracy of this methodology only are limited by the characteristics of the applied measurement devices. As such, high-precision data acquisition and corresponding analysis of timely behaviour of the test object are feasible.
- The test specification shall be independent from the test object and the applied specific test-system implementation. The test methodology shall be defined in an abstract way. The test methodology describes, among others, which interfaces of the test object must be served and whether there is a single test object only or

a multitude of networked test objects. The tests themselves shall be subdivided into single abstract test steps. A test step, for instance, specifies the generation of a specific test stimulus or the check for a specific response of the test object.
- It shall be feasible to organize the individual TCs into groups which are oriented on the features of the test object. It is recommended that a test suite shall allow the selection of single tests or groups of tests depending on the test object.
- Clear and detailed specified test sequences in conjunction with non-ambiguous evaluation criteria assure comparable and repeatable tests. Tests always must start at a well-defined initial state. The kind of tests must be identical for all IUTs. A repetition of tests must issue the same results.
- Any specific capability (option) of a test object must be checked whether it is implemented and if so its functionality must be tested. Any capability which, according to the supplier, is supported by the test object must be checked by the conformance test. Options which are not supported are not tested and consequently they are not confirmed in the test report. None of any specified specific options shall be evaluated as "not supported" retrospectively if they do not meet their functional requirements in the prior test. Any of the obligatory features and the ones which are assured by the supplier shall be checked. In case of a failure of a single test, the overall test verdict is "fail". The individually failed tests must be listed in the test report. The supplier of a test object specifies which functionality is supported by his implementation. The supplier declares the conformance of his implementation to the components specification in the so-called conformance declaration. The conformance declaration quite often is structured in the form of a questionnaire listing all the options which are specified by the standard. The conformance test correspondingly confirms or contradicts the conformance declaration.
- All TCs shall be specified in an abstract and configurable way in order to ease testing of test objects options. This allows, for instance, optional signals to be stimulated, monitored and evaluated only if they had been implemented in the specific test object. As such, the same tests are suitable for test objects with minimal and extended functionality.
- The chosen test method and the process of controlling the interfaces of the test object must adequately correspond to the technical conditions.
- Test results must be non-ambiguous and reproducible. The test verdict shall be limited to "pass" or "fail" only. No other verdict grades are allowed. Reproducibility includes the existence of a human-readable log file, which memorizes the exact test sequence with its stimulation and reaction of the test object.
- When testing according to ISO 9646, if possible, each single protocol layer shall be separately handled from its adjacent layers in order to limit the efforts to a reasonably low complexity. Insofar, test systems quite often are optimized and adapted to a specific component or protocol layer of a communication system correspondingly.

6.1.3.1 Testing Technologies and Their Implementation

Any actions which are to be performed by the UT and LT must be predefined in the test description concerning their performance and timely behaviour. The test coordination fully depends on the test description and its resulting execution. In some cases, a simple sequence of the specified actions is sufficient. Time-critical actions obviously require the specification of the points of time by when those test steps are to be executed.

In general, there are various ways to perform tests. Typically there is the execution of the test description at runtime as well as its separation into a preparation phase, test runtime phase and a post-processing phase. Both of these methods have the characteristic advantages and disadvantages. The runtime execution, for example, allows specific reactions on the behaviour of the test object—decisions can be taken at runtime. However, the second method with its preparation phase is the only one which is real time capable. In the preparation phase, the complete TC is read and all of the participating entities of the test system are initialized and configured according to the test sequence. After the test starts, the test runtime is executed in real time. But in this case, the correct behaviour of the test object must be entirely predictable, because typically a dynamic reaction on the behaviour of the test object is not feasible. In most cases, this is not required, because according to the specification of the test object, its behaviour on a specific stimulus is predefined. TCs must be defined in such a way that only a unique reaction on a test scenario is the correct one. Whether a test system, in general, can react dynamically or whether it is real time capable does not depend so much on the choice of the test description language but more on how the test execution is performed. Presumably, nevertheless, response time and the sequential order of reactions monitored at various interfaces of the test object can be of interest. Therefore, the application of a non-dynamic but real time capable and thus fully deterministic test method may be recommended.

6.1.4 Functional Tests, Timing Behaviour and Performance

Conformance tests are functional tests. As such, the false conclusion seems to be quite likely that in such tests only checks on the expected reactions on distinct stimuli are sufficient. However, specifications quite often specify timing relations too, which must be met as well by conforming implementations. As such, reactions are specified and to be checked not only with regard to their content but also with regard to their timing behaviour.

Distributed systems require synchronous actions although their various members have different time bases. These synchronous actions, for instance, are required for distributed control execution and for a synchronous state change of the whole network. Otherwise, if there are problems with regard to the synchronous behaviour, for instance, the stability of the distributed application may be disturbed and subsystems may drop out or a desired state cannot be reached, such as a sleep or power-down state. The reasons for such failures may come from any one of the hardware or software layers of the communication system. Therefore, testing of all involved components is a condition sine qua non. Especially, non-widely applied and non-standardized software may lead to problems because independent tests do not exist which may have discovered any misinterpretation of the original standard or the specification correspondingly. Quite often, the problems are due to the design and not (only) due to the implementation. If a supplier performs tests of his implementation against his design on its own, the supplier is not being able to discover such kinds of problems, because he or she performs tests on the implementation but not on the design specification. Standardized solutions are less vulnerable, as more commonly conformance tests are applied for quality assurance.

Synchronization between UT and LT is required when timing measurements are to be performed. Furthermore, more precise and complex test scenarios can be performed, if a common time base at all interfaces of the test object is available. Without that common time base certain features even cannot be tested. Tests executed on the target hardware give a clear example for that situation, because actions onto the software interface of the test object are required to be correlated with test messages on the bus.

Particularly the software of the driver layers always must be tested in conjunction with the corresponding target hardware. There are evident reasons for that. The interface between driver and target hardware is not standardized and as such would require complex efforts for interfacing the IUT to any one of the test systems. However, the interfaces of the drivers to the higher software layers (the offered services) as well as the bus itself represent the nearest standardized and adjacent accessible interfaces which are independent from any corresponding target hardware.

For synchronized test scenarios, it quite often turns out to be a good solution to install the common time base for UT and LT within the LT itself. The LT provides the well-known and constant performance as compared to the host microprocessor of the target platform on which the UT is executed. For highly accurate measurement tasks, the UT can be stimulated by distinct trigger signal to execute actions at predefined points of time. However, as mentioned above, this requires the predefined specification of the desired actions at the occurrence of the next trigger event which causes problems for the interpretation of the TC description at runtime. Fully deterministic, sequentially and timely predictable test scenarios can only be achieved by a test runtime phase in real time as well as corresponding preparation and post-processing phases. In the preparation phase, the corresponding measurement devices are initialized and the desired message and control data files are set up, which are executed interrupt-free at test runtime and thus at distinct speed.

Adaptation efforts of a test system to all test candidates of a kind are the same if those test candidates provide unique and standardized interfaces. As such, tests not only require comparably low set-up efforts, but also require top benchmarks, e.g. data throughput over time and reaction times can be explicitly measured. Thereby, implementations of various suppliers can be compared.

6.1.5 Verification and Validation of Test Systems

Test systems are implemented following the requirements given by the tests to be executed as well as by general requirements specified for the environment of the test object. From these requirements, functions or features of the test system are derived which must be implemented correspondingly. Most important for trustworthiness of test-system implementations is that all functions of the test systems are checked on functional correctness, such as communication with the test object, execution of TCs, derivation of the test verdict and logging of those data.

This is done in the verification phase of the test system, which should be formal and carefully planned as well as correspondingly well documented. Any one of the functions should be tested individually. Especially the decision thresholds for automatic evaluation methods must be checked, because they directly influence the test result. However, the generation of individual stimuli for either software or hardware interfaces also must exactly match the expectations. The verification process proves that all test means, all "test tools" operate reliably and as expected. Therefore verification is the precondition before applying a (verified) test tool on a component/module/system which is to be tested.

Not only is the test system itself requested to operate correctly and to accomplish its intended purpose, but also the TCs, which are executed on the test system, must be applied appropriately. At this point an important issue follows, which is the check whether the test system as well as the implemented tests together are well suited for their operational range. Along with its verification it was proven that the functions of the test system work as anticipated. Now the focus is on its intrinsic value, plausibility and applicability. These features are checked in a validation process.

As widely known TCs result from appropriate combinations of various test functions for stimuli generation, monitoring and evaluation. It must be proven that this is the case for any one of the implemented TCs. In particular it must be checked whether the combined test functions are applied in an appropriate order in conjunction with the correct parameters and the correct timely context, in order to implement the given test scenarios of the test specification. The correct stimuli must be applied at the correct point of time to the correct interfaces of the test object, as well as all of its relevant interfaces must be monitored meanwhile in the predefined time window. Although, known from the test scenario and component specification, the correct reaction of a conforming implementation always is predictable, but as however the chances for erroneous behaviour practically are indefinitely high, test runs must be independent from the correct reaction of the test object or, more generally, test runs must not be dependent on how the test object may react. Otherwise, in case of a failure this could result in blockades and thus disturb the test process.

A test is validated if it is suited to prove that testing of the requirements of the test object is accomplished. A sequence of test functions definitely may be inappropriate to prove a distinct requirement, equivalently to a behaviour. As such, a test is useless if, e.g. for checking the error-detection feature of a test object, the timely sequence of test steps reads the error memory of the test object just before the error is stored. According to the component standard, the test object ought to store the error in the error memory, but this cannot be accomplished anticipatorily. Obviously, first the error must be stimulated and then subsequently the error memory can be read in order to perform the error-detection check. Additionally the entry must not be persistent.

A simple trial-and-error process to check whether a TC with a test pattern can be successfully applied is not suited for a validation process. If, as described above, a test does not adhere to an appropriate sequence of test steps or initial states are not checked, tests erroneously may lead to a "fail" or "pass" verdict. Referring to the example, a standard conforming implementation erroneously would result a "fail" only because the error memory would not yet have stored the expected error when reading the error memory. Vice versa, there are cases in which wrong implementations may erroneously be judged as conforming implementations. Referring to the above example, this would be the case, if the erroneous test object would permanently indicate an error and the tester would not check the—wrong—initial state.

A validation must prove that such kinds of errors do not exist and that only a conforming behaviour of a test object results in a "pass" verdict. In order to meet the process requirements of ISO/IEC 17025 (see [ISO17025]), a technically appropriate and fully documented validation of a test system is a condition sine qua non, which nevertheless corresponds to a rather high quality level.

6.2 CAN Transceiver Conformance Tests

This section addresses the testing of controller area network (CAN) transceiver. They are a part of the CAN physical layer. The general standardization of CAN has already been described in earlier sections. In Sect. 6.2.1, first the standardization of physical layer of CAN will be discussed. Section 6.2.2 gives an introduction to why it is necessary to submit systematic tests to the CAN transceiver before their use in real applications. Subsequently, the test idea underlying the tests is presented in terms of test method and test principle in Sect. 6.2.3. The structure of the test system is described in Sect. 6.2.4. At the end of this section, the TCs and their focal points for different CAN transceiver implementations are described in Sect. 6.2.5.

6.2.1 Standardization of the Physical Layer of CAN

For a general history and standardization of CAN, see the sections before.

The parts 2, 3 and 5 of the ISO 11 898 are the relevant standards of the CAN physical layer and are briefly introduced below.

ISO 11898-2: *High-speed, medium-access unit* The standard ISO 11898-2 is the most implemented standard of CAN physical layer. It describes the functional[1] and physical[2] interface to the transmission medium. Also, the specification of the

[1] Functional interface: Medium-Dependent Interface (MDI).

[2] Physical (electrical, optical) and mechanical interface: Physical Medium Attachment (PMA).

physical transmission medium (bus) is part of the standard. The defined data rate is a bit rate of up to 1 Mbit/s.

ISO 11898-3: *Low-speed, fault-tolerant, medium-dependent interface* This standard was developed as part of Generalized Interoperable Fault-tolerant CAN transceiver (GIFT)/International transceiver conformance test (ICT) projects (see 9.2.3) and also defines the functional and physical interfaces to the transmission medium as well as the physical transmission medium. The data rate is defined up to a maximum of 125 kbit/s. The standard also specifies a fault management to implement a fault-tolerant behaviour.

ISO 11898-5: *High-speed, medium-access unit with low-power mode* The standard ISO 11898-5 is an extension of the standard ISO 11898-2.

The extension consists mainly in the description of the behaviour in the low-power mode, or in the description of transitions between the possible modes of operation. Implementations according to this standard (ISO 11898–5) can be used in conjunction with implementations according to the standard ISO 11898-2 in one network.

6.2.2 The Need for CAN Transceiver Testing

Section 6.1 has already given an introduction to the topic of testing. In this section, this is specifically discussed and why it is necessary to test CAN modules and—as part of them—CAN transceiver.

6.2.2.1 Use of CAN in Complex Bus Systems

Once bus systems were introduced in luxury automobiles, they have made their way into all classes of vehicles in the automotive industry. Here CAN is the most important and most widely used serial bus system. Almost every car today has at least one CAN network on board.

CAN is used not only in automobiles but also in industrial control systems, control systems of ships, trains and airplanes, in plants of agriculture technology and building automation, medical technology and renewable-energy systems.

The bus system is thus used in areas where high reliability and fault tolerance are required. Within the bus system, CAN bus modules with different functions and from different manufacturers are linked together via a common data line.

Those CAN bus modules have to work properly due to the application in safety-relevant areas and meet all requirements which are placed on them. The requirements are defined in detail in the standard ISO 11898 (see Sect. 6.2.1). Core of the specification is the data exchange between modules.

If each CAN module—irrespective of the manufacturer—which is involved in a CAN network is compliant with the standard, it can therefore be assumed that the components work together void of errors and the network as a whole meets its duties.

The first step on the way to a desired interoperability of CAN modules from different manufacturers in a CAN bus system was so taken with the publication of the standard ISO 11898.

6.2.2.2 Risks of Complex Bus Systems

Due to the use of CAN in exceedingly complex and safety-related systems, a use of CAN modules without prior checking is not feasible. Several risk factors can restrict the proper communication between the CAN modules in such a significant way that the entire bus system is no longer able to fulfil its function:

- *Faults in the implementation:* Standardized interfaces are only the "outside", but not the realization of the implementation. By implementing this freedom, can faults be built into a device, at which the blur of interpretation[3] has an important part? Especially those errors which lie in the blur of interpretation may not be recognized in the accompanying development tests. Errors in the implementation are often only seen when the module is embedded in its real environment with all its influences and interactions.
- *Interdependence of a module with its environment: The interdependence of a module with its environment is one of the most common sources of faults in a bus system.* The networks themselves and their tasks are characterized by a very large and still growing complexity. Therefore runs the number of possible events, event sequences and combinations of events that result from the interaction of the various components and the effects of component and network environment, gets very large indeed. Thus, the probability is very low that in the specification of the module all requirements are defined, which result from later applications.
- *Gaps or errors in the underlying implementation of a standard:* Information about the first publications of the ISO 11898 standards can be found at [CIA10] the citation "[….] the ISO standard 11898 for CAN was published in November of 1993. [….] Unfortunately, all published CAN specifications and standardizations contained errors or were incomplete". These early times are long gone, yet gaps can occur in a standardized specification or subsequently become relevant. During the development of a specification, the boundary conditions of the subsequent use of the object of specification are often not considered or considered just insufficiently. The reason for this lies mainly in the fact that not all later applications are known and can be known. The type and depth of the aforementioned interdependence of a module with its environment are often ignored when creating the specification. The development of a test specification is therefore always a kind of "test for completeness" of the relevant product specification.

[3] According to Sect. 6.1, *blur of interpretation* means: in the interpretation of a specification for the purpose of implementation, small deviations between the intended content of specification (goal of the author) and implementation (interpretation of converter ligands) can occur.

6.2.2.3 Error-Poor Systems by Testing

In considering the above risk factors for buses, it is clear that errors in newly developed CAN modules are very probable, despite the underlying ISO standard 11898.

This results in a high risk potential with possibly dramatic consequences. This risk potential can be reduced by detecting the errors before the modules are used in their real environment.

The solution is therefore *testing*: The indispensable method for locating errors in the modules is the implementation of systematic and elaborate tests.

A CAN module consists of several layers in the sense of the ISO/Open Systems Interconnection (OSI) model (see Sect. 6.2.1). To determine the origin of a detected error, the layers are each considered and tested in isolation.

This section describes the CAN transceiver as part of the physical layer of the CAN modules. Sought are a test idea and a test system with which implementations of the CAN transceiver can be tested to ensure that as many errors as possible are found: "Testing is the execution of a program with the intent of finding errors".

6.2.3 The Concept of Testing the CAN Transceiver

The need of testing the CAN transceiver is indisputable according to the statements of the previous sections. However, with an increase of only representative tests, following an appropriate test concept, the probability of error-poor systems increases. Test methods and test principles must be chosen in a way that the implementation of the tests minimizes the effects of the risk factors listed in Sect. 6.2.2. In 1999, a group of semiconductor and automobile manufacturers with partners started projects in the field of the physical layer of CAN to develop, among other things, an appropriate test system for the physical layer of CAN.

6.2.3.1 GIFT-Project and ICT-Project

The use of bus systems in fields where high reliability and fault tolerance are required requires as a solution not only the redundancy[4] which is not always feasible, but also, in particular, the fault tolerance. A fault-tolerant product fulfils its basic functionality with reduced performance in case an error occurs.

The lowest layer according to the ISO/OSI model is especially vulnerable. Cable breaks and short circuits can have a momentous effect on the functionality of the lowest layer and ultimately on the whole system.

The need for a fault-tolerant bus system led to the development of a low-speed CAN transceiver that includes a fault-tolerant concept and is able to automatically

[4] The principle of redundancy is the following: If a system fails, another one is available that takes the tasks of the failed system.

Table 6.2 GIFT/ICT: Partner

Automotive manufacturer		Semiconductor manufacturer	Chairman
Audi	BMW	Freescale	C&S group (Testhouse)
Daimler	Ford	Infineon	
PSA	Volkswagen	Philips[a]	
		STMicroelectronics	

[a] Today: NXP B.V.

match line faults. Thus, together with other CAN controllers, fault-tolerant bus systems are built. The concept of this transceiver was well received.

A working group was set up, which had set itself the task of developing in two projects a general specification of a fault-tolerant transceiver and the tests for this transceiver. The working group is composed of automotive and semiconductor manufacturers. Chairman of the working group is still the C&S group (see Table 6.2).

- The project GIFT had the objective of specifying and standardizing a fault-tolerant CAN low-speed transceiver to ensure interoperability of CAN modules with transceivers from different manufacturers. The specification is now published as ISO 11898-3 standard (see Sect. 6.2.1).
- Within the project ICT, the conformance tests[5] for this standard were developed. The C&S group,[6] one, at this time, already-established test house for CAN chips, in cooperation with the GIFT/ ICT partners, was responsible for the specification of TCs and their implementation in a test system. The test specification was published in 2001 under the title *International CAN Low-Speed Transceiver Conformance Test.*

The group continued to work after completion of the projects, and expanded the standard ISO 11898-2 to ISO 11898-5 (see Sect. 6.2.1).

In this standard also, the C&S group was responsible for the specification and implementation of TCs, in cooperation with the GIFT/ICT partners. The specification is titled *CAN High-Speed Transceiver Conformance Test.*

Test Method and Test Principles

At the beginning of the development of a test specification and a test system, decisions must be made about the principle way of implementation in terms of test method, test setup and determination of the TCs.

[5] A conformance test verifies the conformity of the features of a test object with the requirements that are put to the test object. These requirements are described in the specification of the test object.

[6] At that time, a group within the University of Applied Sciences Braunschweig/Wolfenbuettel; today: C&S group GmbH.

6.2.3.2 Conformance Test

The CAN transceiver tests are implemented as a conformance test (see Sect. 6.1). A CAN transceiver is part of a CAN communication module and is therefore working inherently together with transceivers from other CAN communication modules, in one CAN network. Each CAN transceiver is at any point in a particular state and changes in reaction to network events in a different state. The CAN transceiver has to go through a change of state or a sequence of changes according to the underlying part of the standard ISO 11898.

Whether this change of state occurs, and whether it occurs at all CAN transceivers at the right time window, is checked by conformance testing.

The CAN transceiver is initialized and stimulated according to the possible states and events in a CAN network. The reactions of the CAN transceiver are observed and the observed change of state is compared with the expected change of state. A match of observed and expected change of state indicates conformance of the CAN transceiver with its specification for the item tested with the particular TC. Conformance tests are aimed at minimizing the risk factor, *Error in the implementation* mentioned in Sect. 6.2.2 (see also Sect. 6.1).

6.2.3.3 Black-Box Test

The standard ISO/IEC 9646 *Information Technology—Open System Interconnection—Conformance Testing Methodology and Framework* describes a general approach to test an implementation for conformance to the specification. Conformance testing by ISO/IEC 9646 aims to increase the probability that different implementations of the protocols and interface services of the ISO/OSI layers can work together. They should be "interoperable".

The internal structure of the implementations is not important as long as they meet the standardized "outside" requirements.

The tests of CAN transceivers are therefore performed as so-called black-box testing: The transceiver is seen as a "black box"; the internal structure is not known and is not considered.

Whether the interfaces to the outside meet the requirements on the functionality of the CAN transceiver, which are specified in the standard ISO 11898, is tested.

6.2.3.4 Network Test

The CAN transceiver is a communication module and in the standards for the CAN transceiver, a minimum number of CAN modules for each CAN network is defined. Hence, the focus in the design of the test is the consideration of the common communication behaviour of several CAN transceivers in a CAN network.

For that reason, the conformance testing of the CAN transceivers is performed in a network that consists of a defined number of CAN nodes[7]. Thereby, the behaviour of each individual CAN transceiver is observed and compared with the behaviour expected of it. At the same time, however, the behaviour of the CAN transceiver in its entirety is considered in a network. The TCs are determined such that during the testing each CAN transceiver communicates with the other.

The respective maximum specified in ISO 11898 is chosen as the number of considered CAN transceiver.

The structure of the network is specified in detail as a so-called default network in the test specifications for CAN transceiver. The definition of the standard network was developed under the ICT project and takes into consideration the realistic and relevant conditions of use for CAN modules.

There are two types of network tests.

- *Homogeneous network:* For homogeneous network tests, the standard network is equipped with CAN transceivers of only one manufacturer.
- *Heterogeneous Network:* In a CAN network, CAN transceivers from different manufacturers must communicate as error poor as possible. By that reason, for heterogeneous network tests CAN transceiver from different manufacturers are considered in their joint behaviour. When equipping the standard network, the positions for the CAN transceivers from the different manufacturers are defined in the test specification.

The decision to carry out conformity tests as network tests, in conjunction with a systematic TC determination, impacts a significant reduction of the risk factors listed in Sect. 6.2.2, *entanglement of a module with its environment.*

6.2.3.5 Test Case Determination

It has already been mentioned that a test object, must not only demonstrate that it meets the functionality defined in its specification, but also show that it meets the specified functionality even under conditions that correspond to its later field of use.

Ideally, conditions and events that describe the environment of a test object in a broader sense are listed in the specification. In practice, this is unfortunately often not the case. The specification of the CAN transceiver and the associated test specifications, however, were developed or revised in cooperation with users (automobile manufacturers), manufacturers (semiconductor manufacturers) and a test house. Therefore, it can be assumed that the determination and compliance with real application conditions is well done, because the group members have different perspectives and at the same time a wealth of experience.

In practice, the test of a product can generally only be a selection from a very large amount of possible TCs. This clearly shows that a systematic test case deter-

[7] Another reason is the difficulty to satisfactorily emulate the bus–side interface of the CAN transceiver with its complex dynamic impedance and time response by appropriate generators.

Table 6.3 Vectors and subspace vectors of the system operation vector space for CAN transceiver

SOVS for CAN low-speed transceiver conformance tests:

{system configuration}x{communication}x{power supply}x{ground shift}x{failure}x{operational mode}

System operation vector	Under vectors
{System configuration}	{Baudrate}, {termination}, {topology}, {number of nodes}, {composition}, {environmental conditions}
{Communication}	{Bus communication}, {identifier}, {data}
{Power supply}	(Parameter in vector *power supply*)
{Ground shift}	(Parameter in vector *ground shift*)
{Failure}	{Single bus failure}, {double failure recovery}, defined failure position
{Operational mode}	(Parameter in vector *operational mode*)

mination as a basis for a relevant selection of TCs is of major importance. The basis for such a systematic TC determination is the method of the system operation vector space (SOVS). The SOVS goes—among others—back to an idea of Prof. Dr. W. Lawrenz, and forms the basis for an effective TC determination. In the SOVS, each condition and each event are represented by a vector, which must be independent. With this kind of representation, all possible TCs can be determined if the associated parameters are assigned to the vectors and if then appropriate variations between the parameters are made—assuming that all conditions, events and parameters were recognized.

The subsequent selection of relevant TCs from the set of all possible TCs uses techniques such as equivalence class analysis[8] and boundary value analysis[9] and logical thinking. A very important tool is the insertion of experimental values.

Exemplary Table 6.3 shows excerpts from the SOVS for the CAN low-speed transceiver. The parameter assignment can be found in the test specification for each CAN transceiver.

The systematic TC determination is of importance for minimizing the risk factors which are listed in Sect. 6.2.2. In applying the SOVS method, the environment of the object under test is considered in real-use situations. Therein lays the great opportunity to discover, even gaps and errors in the specification of the object.

6.2.4 *The Test System for CAN Transceiver*

It was both realized: a test system for CAN low-speed transceiver as well as a test system for CAN high-speed transceiver. The basic design of these test systems,

[8] The stimuli, to which a test object is to be exposed, are divided into a finite number of classes. From the values within each class, it is believed that they deliver an equivalent result, and therefore, only one value from each class must be considered.

[9] Extension of the equivalence class analysis: the limits of the classes and the values directly above and directly below the limits are considered.

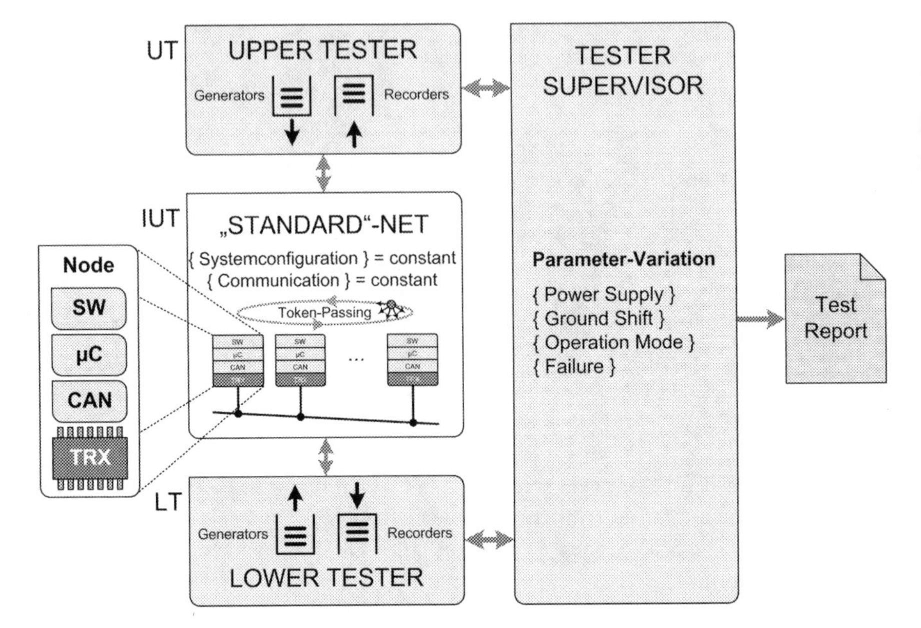

Fig. 6.5 Network tests of the CAN low-speed transceiver

however, is very similar. Therefore, in the following description of the test system it will be spoken from the viewpoint of the test system only. The differences are mainly in the parameterization of the individual components of the test systems. The parameterization can, in detail, be found in the test specifications (see Sect. 6.2.5).

Following the test methods defined in the standard ISO/IEC 9646 (see Sect. 6.1) as the Local Test Method is the method of choice for the testing of the CAN transceiver. The implementation of this method is illustrated in Fig. 6.5 by the example of the CAN low-speed transceiver test system.

Using the standard terminology of the ISO 9646, the components of the test system can be described as follows:

6.2.4.1 Standard Network (IUT)

The tested object is called IUT and is embedded in the SUT. Considered as the IUT, it is the entirety of the CAN transceiver in a standard network. In this entirety, however, simultaneously each single CAN transceiver is stimulated and observed. The properties of the standard network are specified in detail in the test specification for each CAN transceiver.

The standard network consists each of a defined number of CAN nodes.[10] Each node is designed the same way and is in particular concerning bus interface and data

[10] The number of CAN nodes in the standard network is different for the testing of CAN low-speed transceivers and high-speed transceivers.

communications defined exactly. In addition, the structure of the standard network is specified by further definitions:

- The types of CAN lines, topology and cable lengths are defined.
- During testing, the standard network nodes communicate with each other. In this, each node communicates with each node. The type of communications and content of the messages communicated are defined the same way as the order and the time delay, in which each node sends messages.
- During several tests, errors are generated and the behaviour of the transceivers is checked out. The position of the errors in the default network is defined exactly.
- In some tests, a ground shift must be generated; realization, position and value of the ground shift are defined.

Details of these definitions can be found in the test specification for each CAN transceiver. Figure 6.6 shows schematically the structure of the standard network using the example of the CAN low-speed CAN transceiver testing.

6.2.4.2 Upper Tester

Via the—according to the layers of the ISO/OSI model—"upper" interface of the standard network (IUT), the UT manages and controls the standard network with the therein-contained nodes and thus has the function of the next higher layer.

The UT offers services[11] that allow the selection and initiation of different types of ring communication or of single messages. The status of communication and the standard network can be queried. The UT initiates and monitors changes between the operating modes of the CAN transceivers and allows the readout of these modes.

6.2.4.3 Lower Tester

At the "bottom" interface of the standard network the LT undertakes the function of the next lower layer. The LT controls the technical peripherals within the test system. The services of the LT allow the generation and control of various defined errors and of the ground shift. The supply voltage can be controlled and, for example, a resistor decade for the defined application of ohmic bus loads can be switched. Also the control of the measuring devices takes place via the services of the LT.

6.2.4.4 Supervisor

UT and LT are controlled and coordinated by the TCP (see Fig. 6.4). The TCP is in the test system for CAN transceiver realized as the so-called *Supervisor (SV)*.

[11] Details of these services can be found in the test specification for each CAN transceiver.

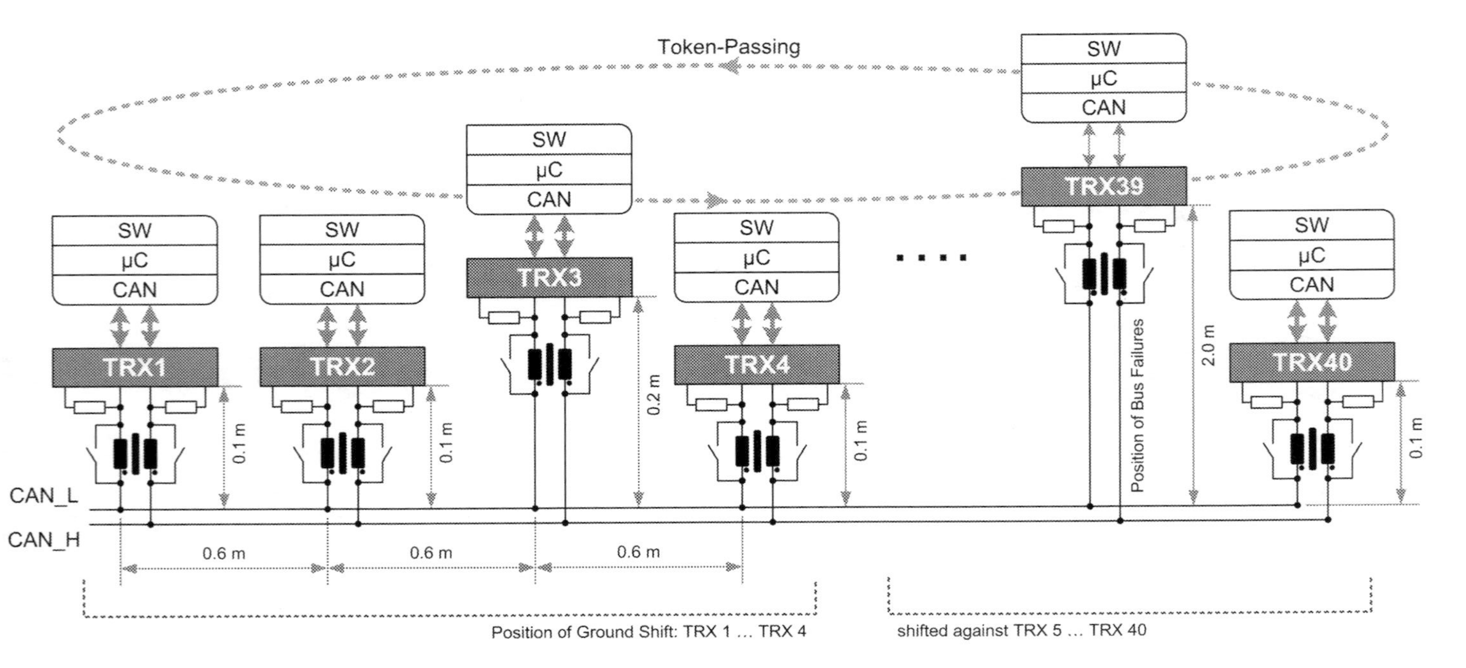

Fig. 6.6 Standard network for CAN low-speed transceiver conformance tests

The SV coordinates the entire test sequence from the initialization of the test system to the generation of the test report. The individual testing procedures are controlled by using the LT and UT and the measured test data are stored. The Supervisor compares the stored test data with the expected values and generates a list of errors and deviations and at last the test report with the relevant information and the corresponding test result.

6.2.5 *The Test Cases of CAN Transceiver Tests*

This chapter describes the test flow followed by an overview about performed tests in CAN Low-Speed and CAN High-Speed area. Afterwards, the so-called CAN ISO Tests will be mentioned.

Besides, a detailed test report and also result files concerning communication behaviour, measured values and diagrams can be provided. In case of fail behaviour it will be separately documented in the *Problem History* of the report. In addition, the customer will receive the *Authentication* sheet: one hard-copy document containing the scope of tests and the rating on one page.

6.2.5.1 The Test Flow

The test execution is defined in the test specification and is always the same:

- *Initialization:* The standard network and the CAN transceiver under test will be set to a defined state according to the test specification.
- *Stimulation:* The standard network and the CAN transceiver under test will be exposed to special events/stimuli according to the test specification.
- *Recording and evaluation/verification:* The behaviour of the standard network and the transceiver under test will be recorded. After test completion, the result data will be compared with the expected behaviour and evaluated.
  - *PASS*: The observed and recorded behaviour matches the expected behaviour.
  - *FAIL*: The observed and recorded behaviour does not match the expected behaviour.

The test execution is fully automated which gives some advantages:

- At the test execution and evaluation, the human error will be reduced.
- Because of the automatic test settings, each TC can be repeated in a defined way and is independent of test executor and test conditions by accident.

Each TC performed within these CAN transceiver tests is autonomous and can be performed as single TC.

6.2.5.2 Test Cases of CAN Low-Speed Transceiver Tests

The CAN Low-Speed Transceiver tests are based on the following specifications:

- Transceiver specification:
  - ISO 11898-3:2006 Road vehicles—Controller area network (CAN)—Part 3: Low-speed, fault tolerant, medium-dependent interface
- Test specification:
  - International CAN Low-Speed Transceiver Conformance Test, Test Specification V1.5

6.2.5.3 Behaviour at Short Circuits

The main part at CAN Low-Speed transceiver tests consists of short circuit tests applied to the bus lines. The reason for that is the *Fault Tolerance* of this transceiver type: Every time the short circuit voltage exceeds a threshold, the differential 2-line signalling has to switch to 1-line bus communication. Despite the loss of differential noise compensation, it is a sufficient compromise for a moment (as a "cheaper" CAN implementation, there exists also a *Single Wire CAN* with only one bus line).

The short circuit voltages are in the range from −1.5 V up to 16 V and will be applied to either CAN_High or CAN_Low line (under consideration of ground shift effects).

According to the deviations in a 12V-standard on-board power supply, the tests will be performed with three supply voltages:

- 7.5 V for case of under-voltages
- 12 V nominal voltage
- 27 V for case of static over-voltages

Main reasons for this are the supply voltage-dependent thresholds for fault detection in many devices. Especially in mixed networks, this results in non-compatible 2-/1-line subnet areas.

In a car, the technically given voltages for shorts are:

- Battery voltage (12 V)
- Generated supply voltage for the electronic components (3.3–5 V)
- Ground (0 V)

When the short circuit detection with 2-line/1-line switching occurs, this will be stabilized by switching off the terminations of each transceiver: One device detects the short circuit first and disables its nominal termination; thus the bus line will be shifted a little more to the failure voltage, resulting in an avalanche effect and (all) other devices of the same type will switch to the new state. In addition, the tests will be performed with some devices under ground shift, because ground shift represents no separate failure state but should be included within the normal fault tolerance handling.

For an accurate determination of the switching level, the short circuit voltage will be incremented in 0.1 V steps. Thus, also singular implemented deadlocks can be detected. The short circuit voltages will be applied via resistors to the bus lines; 0 Ω up to 50 kΩ are available, so the reaction to all possible voltage-resistor combinations can be evaluated.

Due to the fact that at the switching edges the communication may be not totally complete, the car manufacturer defined *valid areas* enclosing the three main short circuit voltage-resistance areas. There the communication has to work properly. The areas had been fixed in this way that the switching levels of the standard transceiver at that time are outside the area limits. This was valid for the homogeneous standard network and also for the heterogeneous network (see Chap. 6.2.3), which consists of a defined composition of the "standard devices" for comparison[12].

Due to the different implementations of the components in the mixed network, the resulting common "negative" edges are close to the allowed limits, so each additional device under test has to sufficiently conform to the others and not to prolong the edges into the *valid area* which would cause a failed test.

6.2.5.4 Behaviour of the Transceiver States and Their Mode Changes

Besides the behaviour in *Normal Mode* state the *Low-Power mode* with its low current consumption is to be examined, which is very important for a working battery, e.g. 'next morning'. The Low-Power mode has to be reachable also under failure conditions and should be stable at not allowed wake-up events. The only way to wake a node should be the Wake-Pin for local wake-up or a message on the bus to wake all nodes of the network.

6.2.5.5 Ground Shift

- It will be applied at all *Normal Mode* tests with 12 V supply voltage: Four of the network nodes are shifted with 1.4 V against the other 36 nodes.
- At one node the ground will be shifted until shutdown of the test board to determine the max possible shift for fault-free communication of the remaining network.

6.2.5.6 Recovery from Double Failures

120 different combinations of the following will be tested: applying two bus failures and removing one. Afterwards, the devices have to work under the one remaining failure. Due to possible different state changes of the devices, groups of transceiv-

[12] Manufacturer: Philips, Infineon, Motorola, STM.

ers may switch off the "wrong" line resulting in stable subnets which can no longer communicate with the "other" nodes.

6.2.5.7 Loss of Power/Ground

When a node has a disconnected Power or Ground line, it is not allowed to disturb the communication of other nodes, e.g. by corrupted frame transmission caused by backward-supplying from the bus lines.

6.2.5.8 Signal Propagation Delays

In a dedicated network the transmission time of signals will be measured. The "digital signal" will be applied to the transmit-pin (TxD) of the transceiver; there it will be transformed into the "physical" bus signal. Afterwards, it can be received by the other nodes and also by the sending node. While receiving, the transceiver will retransform it to a "digital" signal and puts it out at the receive-pin (RxD). The allowed delays between TxD of the sending node and RxD of the sending and another receiving node are limited. Also the delays under ground shift and bus failure conditions will be evaluated.

6.2.5.9 Verification of the Functionality of Further Essential Mechanisms

For the working of the correct state changes required by *fault tolerance*, some special implementations are essential. So, e.g. the transceiver has to distinguish between "hard" and "weak" short circuits at the bus line because the differing bus levels in Low-Power mode "behave" like weak shorts.

6.2.5.10 Test Cases of CAN High-Speed Transceiver Tests

The CAN High-Speed Transceiver tests are based on the following specifications:

- Transceiver specification:
  - ISO 11898-2:2003 Road vehicles—Controller area network *(CAN)—Part 2: High-speed medium access unit*
  - ISO 11898-5:2007 Road vehicles—Controller area network *(CAN)—Part 5: High-speed medium access unit with low-power mode*
- Test specification:
  - CAN High Speed Transceiver Conformance Test, Test Specification V1.0

Due to the fact that the CAN High-Speed transceiver owns no "active fault tolerance" (auto-switching into one or two line mode, depending on failure occurrence/recovery), at that test only "hard" failures will be applied. At some failures, the communication may be possible, at others, never. So, the *Recovery* behaviour will be mostly tested: After application of a failure for 10 s followed by removal, the transceiver still has to work properly and must not be damaged anyhow.

Transceiver with Low-Power mode must be able to switch into this mode under all failure conditions. So at least, no unwanted current consumption would be possible.

Ground Shift and Loss of Power/Ground failures will be tested at all nodes, one after the other.

6.2.5.11 CAN-ISO-Tests for CAN Transceiver

In the network tests described above, the electrical parameters of the transceivers as bus levels and thresholds will be tested only implicitly on the basis of the communication behaviour. These parameters should be stated reliably in the data sheets, but nevertheless some car manufacturers require them to be tested. So it has to be evaluated if the exact limiting values will be met as they are listed in the corresponding specifications of ISO 11898 standard. To fulfil the requirements over the whole temperature range, the tests will be also performed at low and high temperature (−40°C/+125°C).

6.2.6 Conclusion

CAN transceiver tests and, in particular, the presentation of the test idea on which these tests are based were subject of this chapter. These test ideas combine appropriate test methods and test principles so that the risk factors that are critical for the proper functioning of a CAN network can be minimized.

These test ideas were realized in the test systems for CAN transceiver tests; their functionality and basic structure were also described in this chapter.

The test systems are located at C&S group GmbH, Wolfenbuettel. Since 2001, newly developed CAN transceivers are tested there. Statistics show that the procedure of CAN transceiver testing is reasonable and of utmost importance, because not all of the tested transceivers were without faults in the first run.

If the manufacturer permits the publication, the name of the respective tested CAN transceiver together with the manufacturer's name is included on a list. This list is available on the website of C&S group GmbH.

The test system is modular. This makes it possible to adapt the proven test idea to new requirements resulting from revisions and extensions of the CAN specifications.

The test specifications mentioned in this chapter are public and can be viewed on the website of C&S group GmbH. Beyond this, the website provides also further information about the testing activities of the group.

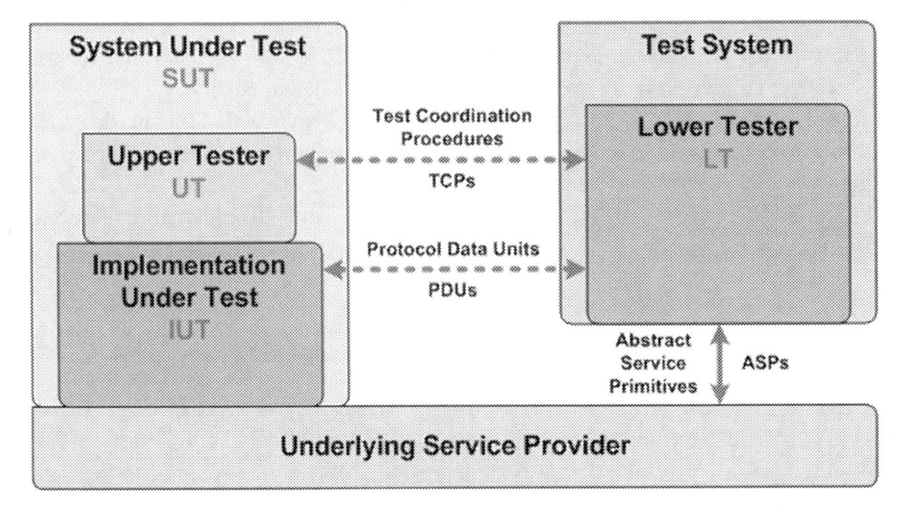

Fig. 6.7 General test method according to the ISO 9646

6.3 CAN Data Link Layer Conformance Testing

The CAN Conformance Test plan is described in the ISO 16845. It provides the methodology and abstract test suite necessary for checking the conformance of any CAN implementation specified in ISO 11898-1.

The empirical findings in this area have proven that additional efforts are needed in order to obtain a better test coverage. Therefore, C&S Group had extended the ISO CAN Data Link Layer (DLL) Conformance Tests, based on the experience gained while having performed such tests for many emulated and silicon CAN implementations. Due to the fact that the extended tests include the standard tests, the extended CAN Conformance Tests can be looked at as a super set of the ISO standard, giving a much greater certainty that a CAN device under test meets its standard specification requirements.

6.3.1 Architecture and Implementation of the Test Environment

The test system is implemented based on the architecture described in the ISO 964, presented in the preceding sections.

The test platform presented in Fig. 6.7 consists of the "SUT", usually a micro-controller evaluation system running a specific software module named *UT*. The UT is normally embedded in a microcontroller and it is linked to the stand-alone CAN implementation (IUT). In case that the CAN implementation is integrated on a microcontroller, the microcontroller will be used as UT host. The *Underlying Service Provider* consists of the CAN receive (CAN RX) and CAN transmit (CAN TX)

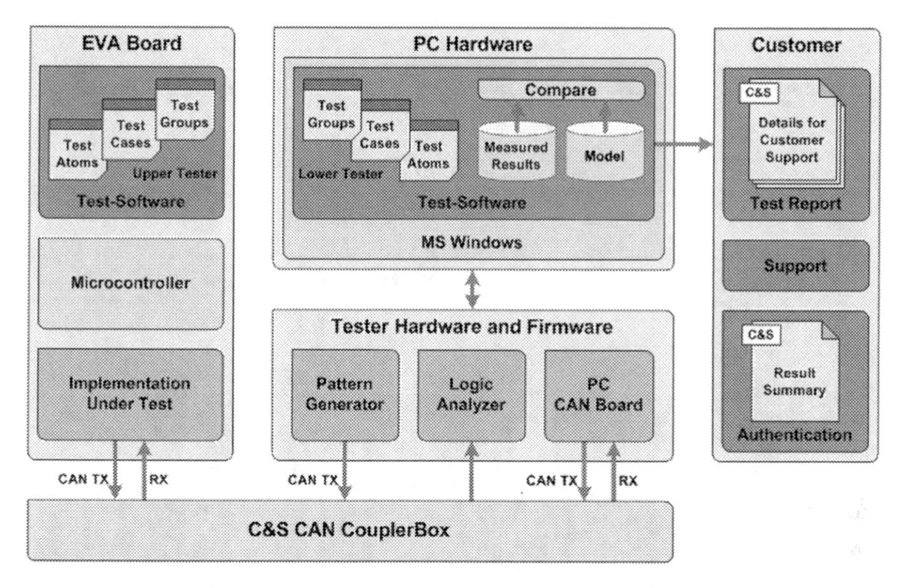

Fig. 6.8 C&S CAN Conformance DLL Tester architecture

signals. These signals are directly connected to the test system avoiding the use of transceivers in order to obtain better signal manipulation and precise timing.

The UT is remote controlled—via CAN bus—from the *LT*, a specially equipped PC. The test coordination is performed by a program executed on this PC.

The TCs are defined as ASCII script files. The script allows defining the frames that will be sent by the LT and the expected IUT reaction. The script language permits the complete manipulation of the CAN frames at sub-TQ level, allowing the test engineer to debug the IUT and generate appropriates test reports.

As it is described in Fig. 6.8, the test system consists of:

- A PC running on MS Windows®.
- The test controller software (C&S CAN CONFORMANCE TESTER v3.0).
- The test scripts.
- A logic analysis system equipped with a pattern generator. They are used to measure and generate the CAN frames respectively with high accuracy and reproducibility (see Fig. 6.9).
- A coupler box that facilitates the signal interface with the IUT.

6.3.2 *Main Test Types*

The C&S Group classifies three main test types to be executed to ensure a sufficient test coverage of a CAN node. These main test types are:

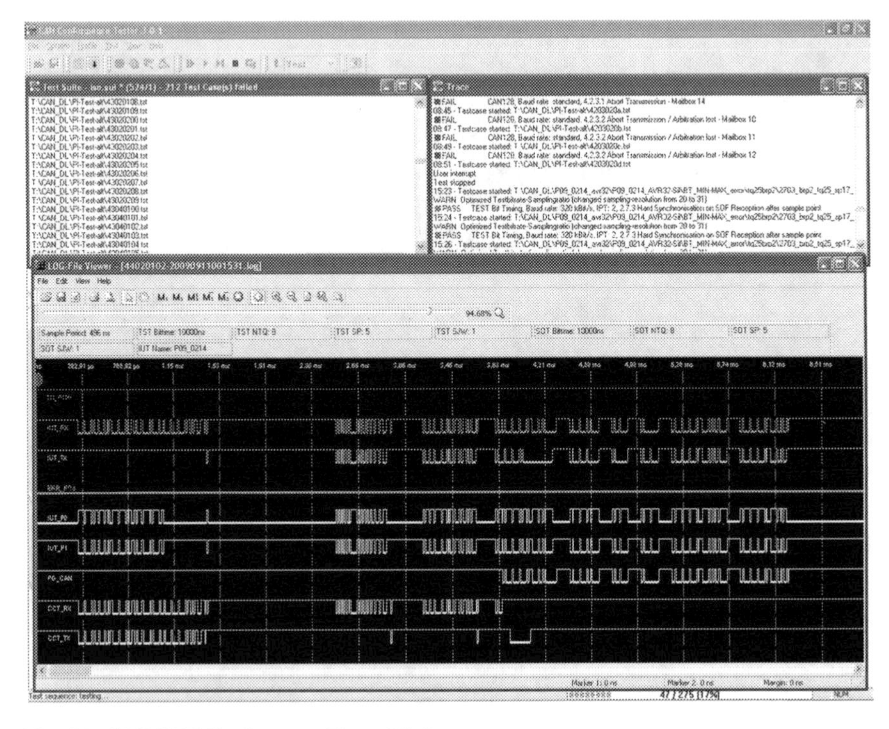

Fig. 6.9 C&S CAN Conformance Tester V3.0

- ISO test types (ISO16845 + C&S add-on tests)
- Processor interface test types
- Robustness test types

Some CAN implementations present advanced features leading to customer-specific tests:

- Gateway tests
- DMA tests, etc.

6.3.3 Test Types, Classes and Test Cases

According to ISO Conformance Test specification the main test types are:

- Receiver frame type,
- Transmitter frame type or
- Bidirectional frame type.

A TC normally belongs to one of the test classes:

- Valid frame format,
- Error detection,

- Active error management,
- Overload frame management,
- Passive error state and bus-off,
- Error counters management or
- Bit timing.

A TC may contain several elementary tests in case the TC itself depends on parameters. Each elementary TC is realized in one test script. An automatic test script generator ensures the proper creation of the different script variations.

The valid frame format TCs have a very limited scope; they are used to determine whether the basic functions, like send and receive a CAN frame, work correctly. Sufficient basic functions are the prerequisite for further and more deepened testing. The basic function TCs are performed at the initial stage of the test session.

The ISO test suite consists of about 500 TCs. Depending on the functions implemented in the device, the needed TCs are selected from the test suite and form the so-called selected test suite, against which the implementation will be checked.

The ***Processor Interface*** group represents a particular case, since each CAN implementation provides different interfaces towards the central processing unit (CPU); therefore, implementation-specific TCs must be defined by the test engineer.

Testing experience showed that sporadic errors occurring under particular conditions are covered neither by short deterministic TCs nor by simulation tests. These problems are usually caused by clock skew or a wrong internal variable whose values are not explicitly tested in simulation and they lead, only after a complex sequence of events, to a visible error. Being aware of the problem that a fair number of short deterministic TCs cannot reach 100 % coverage, the so-called random TCs have been developed. The *Robustness Test* is based on the simultaneous exchange of pseudorandom messages at 100 % bus load between the LT and the IUT.

6.3.3.1 The ISO Test Types

The *ISO 16845:2004 Road vehicles—Controller area network (CAN)—Conformance test plan* defines and describes all the TCs of the ISO test type. The C&S group offers a specific test suite covering all the ISO Tests mentioned by *ISO 16845* and additional TCs defined by C&S.

Scope of the ISO CAN Conformance Tests

The CAN specification covers the ISO OSI layers 2 (DLL) and 1 (physical layer) as shown in Fig. 6.10.

Nowadays, the CAN network interfaces are usually made from two separate chips: One chip covers the upper part of the physical layer—physical signalling sublayer (PLS)—and another chip for the DLL including the full logical link control (LLC) part, which is representing the final interface to the processor.

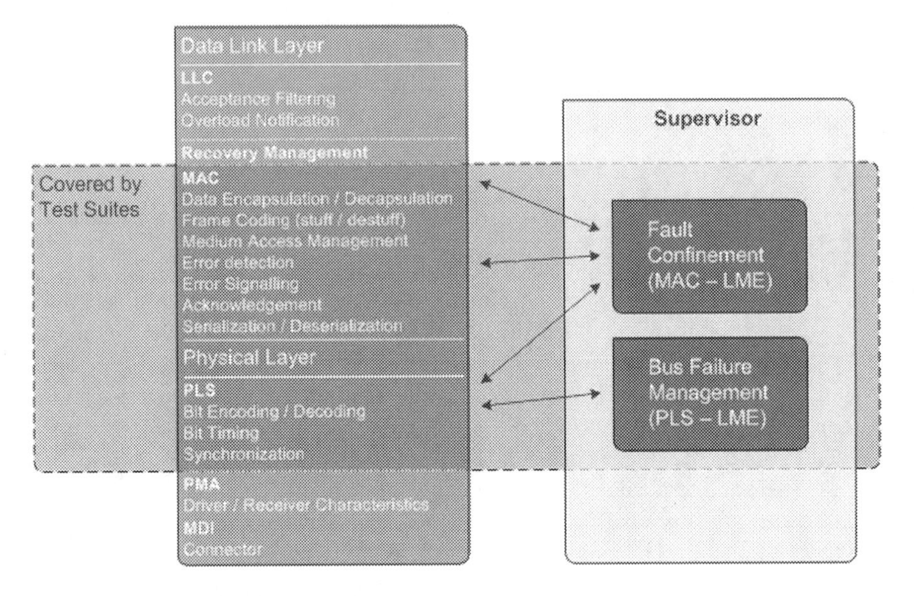

Fig. 6.10 ISO CAN Conformance Test coverage

The ISO conformance test is covering the lower part of the DLL and the upper part of the physical layer. The access points for the tests are given by the TX/RX lines from the physical layer side. For the DLL side, the access points for the tests are given by a unified virtual receive and transmit register interface. This simplification was necessary, because real CAN chip solutions differ very much in the register architecture, representing the processor interface. As no such *standard* interface is available, a virtual *standard* register interface was artificially defined. When performing these tests on the tester, the virtual registers must be mapped by accordingly tuned test software onto the actual real register interface.

As mentioned above, the ISO standard conformance test only covers the so-called CAN cell or core. The standard tests neither cover the individual processor interface nor the line driver part. However, there are extended conformance test solutions as described in the next sections.

6.3.3.2 The Processor Interface Test Types

Standard Conformance tests are performed on the assumption of a standardized simplified received and transmitted register interface between the CAN device under test and the UT. This simplification never corresponds to a real CAN implementation. However, obviously the real CAN register implementation with their corresponding control bits have a great influence on the proper function of a CAN device. As a result a set of additional "de facto" standard register TCs was developed. Each of these extended register TCs is dedicated to perform specific checks on special characteristics of CAN registers.

For solutions comparable to the basic CAN architecture, there are different implementations to be checked such as transmit buffer registers with/without internal prioritization, different write capabilities onto the transmit register queue which influence heavily the overall arbitration process and thus the latency times of message transfers, etc. Concerning CAN solutions comparable to the full CAN architecture, specifically adapted checks must be executed to test the individual solution of structure and number of transmit and receive registers, the kind of masking capabilities, the access techniques to the registers, etc. There are various solutions for the status and control registers which must be checked such as receive/transmit error counters, time stamps, transmission success checks, interrupts, non-unified position and definition of status/control bits and so on.

6.3.3.3 The Robustness Test Types

According to the ISO 9646 standard, a test suite usually consists of short deterministic TCs whereas each TC focuses only on a particular function of the protocol. This approach neglects the complex interaction between various state machines of the implementation. When checking a state machine, it is not sufficient to check each of the various paths individually only. The behaviour of a state machine is dependent on the sequential order of the paths too. Therefore, these dependencies ought to be checked, also in CAN devices, in order to ensure their compliance to the CAN standard.

Therefore, C&S Group has done many efforts to develop generic robustness tests verifying the correctness of the implementation under various bus loads and during a long period of time. As the number of combinations of sequential paths in CAN is very large, these test sequences are generated randomly. These robustness tests cover implementation problems like clock skew due to critical timing, sporadic errors occurring only under very particular conditions and other protocol or processor-interface problems not covered by the limited scope of the deterministic TCs.

The test sequences typically run for hours up to a couple of days, depending on the complexity of the CAN device and the number of correspondingly required tests cases. There are various parameters modified while producing the sequential order of "standard" and "extended" TCs such as random generation of identifiers with standard or extended length and coding, random length and random content of data, various baud rates, very high bus load such as ~100 % and ≥50 %, with/without errors, etc. All the tests are executed in real time. The verification of the tests is done in real time at message level.

6.3.3.4 Characteristics of the Robustness Test

- Bus Load

Each robustness test is executed at ≤100 % bus load. The maximal baud rate used to run the tests at 100 % bus load depends on the computing power of the UT. The UT must be able to calculate the random values and generate the random messages to be transmitted guaranteeing a bus load of minimum 50 %. In parallel, the UT must generate the random messages expected to be received from the LT and to check them before the next reception.

As both, LT and UT try to send at 100 % bus load, arbitration occurs at each transmission. At lower bus load, the transmitter part of the UT is delayed by hardware limitations and LT is delayed by test configuration.

- Error generation

The robustness tests can be classified in two main groups:

- Robustness tests without error
- Robustness tests with errors

The errors are generated by the test environment (Pattern Generator) as follows. The test environment forces a bit sequence of the frame transmitted by the IUT, to dominant. Depending on the error sequence, the IUT will stay in Active Error State, reach the Passive Error State or even go to BUS OFF. Whether Active, Passive or even BUS OFF will be reached depends on the error sequence frequently generated, possibly disabling the IUT to recover its error counters between subsequent error bursts.

- CAN Frame generation

The messages transmitted by the LT and the UT during the robustness tests are randomly generated. For each message, identifier, data length code and data are generated with a 16-bit pseudorandom generator.

The same random generator is implemented on the LT and on the UT, but different start values are used for the random generator leading to various random values, hence various CAN messages.

To avoid errors caused by the simultaneous transmission of the same identifier but different data by the LT and the UT, during one test odd identifiers will be used by one component only (e.g. LT) and even identifiers by the other component. Both configurations will be used to ensure the IUT is able to handle the whole range of IDs the random generator could provide.

- Verification

To verify the correct behaviour of the IUT during the robustness test, checks are performed at processor interface level.

At the processor interface level, the verification is performed by the LT and the UT. The LT checks at runtime that every message sent by the IUT matches the expected random message. This verifies that the IUT sends all messages to the LT correctly and in the right order.

The UT checks at runtime whether every message received by the IUT matches the expected random message. This verifies that the IUT receives all messages from the LT correctly. To ensure this verification process, each component must calculate the random messages it has to transmit, and in addition, it expects to receive the random messages in the right sequence.

The logic analyser stores the last messages transmitted before an error has been detected by the LT or the UT. These data on the logic analyser are necessary to reconstitute and investigate the sequence which has led to the faulty behaviour.

6.3.4 Conformance Test Results

The experience from more than 15-year testing of CAN cores has shown that almost all the new implementations have not passed the conformance test in the first attempt. Problems with bit timing and synchronization are the usual cause of the failures. Thanks to the actual field-programmable gate array (FPGA) technologies, the CAN implementers are able to test the cores during the development phase, avoiding finding unexpected errors in the silicon version of the device.

Consolidated CAN cores are usually adapted to new families of microcontrollers. The statistics show that in approximately 30 % of the cases, issues are found during the processor interface or the robustness tests.

6.4 CAN Software Testing

The term *CAN-software* comprises communication functions which make use of the CAN hardware for data exchange between electronic control units in order to enable the operation of application and network management (NM). CAN-software is required on each control unit, for instance, in order to configure and control the corresponding CAN controller(s) and bus drivers, or to compile and transmit data packets when requested by the application, as well as to provide the packet contents at the receiver - Communication layer (COM), or as well to monitor and control the state of the network (NM). For classification purposes, quite often CAN-software is subdivided into layers, for which the higher layer always makes use of the services offered by the lower layer. The term *CAN-Software-Stack* is applied synonymously.

6.4.1 Test Objects in Software Tests

The individual functional units of a CAN-Software-Stack can be subdivided into layers depending on their individual objectives. Lower layers are directly hardware dependent while higher layers can be designated as hardware independent—at least from a logical point of view. This becomes more plausible when considering the tasks of the individual layers.

The interface for configuration and control of a CAN controller, the message buffer structure as well as the connection and the control of a bus driver to a microcontroller are all specific to their supplier or to the component or Electronic Control Unit (ECU) respectively. As such, addresses, assignment and functionality of

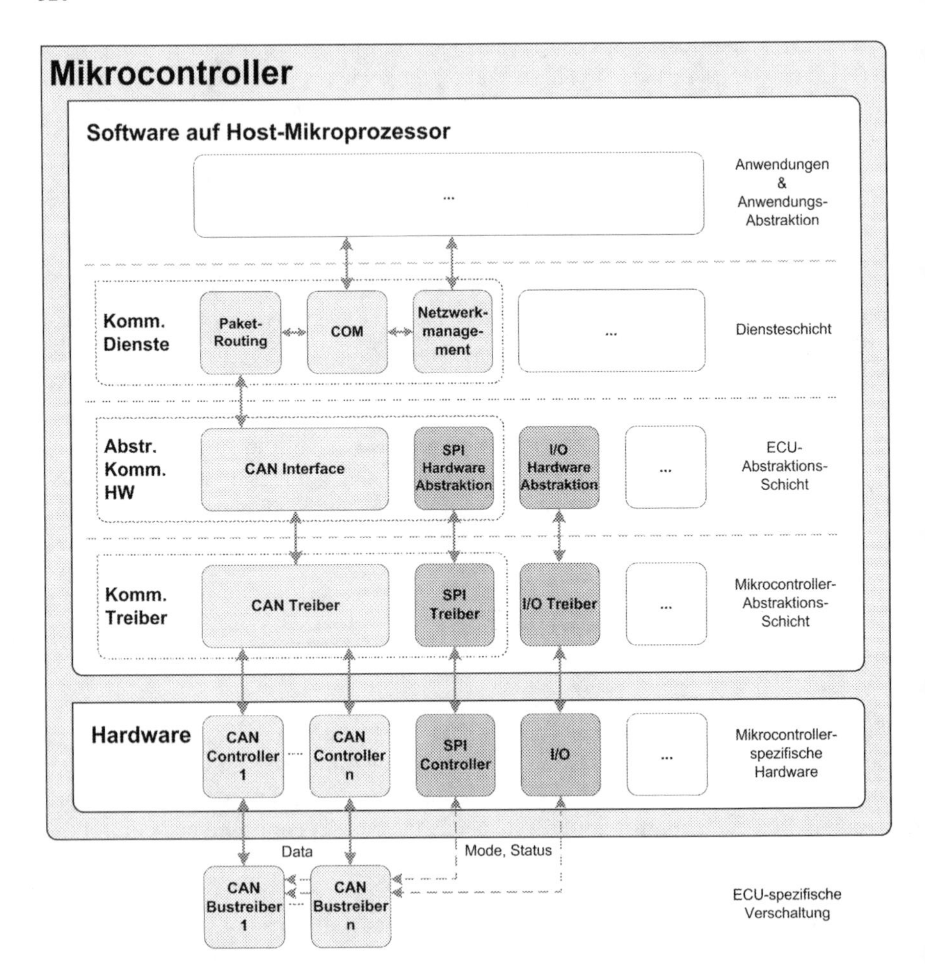

Fig. 6.11 Example of a CAN-Software-Stack with objectives-oriented layers

configuration registers of a CAN controller all are different as well as, for example, connection mode and applied ports of the bus drivers. Depending on the implementation they can be either controlled by digital control lines or by serial buses such as Serial Peripheral Interface (SPI). The abstraction from the underlying communication hardware and hardware driver layers depends on the hardware itself.

Any one of the higher layers for management, compilation of data packets, segmentation of long messages, etc. are functionally hardware independent. However, any software nodule on these higher layers becomes hardware specific whenever it is compiled down into machine language for a specific target hardware. However, any original source code written in a higher level programming language can be reused if no ECU-specific functions had been applied. In conclusion, this implies preferably to standardize any hardware-independent software layers, to specify clearly their functions and interfaces. As such, no further adaptation is required for any individual application purpose. Merely software for ECU abstraction layers and driver layers cannot be fully standardized due to the differences in the hardware to be controlled and the varying functionalities and options of it. Nevertheless, the basic functionality as well as the minimum interface towards the higher software layers can be standardized (see Fig. 6.11).

6.4.2 Trend Towards Standardization

In the past, software for communication functions often was not developed from standards but often from scratch always for each new application. The reason for that certainly was the need to save limited resources of the applied micro-controller as well as to differentiate from competitors. The trend started from non-standardized solutions, then moving from partial standardization of specific components (e.g. OSEK/VDX COM and NM, where OSEK/VDX stands for "Offene Systeme und deren Schnittstellen für die Elektronik im Kraftfahrzeug / Vehicle Distributed eXecutive") and customer-specific guidelines as well as quasi-standards of automotive manufacturers (e.g., BMW-kernel) towards open and customer-independent standards (e.g., Automotive Open System Architecture, AUTOSAR) in the future.

Many benefits arise from standardization such as reuse of once implemented components, substitution ability of standardized software modules by different suppliers, comparability and long-term reduction of development time and cost. The generalized applicability of a software module in conjunction with its included partially complex functions which may not even be needed in specific applications, however, implies the drawback of increased resources consumption, bigger complexity and higher development cost.

Obviously standardization of components only pays off if those components provide a sufficiently high quality and thus a broad market acceptance and long lifetime. A solution, which due to its high complexity hardly can be handled, and which despite of the application of standards does not meet the goals of exchangeability and comparability or which frequently fails, will not prevail the market, which is characterized by an enormous cost pressure and high-quality requirements.

Therefore, the key is quality of the generated standard software modules as well as their exchangeability, correctness, performance and comparability. In order to safeguard these characteristics standardized tests are required which must provide a preferably high level of quality at reasonable costs. Only this prevents a complex system, consisting of various software modules, which typically are sourced by different suppliers, from having a correspondingly high risk of failures which, after the integration phase, are even difficult to be reproduced and localized. This finally turns out to be a very costly task if and even if the erroneous module(s) can be identified at all.

6.4.3 Requirements on Tests and Test Systems

Standards require conformance tests checking whether an implementation is compliant to its corresponding standard. This does not imply that all such implementations must provide the same source code. However, they must match the given specification, e.g. their interfaces, functionality and configurability if specified so. As such, e.g. software modules for communication systems such as CAN must match some requirements which are explained subsequently.

6.4.3.1 Hardware-Dependency, Test Support on Target Platform

Some parts of the CAN-Software-Stack are hardware dependent. Therefore, drivers for CAN controllers are significantly different with regard to their interfaces to the hardware while the interface towards the higher software layers can be standardized, as it is the case, for example, in AUTOSAR. Functionality and performance of a CAN driver can only be reasonably tested in conjunction with the target hardware which, in most cases, is an embedded CAN controller. Finally, this is the only way to check whether the control function of the driver achieves the desired behaviour of the CAN controller towards the bus. A CAN driver applied onto the wrong CAN hardware is not functional. This implies that the real hardware cannot be substituted.

Under the assumption that the applied CAN controller is proven to be conformant, any potential failure must come from the software when performing further tests. The standardized and accessible interfaces to the test candidate, therefore, are the software interfaces towards the next higher software layers as well as the digital signals and, in conjunction with a CAN bus driver, the analog bus signals. Insofar, the lower layer of the CAN controller software is accessed while passing through the CAN controller. This is a typical method which is applied if the direct interface of a test candidate is either not standardized or not directly accessible. Test descriptions and test system must consider this kind of structure. From an abstract point of view the addressed protocol layers below the tested layer belong to the service provider which only acts as a transfer station for test messages or stimuli. AUTOSAR identifies the combined tests of software modules as “class B” tests.

6.4.3.2 Test of Hardware Independent Software Module in Absence of a Target Platform

The said class B tests are neither new nor unusual. Software such as OSEK/VDX NM and implementations of customer specific quasi-standards of automotive manufacturers quite often are tested in conjunction with the target hardware or target platform correspondingly. This is performed in most cases while applying a corresponding evaluation platform and specific test solutions. This is a reasonable process, because the software code is developed, applied and optimized for specific micro-controllers. Class B tests therefore are not new, but very common. The AUTOSAR “Class A” tests for hardware independent PC-based modules are only feasible because of their hardware independency and corresponding programming on PCs. This is quite a new approach. However, this makes it even more difficult for software-module suppliers to generate highly optimized code, which contradicts the idea of a generalized applicability. Nevertheless, this method allows the test of embedded software while not requiring any specific target hardware. In order to execute the embedded code on a PC the object code must be generated for PC execution. The resulting object code and its timing behaviour obviously are not comparable with a code dedicated for an embedded system application.

6.4.3.3 Support of a Script Language for Test Implementation

All up to now widely spread solutions, which are based on a partial standard or a quasi-standard of some automotive manufacturers, typically provide a textual document form for test specification which finally must be transferred into an executable code. Typically, this is achieved by a non-standardized script language, which is adapted to the specific test object and which is human readable. Any one of these script languages finally provide well-defined services, which can be combined in different ways, configured and parameterized in order to specify a specific TC. Calling these services implements the access to the functions of the test system such as to the measurement tools, signal generators, log files and the evaluation.

6.4.3.4 Support of TTCN-3 Test Descriptions

AUTOSAR defines TTCN-3 as the notation form for any of its standardized test specifications for conformance tests of all specified AUTOSAR software modules. This also includes all of those software modules, which are applied in an AUTOSAR CAN software stack. Applying a compiler or interpreter on the TCs of this test description language makes them executable. As such, this human-readable test code is a test specification and, at the same time, an implementation of it. Any one of the newly implemented AUTOSAR software modules must be checked for conformance by well-defined, module-specific conformance tests in the future. For that reason, a test system is required which executes the TTCN-3 conformance test specification (abbrev. "CT-Spec"). This is to be applied for the so-called class B tests in conjunction with the target hardware as well as for the PC based class A tests. It is favourable to have the same, identical solution for both of these classes in order to avoid specific and costly special solutions for one class only or only for just some few software modules.

6.4.3.5 Adaptation of Test Candidates, Connection Between Test Object and Test System

Neither tests written in TTCN-3 nor tests written in a script language can be applied without any further efforts to test a specific test candidate. Any test candidate requires a certain adaptation to the software/hardware interfaces offered by the test system. A software interface, e.g. may differ from others in the kind and coding of the data types. A hardware interface, e.g. may have distinguished signal voltages, pinning, kind of connections, etc.

For any specific interface of the test system, a link is required to the abstractly defined test steps. The link abstracts the specific test object. The communication towards the test description and test evaluation is implemented by communication channels and—if required—also by appropriate electrical interface hardware. This kind of connecting tests is applied for embedded platforms (typical, AUTOSAR class B) as well as for PCs (AUTOSAR class A), while in the latter case inter-process communication is applied instead of electrical communication media.

6.4.3.6 Influences on the Test Result, Independence of the Adaptation and Execution

A kind of test candidates requires only a one-time adaptation to the test system, if all its interfaces are standard. In that case, any other test candidate of this kind makes use of the same generated links. Obviously the more open a specification is, the more a deviation in the functionality and in the interfaces of the corresponding implementation arises. This implies the need for adaptations. It is needless to say that adaptations of the link between test candidate and test system, especially with regard to the TC description and the test evaluation, may directly influence the test result. As such, this must be performed with special care. With respect to public test specifications and the comparability of solutions by different providers, it is recommended that an independent entity performs checks of the test candidate adaptation as well as the test execution.

6.4.3.7 Extensibility, Adaptation Ability, Flexibility

In the past, quite often existing standards were extended, revised and altered by software suppliers and automotive manufacturers for various reasons. Some of the reasonable examples for these efforts are saving of micro-controller resources, making use of special functions or special features of the applied hardware or optimizing of inappropriately specified mechanisms. A standard is obviously stable only as long as there is no good reason to modify it.

Consequently, a corresponding test must be adapted or extended. There is a need for cost effective and fast definition of customer-specific adapted tests, but which must be clearly distinguished from the official standard tests. A flexible system with an extension potential and a comprehensible, easy to apply test description language is required. Comparably, minor modifications of the test should not require a re-design of the test implementation. This risk quite often is underestimated for the sake of an alleged better solution. Therefore, the design of a test system shall consider that none of its components is fully loaded by the already-known tests.

6.4.3.8 Test of Time Behaviour, Performance, Throughput

CAN software modules are expected to provide sufficient throughput to handle all message traffic even at high bus loads. All this is expected from any of the software layers of the CAN software stack; otherwise buffer overruns, loss of data must be encountered and finally predictability and application functionality would fade away. Besides this worst-case scenario and because of competition reasons, it makes sense to specify, require and measure throughput as well as performance. A comparably excessively slow implementation will not prevail in the market, as well as a faulty or resources-wasting one—all this is for the benefit of the applicants. Therefore, it is highly recommended to specify appropriate benchmarks for standardized software modules which evaluate and compare the performance of implementations. Such kinds of tests focus directly on the timing behaviour of software modules and their target systems. Therefore, corresponding methods for test execution in real time, timing control and synchronization of test-system entities are required. Up to now these kinds of tests have not been discussed so much and they are not widely applied. Insofar the customers—mostly automotive manufacturers—must rely on corresponding statements of their suppliers. However, nota bene, standardization is an important prerequisite for comparability. It is recommended to do performance-comparing tests when interfaces and functionality are the same.

6.4.4 Test-System Architecture for Embedded Software Tests

Assuming a modular design, software can be tested module by module. As such, software is comparable to layers of the ISO/OSI-layered model. The basic principles of ISO 9646 standard are applicable as well. All accessible interfaces of the software module are surrounded by the tester, whereas the logical assignment of LT and UT to higher or lower software layers can be done either way, depending from the orientation of the interfaces.

Interfaces of neighbour modules of the same layer are served as well by the tester. In order to avoid conflicts coming from the assignment of either UT or LT—this problem cannot be resolved sufficiently clearly due to the ambiguous orientation to either the higher or lower logical layer—temporarily the term "lateral tester" can help out, which addresses those interfaces.

6.4.4.1 The Residual System, Tests in Conjunction with the Application

Quite often, a software module cannot be fully separated from its surrounding system, for instance, if it requires a specific operating system or if it is highly interwoven with the application. These parts of the surrounding system must remain on the target platform and can be called "residual system" (RS). As the RS has an impact on the test, the specific characteristics of the RS must be taken into account when testing. This is the reason why the CPU of a micro-controller cannot be exclusively assigned only to the test execution, but this assignment nevertheless should have a high priority. If the application cannot be completely substituted by the test code, the application typically limits the applicable test scenarios. In addition quite significant adaptation efforts are required in order to achieve a distinct behaviour. In other words the test object is quasi-shielded by the surrounding RS. There is only an indirect and fuzzy visibility and accessibility onto the test object.

The characteristics of the RS are mostly not standardized. Insofar, standard tests are not applicable. With regard to their distinctions, applications react completely different on specific bus messages or failure scenarios. They ignore, for instance, any messages which are not addressed to them. There are limited possibilities to influence and control applications and very detailed knowledge of the surrounding components is required in order to test the covered test object sufficiently. If this is not the case, it is an encapsulated test object scenario which more or less modifies or filters all input and output signals. For test execution, the impact of the encapsulation must be fully known and the test steps must be modified correspondingly in order to get the desired signals provided at the test object itself and to purify the output signals from their encapsulation effect. This process, in general, can be automated but not automatically generated. Ideally, all components surrounding the test object are substituted by well-defined test code in order to exclude any non-desired influence.

6.4.4.2 Implementation and Synchronization of Upper and Lower Tester

UT and if required lateral tester for testing of software modules of a CAN stack are implemented in software, whereas the LT applied for the above-mentioned combined tests of driver software together with the underlying hardware can also be implemented using measurement tools and electrical signal generators. Synchronization signals are provided for precise timing control of the time base, which typically is located in the LT, and the driver software in the micro-controller. This enables test scenarios with precisely tuned timing behaviour as well as precise timing measurements of the test object.

All functions for stimulation and control of the software interfaces of the test object usually are pooled within a separate software unit. This unit simultaneously generates the designated stimuli, provides the functions which can be called by the test candidate, logs all actions and reactions and adopts as well the abstraction of specific features of the test object. However, the stimulation and control of electrical signals are mostly implemented by separate real time capable devices which are specialized for their particular purpose.

6.4.4.3 Control of Electrical Signals, Reliable Measurement and Self-Control

Control of all relevant electrical input and output signals is most important for tests on the target platform. This not only allows monitoring the reactions of the test object but also enables control and recording of the applied stimuli. In general, it is recommended also to log all of the applied signals in order to achieve a comprehensive image of the complete test sequence. This enables the self-control and the continuous consistency check as well as the later proof of the correct operation of the test system. When only checking the output signals of the test object, the only assumption can be concluded that all test scenarios had been applied properly, which by no means can be recommended. In order to assure the reliability and availability of the applied control and signal-recording devices, it is recommended to apply calibrated and commercially available measurement technology only. This allows a realistic and pure image of the applied test sequences.

6.4.4.4 Time Resolution, Sampling Rate and Buffers

The sampling rate or the time resolution of all devices shall be chosen so that a significant over-sampling of all electrical and test-relevant signals is achieved. The application of a device with a just sufficient performance can lead to problems if, for example, another recording channel is required or another specific TC with a higher time resolution is required. For any of the software modules of the CAN software stack in this context, based on experience reaction times, time-outs, delay times or intervals in the range of milliseconds are specified. Therefore, a test system is recommended which allows a test coordination with tolerances in the range of some few milliseconds. Of course, the applied measurement tools must be capable to read CAN bus messages with a sufficient over-sampling rate to make them reproducible.

6.4.4.5 Test Control, Important Services and Test Installation in General

A test system must be able to perform various tasks, which are described subsequently.

On one hand, the applied test description method must be supported which is either the execution at runtime or separation into a preparation phase, runtime phase in real time and post-processing phase. An ideal solution is to not only allow the interpretation of the test description at the runtime phase, but also support the preparation and post-processing phases in order to, e.g. initialize the test installation or prepare time-critical phases for easy activation at later times. As such, the benefits of both test-performing solutions are combined, whereas the real time capable interface hardware typically does not allow the preparation of too many sequences which can be either executed if required or even dynamically.

On the other hand, services are required for communication with UT/Lateral/LT over different communication channels, data logging, test evaluation, presentation, management, as well as for the control of various measurement devices, interface hardware and stimuli generators. The aforesaid interface hardware can also serve as part of a communication channel to the test system surrounding the test object, which can be applied, for instance, for communication towards the UT of the host micro-

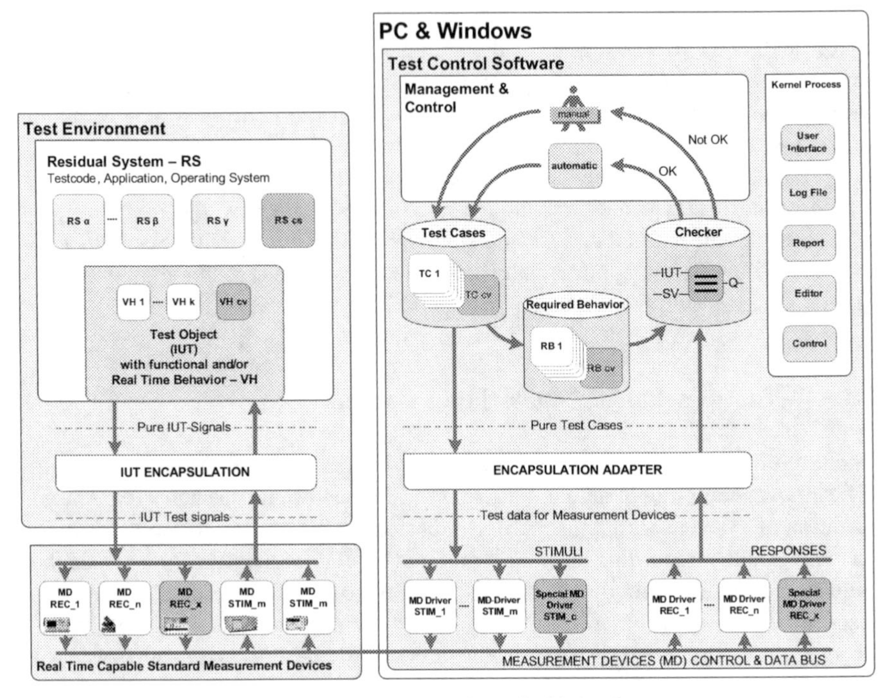

Fig. 6.12 Schematic presentation of a test system for embedded-software tests

processor on the target platform when executing software or hardware combined tests. Interface channels once made available are applied as abstract ones. Therefore, they can be used for tests of various software modules on the same target platform.

6.4.4.6 General Test-System Architectures for Embedded Software Tests

Figure 6.12 depicts the recommended architecture of a test system for communication stack software tests on the target hardware in conjunction with a test object (IUT) together with its enclosed behavioural modules/features (VH = VerHalten = behavior (German)).

On the right hand side is the test-system kernel with its associated tasks such as linking of the TC description, driver pool control of interface hardware and measurement devices, providing communication channels, test evaluation, logging, etc. With regard to test control, test signals to and from the test object are to be prepared correspondingly. Thereby, the test object gets the appropriate signals and thus the test-object reactions can be uniquely reproduced. This job is performed by the encapsulation adapter. Control information is derived from the TCs and transferred through the encapsulation adapter to the interfaces of the measurement devices and the target platform where they are applied to the test environment as IUT test signals. The encapsulation is performed on the test object and also on the RS. Insofar, pure and predictable test signals are applied to the test object. The reaction of the test object is bounced back through correspondingly similar signal paths for further TC evaluation in order to check whether these signals match the required behaviour (RB). The RB data are either derived from an executable model or by a correspondingly generated database. Quite often, there is no comprehensive reference model of the test object which can be applied in a test system. As such, a manual prediction of the desired behaviour for each TC is the only solution. Test results are collected in a report, while any other test relevant information such as generated stimuli, measurement data and evaluation steps are continuously recorded in a log file. As such, at any time a reproduction or a reevaluation of the executed TC is feasible.

On the left hand side in Fig. 6.12, there is the general architecture of a test object on an embedded platform. All the interfaces are surrounded by the RS, which ideally is represented by test code, but which may also contain, an application or additionally an embedded operating system. The RS encapsulates the actual test object. Therefore, the pure test signals applied directly to the test object may not be visible to the outside world. All electrical input and output signals are generated, monitored and controlled by real time capable measurement devices. The test object runs in its intended environment. Its functionality is directly checked together with the applied target platform. The applied real time capable interface hardware and measurement technique can be virtualized and thus is flexible; real devices can be easily substituted by other ones.

This test architecture allows tests of various kinds, from driver modules to higher layer software up to complete software stacks.

6.5 Model-Based Testing

Currently tests are not only executed when real test objects exist. Quite often, customers require that tests must be executed on models which, for example, is the case when performing tests on physical layer models. Automotive manufacturers expect from these early tests a reduction of their development costs and a significant increase of test quality for electronic systems in cars and other applications. Model checking[13] though is in its beginning stages with regard to standardization or harmonization. In the area of networking, there are almost no standards with respect to implementation models or how to analyse these models, for instance, with regard to conformance. That is why up to now only some few manufacturers dared carefully to do their first steps in this area, mostly based on proprietary solutions. Under these circumstances, models, test runs and test results are neither comparable among themselves nor repeatable.

Going into details makes the benefits of simulation for automotive manufacturers become more visible. On one hand, a model bares the opportunity to analyse

[13] This is related to the scope of this section with regard to the physical layer of automotive networked electronics. In other application area, such as mechanical engineering, modeling and model checking are more typical.

borderline cases, which are difficult to implement in the physical world. Such a realistic worst case scenario, for instance, is the network topology layout of networks, whereas one control unit under the hood may operate at +100°C while another one in the trunk may be at −40°C. Different environmental temperatures influence significantly the operation characteristics of the corresponding physical bus drivers. Simulating such a scenario only under the assumption of typical or almost ideal environmental conditions bares a high risk for corrupted signal integrity and finally for the safe operation of the whole end product as well. On the other hand, simulations can easily consider tolerances of the physical bus drivers as well. Topologies, therefore, can be analysed in high-frequency range as well as charged with non-typical loads in order to generate borderline cases. Among others, the number of nodes applied in a topology has great influence, because not only the bus drivers but even the chosen bus lines influence the signal quality.

In order to meet the requirements for high-quality tests correspondingly, appropriate models supplied by the component manufacturers are required. Unfortunately, thereof the first area of problems arises: As there is no standardized validation method, there is no way to check these models for compliance to the requirements of the standard. Furthermore there is no compatibility among the different languages which are applied for model specification. On one hand, there is the traditional language SPICE[14] which unfortunately is not well suited for generating more complex models such as for bus drivers. SPICE quite obviously has its weaknesses when functional or behavioural descriptions are required. On the other hand, non-open specified description languages are applied; one of them is MAST, a proprietary language. That is why, there is a certain interest to create standards for generic and specific requirements on models, requirements on simulation environments and requirements on model test specifications. Only when following this path independently, reproducible and comparable test results and a corresponding quality evaluation on model basis can be achieved.

Subsequently, a model-based test method is presented. This test method not only is well suited for checking transceiver implementations against their existing standards, but also can be applied to check the overall physical network with special focus on the integrity of signals on the communication media. As such, two major aspects with regard to analysis of the signal integrity are met: on one hand the validation of individual components and on the other hand the observation and evaluation of a complex topology.

6.5.1 VHDL-AMS

VHDL-AMS (**V**ery **H**igh Speed Integrated Circuit Hardware **D**escription **L**anguage - **A**nalog and **M**ixed **S**ignals) is an extension of VHDL (see [IEEE1076]) for simulation support of digital, analog and mixed signals systems and technologies. VHDL AMS allows

[14] Simulation Program with Integrated Circuits Emphasis—there are optimized and commercially available versions such as PSPICE or HSPICE.

the description of the continuous behaviour[15] of systems. Furthermore conservative[16] as well as non-conservative[17] systems can be modelled. They can be connected to each other and jointly simulated.

Because VHDL-AMS is the only standardized hardware description language for digital, analog and mixed signal systems, VHDL-AMS is in conjunction with the before mentioned characteristics the preferred means for modelling of automotive networks on analog, physical level. Neither SPICE nor any other proprietary or non-standardized language provides a comparable flexibility. Furthermore SPICE net lists can be translated into VHDL-AMS without any problems. Therefore, integration of existing models is feasible. That is why VHDL-AMS is required by various specifications and standards as the sole description language for automotive network components. Since quite a while work groups with the broad participation of automotive and semiconductor industries and others provide specifically tailored model libraries for the application in the physical layer network area.

6.5.2 *Test Methodology*

The most significant characteristic of this test method is the systematic structuring of the individual components of the test system following the guidelines of ISO 9646 for conformance tests according to the distributed method. The term "model-based testing" only indicates that the test object (IUT) must be provided in the form of a model. There is no further indication on what is outside of the test object and as such there is plenty of freedom for further interpretations and implementations. Subsequently more than the simple alternative of mapping the IUT solely into a model is discussed. The reasoning will be explained.

The modelled IUT is stimulated by distinct test signals while a TC is executed. This correspondingly requires that stimuli must be available as models, too. This is mainly due to the fact that stimuli typically are analog signals within a conservative system description influencing the IUT correspondingly. These stimuli can be classified as host stimuli or bus interface stimuli depending on the model alternative. The same is true for corresponding measurement devices. Furthermore, a model of the environment (e.g. power supply) is required complementing the IUT model. UT, LT and the environment model with its stimuli generators, measurement devices, etc., are fully translated into the VHDL-AMS model description of this virtual protocol test system.

Tests of an executable model description are run on a simulator. Such simulations are performed by either the algebraic solution of a set of equations in a well-defined time period of a closed system or by the description of a fixed system described before the simulation is started correspondingly. In the latter case, the test

[15] Continuous behaviour with regard to time and/or data.

[16] Based on the law of conservation of energy; e.g., Kirchhoff laws.

[17] Signal flow systems for e.g. transfer functions in control systems.

bench[18] cannot actively[19] be either controlled nor its structure be manipulated from outside while the simulation is ongoing. Therefore, the executing simulator is not able to address and control the UT and LT of the test system coded in VHDL-AMS modelling language. This fundamental statement is of important significance when recognizing the continuously ongoing transition from test system for real, physical systems towards model-based testing. The major difference to a non-simulation based test-system implementation is that an additional device level is inserted between the test coordination and the IUT when applying a simulator. This implies that the device models of the UT and LT no longer can be addressed directly. Insofar, also the test coordination—with its main responsibility to perform the test sequence control and which, therefore, is defined as a set of sequences of statements for the devices of the UT and LT—cannot be completely performed from outside of the simulator. That is why test scripts containing TC-specific model parameters must be provided to the test bench. Therefore, a solution is required for the integration of the test sequence control into the model description.

6.5.2.1 Integration into Models

There are various versions and integration levels. In a first version, all the complete test coordination is integrated into the model. This implies that any required procedures which are needed for test coordination must be converted into the model description. In this case enormous efforts are required for the parameterization and the complexity of the model, especially with regard to measurement data evaluation and test-report generation. This is feasible due to the VHDL-AMS feature to read and write external data from and to a host system. Any interfacing to real tests is difficult though, because existing test scripts cannot be applied as they are, they first must be converted into model descriptions. In this version, a test script simply is a configuration in the VHDL-AMS language, which parameterizes the individual, modeled components of the test system and links TC-specific implementations to the interfaces of the components (see Fig. 6.13).

6.5.2.2 Integration Outside the Simulator

In the second version, the measurement data evaluation and test report generation are outside of the model descriptions. As such, validated functionalities of existing test systems can be applied. Interfacing to those physical test systems therefore can be easily done. Furthermore, this integration version allows adopting test from the physical tests scripts for the devices control of UT and LT. This also supports the comparability of virtual and physical tests. On one hand, this may bare the poten-

[18] Top-level model description of a simulation.

[19] By external programs through the simulator or by the simulator itself.

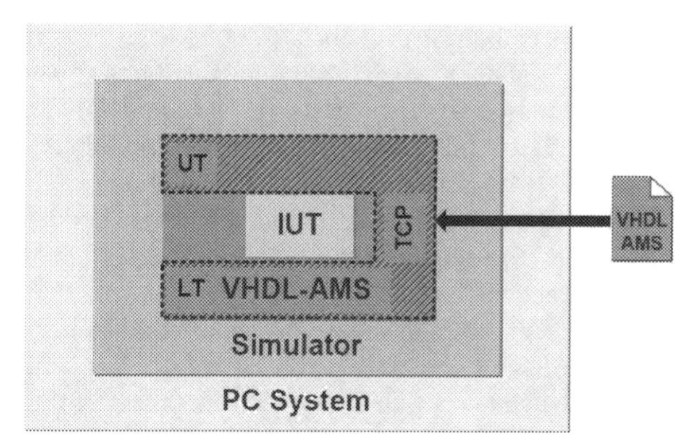

Fig. 6.13 Test environment integration into the model

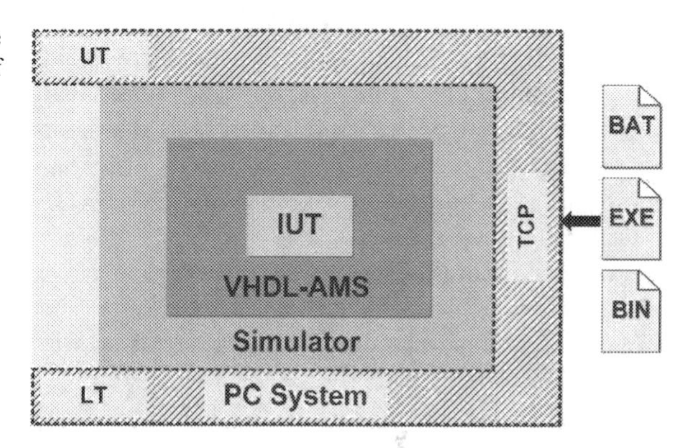

Fig. 6.14 Installation of the test environment outside of the model

tial disadvantage that the device models must provide the implementations of all functionalities for selection which are needed for all TCs; this increases complexity. On the other hand, these device models can be applied in a more versatile way; there is no need for modifications, if the application purpose varies.

For the evaluation and application of textual test scripts by model descriptions in VHDL-AMS, the test bench requires a so-called script interpreter. Unfortunately, the script interpreter again enhances the complexity of all model descriptions. When designing its model-based test system C&S FACTS. Sim,[20] C&S group GmbH decided to choose this version. The focus was on the reusability of test system components of the existing FACTS-based test systems as well as on a high degree of reproducibility and comparability of model-based tests. The complexity of both of the integration versions is in the same range and therefore this should be of minor importance, when voting for one version or the other (see Fig. 6.14).

[20] FACTS stands for Flexible Adaptable Customizable Test System. This is the generic product name of various test products applied at C&S such as FACTS.CAN.COM/NM and FACTS. AUTOSAR.

There is still one aspect open to be discussed: For measurement data evaluation on one hand and for general reproducibility on the other hand outside, it must be assured that those measurement data of the simulator are stored persistently. Because those data are to be recorded from the corresponding measurement device models, the VHDL-AMS feature of reading and writing external data can be applied. Current simulators offer graphical evaluation tools, but these tools are too much different and therefore would increase the complexity of the simulator interfaces between simulator and test system.

Independent from the chosen version the advantages of model-based testing are obvious. Model-based testing allows safeguarding of the whole development cycle of systems while starting with the development of the individual network components up to the complete topology layout. Simulation and model-based test enable a broader spectrum of analysis (e.g. borderline checks and variation of components) and thus can reduce the risks significantly while increasing the test quality, especially when applied in conjunction with real, physical measurements. On top of this, the development costs can be significantly reduced when applying models instead of physical prototypes. The proliferation of simulator tools together with the continuously ongoing enhancements of VHDL-AMS on language coverage and integration abilities—as it is the case, e.g. for C&S FACTS—fosters more and more the application of model-based tests.

Continuous proliferation of simulation tools as well as especially the support of encryption technologies for securing models for better protection of intellectual property when sharing models with other parties will give an additional push to the acceptance and application of model-based testing.

6.6 CAN ECUs Emulation

A first verification step, as described in the previous chapter, is typically used to check whether the nodes in a CAN network meet basic conformance requirements (see Sect. 6.1). This ensures that the controllers, transceivers and physical properties of the respective CAN nodes allow for secure communication between nodes. Now the question is: will the contents of the individual node applications work together? To address this question, the next step checks the logical properties of the applications, individually as well as jointly in an ECU network.

Because CAN plays an important role in the automotive industry, the examples in this section are based on ECU networks used in automobiles. The information provided here also applies, without restriction, to other sectors where devices are networked using CAN.

6.6.1 Overview and Requirements

Modern CAN networks, as used in the current generation of automobiles, are complex systems. The transmission of large numbers of signals in combination with content-based and time-based dependencies can lead to a large number of possible errors, some of which can be difficult to reproduce. In sectors such as the automotive industry, CAN systems are sold in products that are built in large quantities but then used in highly customized ways. For this reason, customers sometimes experience exotic errors, which can result in high costs (e.g. for product recalls). Reputational damage is also a risk, as a number of striking examples from the automotive industry have shown. Beyond the economic impact of undetected errors, safety is also a consideration. With the increase in distributed functions, safety aspects are of increasing importance.

Comprehensive testing is therefore indispensable. The functionality of any CAN network needs to be tested under all the different conditions in which the system will later operate. It is also important to check ECU behaviour in error situations as well as inconsistent states, e.g. when protocol specifications are violated. It is often precisely such theoretically "impossible" cases which lead to particularly unpleasant problems in practical operation. For this reason, it is important that the ECU reacts as defined in all situations.

Typical test scenarios are aimed at stimulating the ECU and then observing and interpreting its reactions. This generally requires at least a partial simulation of the ECU's environment. For instance, it is rarely possible to operate ECUs on a CAN bus without functioning NM. Similar conditions apply to hardware inputs and outputs. Actuators and sensors are partially checked by the ECU. When such a check fails, the ECU enters an error state. Sensor/actuator inputs/outputs therefore need to be addressed correctly in every case.

Errors that are difficult to reproduce are often caused by time-based dependencies. A specific software error may only appear after specific events occur in a very specific sequence. The test system therefore needs to be able to reproduce time sequences and constraints (e.g. cycle times) in exactly the same way as these would occur in reality. However, it should also be possible to create erroneous or unrealistic states in a targeted and reproducible way.

As a broad preventative measure, testing is required from the earliest development phases. This is because it is far more cost-effective to detect, analyse and eliminate errors in earlier development phases than later on in the development process. The large number of possible system states, combined with the need to maximize the utilization of available bandwidth for efficiency reasons, means that tests can be significant in scope. The only efficient way to handle such large-scale testing requirements is to use automated test sequences.

6.6.2 Testing Methods

6.6.2.1 Protocol Tests

Communication at message level is tested using the so-called protocol tests. Such a test does not use the abstraction layers available in the tester (which, e.g. represent signals on messages and ensure cyclical message transmission); rather, it communicates directly with the ECU SUT.

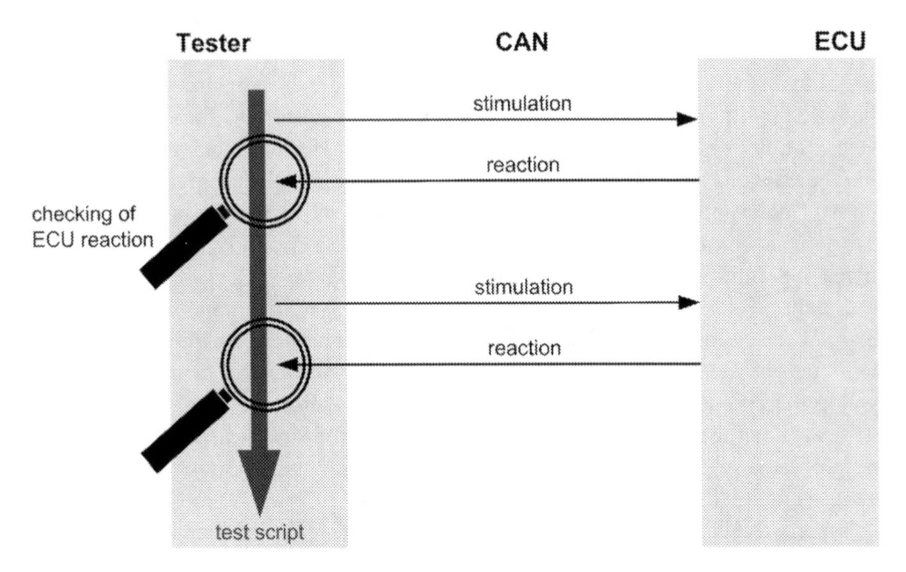

Fig. 6.15 Basic sequence for protocol tests

Protocol tests are generally set up in the following way: The tester stimulates the SUT by sending it a message or message sequence and waiting for the reaction. The test program then checks whether the observed reaction matches the expected one. Depending on the reaction received, the tester then sends another stimulation, waits for a further reaction, checks this reaction, etc.

Individual stimulations and reactions can also consist of input/output (I/O) operations, e.g. activation of an ECU output line. Test sequences can also contain complex conditions and operations. Such a "ping-pong" exchange of actions and reactions, where the tester examines each of the SUT's reactions, is a typical characteristic of this test concept (see Fig. 6.15).

A typical use case for a protocol test is to check the ECU's transport protocol implementation. In this type of test, data items of varying length that are distributed across various messages (CAN frames) are sent to the ECU. The ECU's reaction reveals whether it received and interpreted the message sequences correctly. This also applies in the opposite direction, where the ECU is made to transmit segmented messages. The test system can also send erroneous message sequences in order to interpret the ECU's reaction to these.

Test sequences and check conditions are very tightly interwoven. Test sequences replicate the specified behaviour of the SUT very precisely in order to check whether the SUT behaves in the specified manner. Because the test program itself implements the communication protocol, it can generate exceptional situations and test the SUT's reaction to these. Beyond testing simple protocol sequences, it is also possible to test message contents and timing conditions.

Complete implementation of such protocols within a TC does, however, require a degree of effort when creating the test sequence. Furthermore, the tight linkage

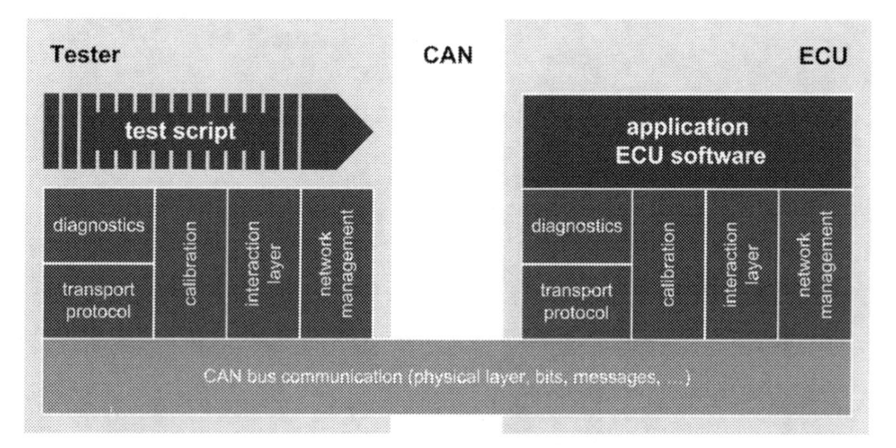

Fig. 6.16 The test program runs on the same abstraction layer as the ECU software being tested

between the tester and the SUT means that any changes made to the SUT have to be taken into consideration.

It is also necessary to know exactly what constitutes correct behaviour on the part of the SUT, even if individual TCs are not always precisely formulated, i.e. when permissible value ranges are used instead of a concrete value. All of this increases the difficulty involved in maintaining and reusing TCs.

6.6.2.2 Application Tests

Application tests are primarily used to test ECU functionality in a complex, networked environment. The objective here is, e.g. to check whether the switch states for light switches, door contacts, control units in roof consoles, light sensors, etc., transmitted via the CAN bus cause the lighting ECU to activate the specified lamps.

To simplify test programs, the tester uses an abstraction layer compatible with the system running on the SUT. The test program thus acts as an ECU application when it communicates with the SUT via a system-compliant communication layer (see Fig. 6.16).

Like a protocol test, an application test also uses a test sequence to stimulate the SUT and then receive and interpret its reactions. Prior to the actual communication, abstraction is implemented via an interaction layer. This means that incoming and outgoing signals are generally read and set in the CAN area. The interaction layer is used to define and implement how and when these signals are sent and received via the CAN bus.

Use of an interaction layer greatly simplifies test sequences, since the alternative is to force the test sequence to replicate all of the properties of the communication protocols precisely. This works on the assumption that the interaction layer is free of errors, which is easier to ascertain by its frequent use. The reduced complexity of the TCs improves their quality and makes it easier to reuse them.

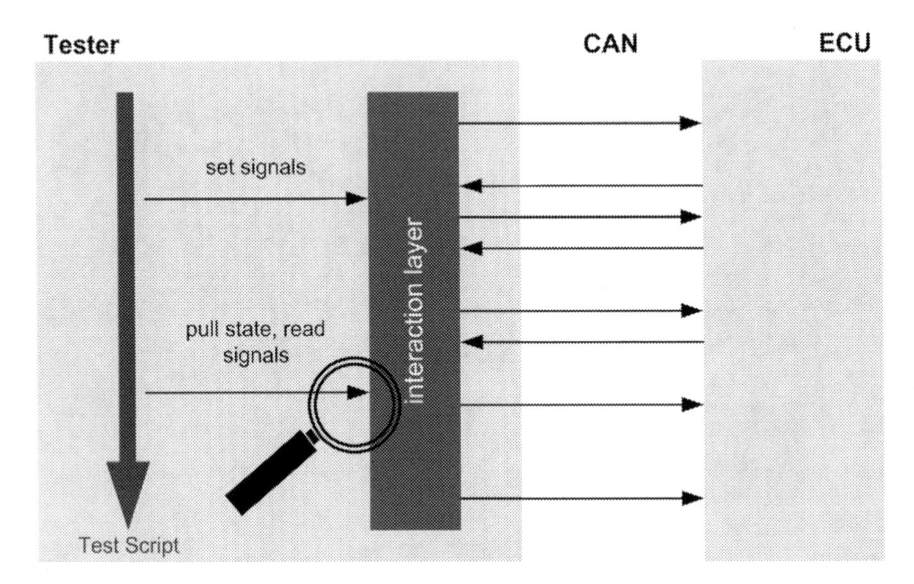

Fig. 6.17 Application tests check for an expected state after a defined interval

Protocol tests are largely event-driven, i.e. the TC reacts directly to perceived events. Such events can include, e.g. receipt of a specific message or measurement of a change in voltage at an ECU port. Application tests, by contrast, tend to be more state based. A test sequence will often wait for a specific minimum interval before it checks whether the expected final state has been reached (see Fig. 6.17). This makes it easy to test complex reactions that involve, e.g. several messages.

It is possible to further increase the degree of abstraction in an application test so that the tester itself no longer actively participates in the communication. In such cases, the test program controls the SUT's simulated environment and interprets the resulting simulation states. This approach is also generally used with hardware-in-the-loop tests. It requires a suitable simulation of the environment and, in particular, of the ECUs on the CAN bus.

6.6.2.3 Invariant Tests

With invariant or observing tests, the SUT is operated under different constraints and its compliance with specified parameters is checked continuously. These parameters are invariants, i.e. conditions that must be adhered to. Such conditions may relate to specific states of the environment or ECU, which means they can be situation dependent.

An example of an invariant condition in a CAN network is the message cycle time. Errors in the cycle time may not necessarily be caused by a software error. They may, under some circumstances, also arise from an overload of the ECU. The cycle time should not be measured while the bus is booting up. An invariant

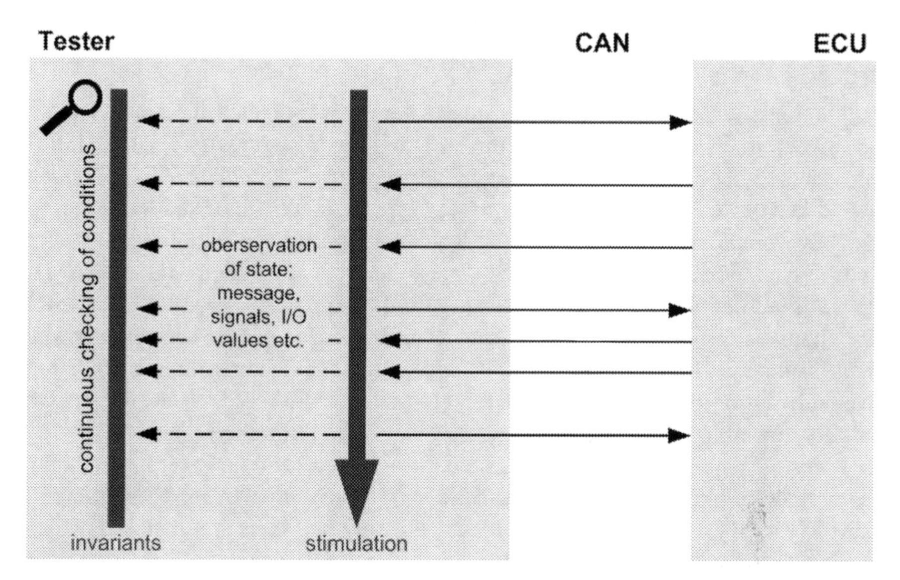

Fig. 6.18 Invariant tests check the communication and behaviour of the SUT

condition can also be made to depend on a specific bus state, and the invariant test can be started at a later point in time if desired. The actual test consists of constant monitoring of the invariant condition while the SUT is subjected to different situations. The latter can be done, e.g. by employing a defined "drive cycle" or by conducting a real test drive. The test is passed when the condition is not violated at any time (see Fig. 6.18). Invariants can be derived from the specification for the SUT, since they essentially describe its properties.

In this testing strategy, stimulation and check conditions are independent of one another and can be developed separately. When developing the stimulation, it is important to run through all the possible types of operational situation as well as all exceptional situations to the fullest extent possible (code coverage, feature coverage, etc.). Developing the stimulation is relatively easy, since the reaction of the ECU being tested does not need to be interpreted. A stimulation sequence can also contain random elements (catchphrase: "playing with the controls"); however, this impacts the repeatability of the test and makes planning test coverage more difficult. On the other hand, random sequences often lead to the identification of errors that may not have been considered by anyone until then. A certain degree of randomness in such tests is therefore advisable, and is certainly unavoidable in real-life environments such as road tests.

Defining sensible invariant conditions can be a tricky task, especially when the goal is to guarantee a wide range of ECU functionality. This is because it is often the case that only a portion of the invariant conditions can be derived from the existing formal specification. For this reason, it may often be preferable to run tests via direct communication when testing complex sequences (see "Application Tests" in Sect. 6.6.2).

6.6.3 Simulating the CAN Network

A functional test for a CAN ECU requires the involvement of the other devices on the bus for two reasons: Firstly, the ECU only works correctly when its infrastructure, e.g. NM, is present and functioning correctly. Secondly, many errors in a networked device can only be observed when it is operating under the same conditions as would apply in a real network. This includes, e.g. the various signals sent cyclically by other ECUs which the SUT must interpret in parallel to the test task. It goes without saying that such loads can impact the SUT's behaviour. In the automotive industry, different vehicle ECU's are typically developed by several suppliers at the same time. For this reason, the ECUs of other manufacturers are generally not available during the development phase. To carry out the necessary testing despite this, the remaining CAN network is usually simulated during this phase.

It makes sense to generate the so-called remaining bus simulation from a suitable network description. The Vector Data Base for CAN communication (DBC) format is widely used for this purpose, as it retains not only messages and signals, but also attributes such as cycle times. In other words, it provides a complete description of the CAN communication, and is often referred to as a database or K-Matrix.[21]

With a remaining bus simulation, it is not necessary to simulate the actual application itself. This means, e.g. that it is sufficient for an ABS ECU in a simulated or partially simulated CAN vehicle network to receive signals relating to wheel revolutions, brake pressure, brake pedal position, etc. and to transmit defined control signals with appropriate values and in compliance with the specified cycle times and protocols. It is not necessary to implement the actual ABS braking algorithm for this purpose.

Remaining bus simulations do not just serve as a simple replacement for CAN network components that are not yet available. They also make it easier to simulate error situations such as the absence or loss of individual messages, which could not be as easily implemented with a real ECU. It is the flexibility of the software-based simulation that is so helpful here, inasmuch as it is easily controlled by the test program. This makes it easy to implement controls for the participating ECUs needed for functional tests in the test program. This is an important prerequisite for test automation.

A remaining bus simulation can also be used prior to running an actual test on a real ECU. The simulation makes it possible to detect possible problems in a CAN network early on and to analyse the impact of changes to the communication definition. The communication definition, in combination with the very illustrative remaining bus simulation, also serves as a specification document for the development of individual CAN ECUs.

[21] K-Matrix stands for communication matrix because the database also describes which signal is sent and received by which node.

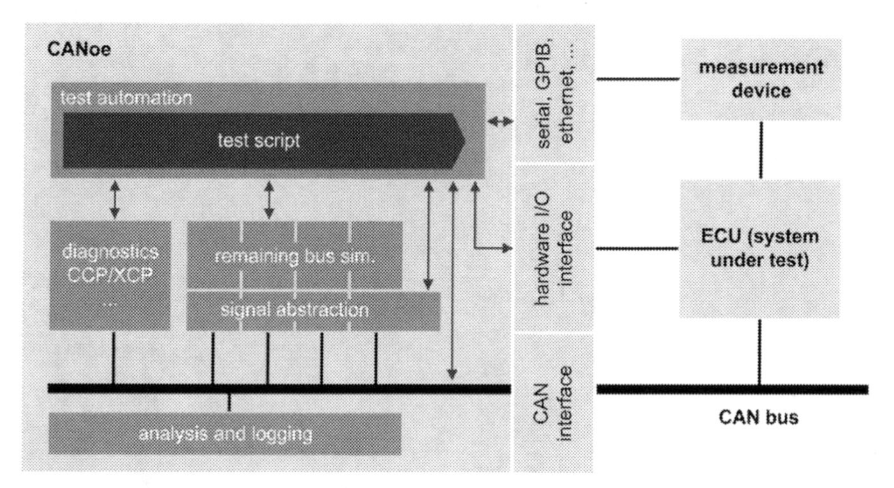

Fig. 6.19 Test cases executed in CANoe can use a range of interfaces to access the SUT

6.6.4 *Testing with Vector CANoe*

A number of programs can be used to develop test systems for CAN ECUs and networks. Here we briefly outline the advantages of automated testing using Vector Informatics widely used CANoe tool suite. CANoe supports the analysis simulation and testing of CAN networks. Because vehicle systems tend to encompass bus systems other than CAN, e.g. MOST for the infotainment system, CANoe is capable of accessing these buses just like it accesses CAN. This enables comprehensive testing of the ECU in question. Various interfaces can also be used to address measurement instruments as well as the hardware interfaces of the SUT and/or the devices in the SUT's environment (e.g. a climatic chamber; see Fig. 6.19).

Test automation is an integral part of CANoe. It enables the execution of test programs that can be formulated in different languages, depending on the application, and can support a wide range of test methods (as described in Sect. 6.6.2). The test program is capable of directly accessing the remaining bus simulation, and can thus control the SUT's environment directly. It can also access protocol layers, such as the transport protocol, NM or signal abstraction, which are integrated as DLLs.

To develop automated tests efficiently it is important to choose the right abstraction layer. Using tried and tested protocol DLLs reduces the likelihood of the TCs themselves containing errors. Protocol tests and the simulation of certain error conditions require direct access to messages, which is also possible in CANoe.

An ECU's diagnostic and calibration interfaces also play an important role in testing. These interfaces are the only way of evaluating the ECU's internal state by means of a test program without modifying the ECU's code. For instance, a diagnostic interface can be used to check whether the ECU correctly interprets an externally generated error pattern.

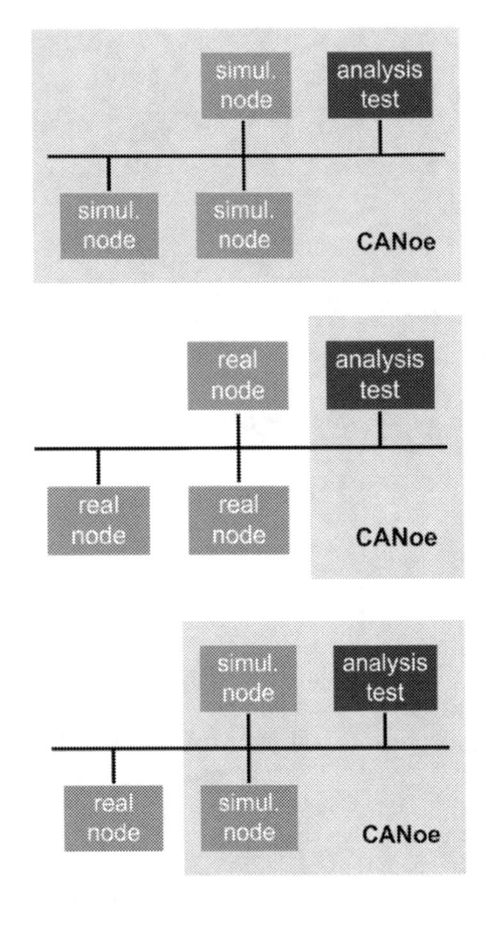

Fig. 6.20 Development of a CAN system in three phases: network design, ECU development and system integration

The diagnostic and calibration protocols used to access the ECU are integrated into CANoe and can thus be readily used in a test program. CANoe interprets the description files (e.g. ODX, ASAP) that define the relevant capabilities of the SUT. The network description is used to automatically generate the remaining bus simulation. Because the transmission behaviour of a CAN system depends on the manufacturer's approach to communications and because various protocols may be used at the same time (transport protocol, NM, etc.), Vector provides manufacturer-specific add-on packages. These contain the components needed to generate a manufacturer-specific remaining bus simulation.

The so-called simulation model builds the basis of the three-phase development of a CAN system in CANoe (see Fig. 6.20). This process enables ECU developers to carry out testing early on, and makes it easy to integrate individual components into a viable overall system. In addition, it enables the use of simulations to solve many problems, which leads to fewer misunderstandings between manufacturers and suppliers.

Phase 1: Network design In the first development phase, the manufacturer specifies the complete CAN network and defines the communication between the individual nodes. This design can be simulated in CANoe in its entirety, which lets developers verify that specifications regarding bus load and latency times are adhered to (See Fig. 6.20 upper section).

Phase 2: ECU development In the second phase, the system description is provided to ECU suppliers as an executable specification. The supplier's developers can now check the communication behaviour of the SUT using the remaining bus simulation (See Fig. 6.20 middle section).

Phase 3: System integration In the third development phase, the manufacturer integrates ECUs from different suppliers one by one to form the overall system. The simulated ECUs are now switched off and CANoe is used to test the real CAN communication (See Fig. 6.20 lower section).

Bibliography

[AUTO01] AUTOSAR GbR, AUTOSAR erreicht ersten Meilenstein—Erste Spezifikationen werden veröffentlicht, Pressemitteilung, 2. Mai 2006, URL: www.autosar.org, Home > Media > Media Releases

[CIA10] CAN in Automation (CiA): URL: http://www.can-cia.org, 2010 Home > Device design > Technology > CAN > CAN history

[CSG10] C&S group GmbH: URL: http://www.cs-group.de, 2010 Home > Conformance Tests > CAN osi-1 > CAN OSI1—Tested Products

[DO178B] Radio Technical Commission for Aeronautics (RTCA), DO-178B Software Considerations in Airborne Systems and Equipment Certification, RTCA, Washington D. C., 1. Dez. 1992

[ETEM02] Berufsgenossenschaft Energie Textil Elektro, Grundsatz für die Prüfung und Zertifizierung von „Bussystemen für die Übertragung sicherheitsrelevanter Nachrichten", Fachausschuss Elektrotechnik, Mai 2002 URL: http://www.bgetem.de/praev/praev_pruefgrundsaetze.html

[HEUR06] Heurung T (2006, 20. März) In-vehicle network design methodology, 2nd IEE Automotive Electronics Conference. The IEE, Savoy Place, London

[HUDI09] Hudi R (2009, 15.–16. Juli) Die E/E Entwicklung im Wandel auf dem Weg zur Elektromobilität, Automobil Elektronik Fachkongress Elektronik, Ludwigsburg

[IEEE1076] IEEE 1076, VHDL Language Reference Manual, IEEE, 2009 ISBN: 978-0-7381-5801-3

IEEE 1076.1, VHDL Analog and Mixed-Signal Extensions, IEEE, 2007 ISBN: 0-7381-5627-2

[ISO9000] DIN EN ISO/IEC 9000 (2005) Qualitätsmanagementsysteme—Grundlagen und Begriffe, ICS 01.040.03, ICS 03.120.10, ISO

[ISO9646] ISO/IEC DIS 9646-1 … 7 (1994–1998) Information technology—Open Systems Interconnection—Conformance testing methodology and framework, ICS 35.100.01, ISO

[ISO11898] ISO, Road vehicles—Interchange of digital information—Controller area network (CAN) for high-speed communication, ISO, 1993 (zurückgezogen, revisioniert durch ISO 11898-1:2003 und ISO 11898-2:2003)

ISO 11898-1 (2003) Road vehicles—Controller area network (CAN)—Part 1: Data link layer and physical signalling, ISO

ISO 11898-2 (2003) Road vehicles—Controller area network (CAN)—Part 2: High-speed medium access unit, ISO

[ISO16845] ISO 16845 (2004) Road vehicles—Controller area network (CAN)—Conformance test plan, ICS 43.040.15, ISO

[ISO17025] DIN EN ISO/IEC 17025 (2005) Allgemeine Anforderungen an die Kompetenz von Prüf- und Kalibrierlaboratorien, ICS 03.120.20, ISO

[KRUE08] Krüger M (2008) Grundlagen der Kraftfahrzeugelektronik—Schaltungstechnik, 2nd edn. Hanser Verlag, München

W. Lawrenz (ed.), *CAN System Engineering,* DOI 10.1007/978-1-4471-5613-0,
© Springer-Verlag London 2013

[LUEB04] Lübke A Car-to-Car Communication—Technologische Herausforderungen. In: VDE-Kongress 2004, Innovationen für Menschen, Band 2: Fachtagungsberichte DGBMT, GMM, GMA, VDE Verlag, Berlin

[MOST06] MOST—Media Oriented Systems Transport, Multimedia and Control Networking Technology, MOST Specification. Revision 2.5, MOST Cooperation, Oktober 2006

[MYER89] Myers GJ Methodisches Testen von Programmen, 3nd edn. Oldenbourg Verlag, München, 1989 ISBN-10: 3-486-21317-2, ISBN-13: 978-3-486-21317-1

[ODVA01] ODVA, CIP Networks Library, Volume 1, Common Industrial Protocol, Edition 3.7, © 2001 bis 2009, ODVA, Inc., November 2009

[ODVA05] ODVA, CIP Networks Library, Volume 5, CIP Safety, Edition 2.2, © 2005 bis 2009, ODVA, Inc., May 2009

[ODVA94] ODVA, CIP Networks Library, Volume 3, DeviceNet Adaptation of CIP, Edition 1.8, © 1994 bis 2009, ODVA, Inc., November 2009

[OEM10] Audi, BMW, Daimler, Porsche, Volkswagen: Hardware Requirements for Partial Networking, Version 1.1, 28.5.2010

[PFEI03] Pfeiffer O et al (2003) Embedded Networking with CAN and CANopen, 1st edn. RTC Books, USA

[SCHI06] Schiffer V (2006) The Common Industrial Protocol (CIP™) and the Family of CIP Networks. Herausgeber: ODVA

[SCHMID] Schmid C ATA26 CAN Application Layer Protocol. SID 2616DD100, Issue 2 (nicht öffentlich)

[SOCAN] URL: http://developer.berlios.de/projects/socketcan

[TIMM09] TIMMO Timing Model, ITEA 2–06005, Version 1.1, 2.9.2009 URL: www.timmo.org

[TIND94] Tindell K, Hansson H, Wellings AJ (1994) Analyzing Real-Time Communications: Controller Area Network (CAN). Fachbereich Computersysteme, Universität Uppsala, Schweden (Tindell, Hansson) & Fachbereich Informatik, Universität York, England (Wellings)

[TIND98] Tindell K, Rajnák A, Casparsson L (1998) A CAN Communications Concept with Guaranteed Message Latencies. SAE 98C050, Dearborn, MI, USA, Convergence

[VDA09] Verband der Automobilindustrie e. V. (ed) (2009) Handeln für den Klimaschutz—CO_2 Reduktion in der Automobilindustrie, 2nd edn, Frankfurt a. M.

[ZELT01] Zeltwanger H (ed) (2001) CANopen. VDE Verlag, Berlin

References to national and international standards ARINC

ARINC 429P1 … P17 Mark 33 Digital Information Transfer System (DITS), Aeronautical Radio Inc.—ARINC, 2004 … 2009

ARINC 600 Air Transport Avionics Equipment Interfaces, Aeronautical Radio Inc., Mai 2010

ARINC 664P1 … P8 Aircraft Data Network, gängige Bezeichnung: AFDX, Aeronautical Radio Inc.—ARINC, 2005 … 2009

ARINC 812 Definition of Standard Data Interfaces For Galley Insert (GAIN) Equipment, CAN Communication, Aeronautical Radio Inc.—ARINC, Dezember 2006

ARINC 825 General Standardization of CAN (Controller Area Network) Bus Protocol for Airborne Use, Aeronautical Radio Inc.—ARINC, Mai 2010

ARINC 826 Software Data Loader Using CAN Interface, Aeronautical Radio Inc.—ARINC, Januar 2009

CISPR

CISPR 16 Specification for radio disturbance and immunity measuring apparatus and methods, siehe auch deutsche Fassung in DIN EN 55016-1-3 Anforderungen an Geräte und Einrichtungen sowie Festlegung der Verfahren zur Messung der hochfrequenten Störaussendung (Funkstörungen) und Störfestigkeit, DIN, Mai 2007

DIN

DIN 19245 PROcess FIeld BUS—PROFIBUS, DIN, 1989 bis 1995

Teil 1: PROFIBUS-FDL (Fieldbus Data Link), DIN, 1989

Teil 2: PROFIBUS-FMS (Fieldbus Message Specification), DIN, 1990

Teil 3: PROFIBUS-DP (Decentralized Peripherals), DIN, 1994

Teil 4: PROFIBUS-PA (Process Automation), DIN, 1995

DIN 41652 Steckverbinder für die Einschubtechnik, trapezförmig, runde Kontakte Ø 1 mm; Maße der Bauform A; Lötanschluss für freie Verdrahtung, DIN, 6/1990

EN

EN 3646 Aerospace series—Connectors, electrical, circular, bayonet coupling, operating temperature 175 °C or 200 °C continuous, Deutsche Version DIN EN 3646: Luft- und Raumfahrt—Elektrische Rundsteckverbinder mit Bajonettkupplung, Betriebstemperatur 175 °C oder 200 °C konstant, Teile 1 bis 10, 2006–2007 (deutsch: 2007–2008)

EN 50090 Home and Building Electronic Systems (HBES)—Elektrische Systemtechnik für Heim und Gebäude (ESHG)

EN 50170 General Purpose Field Communication System—Part 2: PROFIBUS, 1996

EN 50295 Niederspannungsschaltgeräte—Steuerungs- und Geräte-Interface-Systeme—Aktuator Sensor Interface (AS-i), Deutsche Fassung, 1999

EN 50325-4 Industrielles Kommunikationssubsystem basierend auf ISO 11898 (CAN)—Teil 4: CANopen, Deutsche Fassung der EN 50325-4:2002 Industrial communications subsystem based on ISO 11898 (CAN) for controller-device interfaces—Part 4: CANopen, Text in Englisch, DIN, 3/2007

IEC

IEC 61000 Electromagnetic compatibility, ICS 33.100.20

Part 4-2: Testing and measurement techniques—Electrostatic discharge immunity test, Edition 2.0, IEC, 9. Dezember 2008 Hinweis: IEC 61000-4-2 bildet die Basis für die Norm ISO 10605, die fahrzeugspezifische Anforderungen beschreibt

Part 4-4: Testing and measurement techniques—Electrical fast transient/ burst immunity test, Edition 2.0, IEC, 8. Juli 2004

IEC 61508 Functional safety of electrical/electronic/programmable electronic safety-related systems—Deutsche Fassung: DIN EN 61508: Funktionale Sicherheit sicherheitsbezogener elektrischer/elektronischer/programmierbarer Systeme, IEC, Edition 2.0, 4/2010

IEC 61691 Behavioural Languages, IEC, beinhaltet unter anderem Referenzen zu den Sprachen VHDL, VITAL ASIC, VHDL-AMS und SystemC

Part 6: VHDL Analog and Mixed-Signal Extensions, IEC, 14. Dezember 2009

IEC 61800-7 Adjustable speed electrical power drive systems—Part 7-303: Generic interface and use of profiles for power drive systems—Mapping of profile type 3 to network technologies, IEC, Edition 1.0, 11/2007

IEC 61967 Integrated circuits—Measurement of electromagnetic emissions, 150 kHz to 1 GHz, ICS 31.200

Part 4: Measurement of conducted emissions—1 Ω/150 Ω direct coupling method, Edition 1.1, IEC, 27. Juli 2006

IEC 62026-2 Low-voltage switchgear and controlgear—Controller-device interfaces (CDIs)—Part 2: Actuator sensor interface (AS-i), IEC, 29. Januar 2008

IEC 62132 Integrated circuits—Measurement of electromagnetic immunity, 150 kHz to 1 GHz, ICS 31.200

Part 4: Direct RF power injection method, Edition 1.0, IEC, 21. Februar 2006

IEC 62228 Integrated circuits—EMC evaluation of CAN transceivers, ICS 31.200, Edition 1.0, IEC, 16. Februar 2007

IEEE

IEEE 802.3 IEEE Standard for Information technology-specific requirements—Part 3: Carrier Sense Multiple Access with Collision Detection (CMSA/CD) Access Method and Physical Layer Specifications, IEEE, 2008

IEEE 1076.1 IEEE Standard VHDL Analog and Mixed-Signal Extensions, IEEE, 8. September 2007, ersetzt durch Teil 6 der IEC 61691 Behavioural languages

IEEE 1394 IEEE Standard for High Performance Serial Bus, IEEE, 1. Januar 2008

IEEE 1588 IEEE Standard for a Precision Clock Synchronization Protocol for Networked Measurement and Control Systems, IEEE, 24. Juli 2008

ISO

ISO 7498-1 Information technology—Open Systems Interconnection—Basic Reference Model: The Basic Model, ICS 35.100.01, ISO, 1994

ISO 7637 Road vehicles—Electrical disturbances from conduction and coupling, ICS 43.040.10, ICS 01.040.43

Part 1: Definitions and general considerations, ISO, 2002

Part 3: Electrical transient transmission by capacitive and inductive coupling via lines other than supply lines, ISO, 2007

ISO 9000 Quality management systems—Fundamentals and vocabulary, ICS 01.040.03, ICS 03.120.10, ISO, 2005

ISO 9141 Road vehicles—Diagnostic systems, ICS 43.180

Part 2: CARB requirements for interchange of digital information, ISO, 1994

ISO 9646 Information technology—Open Systems Interconnection—Conformance testing methodology and framework, ICS 35.100.01, ISO, 1994–1998

ISO 10605 Road vehicles—Test methods for electrical disturbances from electrostatic discharge, ICS 43.040.10, ISO, 2008
ISO 10731 Information technology—Open Systems Interconnection—Basic Reference Model—Conventions for the definition of OSI services, ICS 35.100.01, ISO, 1994
ISO 11519 Road vehicles—Low-speed serial data communication, ICS 43.040.15, heute mit ISO 11898 zusammengefasst, ISO, 1994
ISO 11898 Road vehicles—Controller area network (CAN), ICS 43.040.15
Part 1: Data link layer and physical signalling, ISO, 2003
Part 2: High-speed medium access unit, ISO, 2003
Part 3: Low-speed, fault-tolerant, medium-dependent interface, ISO, 2006
Part 4: Time-triggered communication, ISO, 2004
Part 5: High-speed medium access unit with low-power mode, ISO, 2007
ISO 13213 Information technology—Microprocessor systems—Control and Status Registers (CSR) Architecture for microcomputer buses, ICS 35.160, ISO, 1994
ISO 13400 Road vehicles—Diagnostic communication between test equipment and vehicle over internet protocol, ICS 43.180, ICS 43.040.10
Part 2: Network and transport layer requirements and services, Committee Draft, in Entwicklung
ISO 14229 Road vehicles—Unified diagnostic services (UDS), ICS 43.180
Part 1: Specification and requirements, ISO, 2006
Part 2: Session layer services, ISO, in Entwicklung (2010-3-28)
Part 3: UDS on controller area network implementation (UDSonCAN), (früher ISO 15765-3), ISO, in Entwicklung (2010-04-18)
ISO 14230 Road vehicles—Diagnostic systems—Keyword Protocol 2000, ICS 43.180
Part 4: Requirements for emission-related systems, ISO, 2000
ISO 14543 Information technology—Home Electronic Systems (HES) Architecture, ICS 35.200, 35.240.99
Part 3-1: Communication layers—Application layer for network based control of HES Class 1, ISO, 2006
Part 3-2: Communication layers—Transport, network and general parts of data link layer for network based control of HES Class 1, ISO, 2006
Part 3-3: User process for network based control of HES Class 1, ISO, 2007
Part 3-4: System management—Management procedures for network based control of HES Class 1, ISO, 2007
Part 3-5: Media and media dependent layers—Power line for network based control of HES Class 1, ISO, 2007
Part 3-6: Media and media dependent layers—Network based on HES Class 1, twisted pair, ISO, 2007
Part 3-7: Media and media dependent layers—Radio frequency for network based control of HES Class 1, ISO, 2007
ISO 15031 Road vehicles—Communication between vehicle and external equipment for emissions-related diagnostics, ICS 43.040.10, ICS 13.040.50
Part 2: Terms, definitions, abbreviations and acronyms, ISO, 2004
Part 5: Emissions-related diagnostic services, ISO, 2006
Part 6: Diagnostic trouble code definitions, ISO, 2005
ISO 15765 Road vehicles—Diagnostics on Controller Area Networks (CAN), ICS 43.040.15
Part 2: Network layer services, ISO, 2004
Part 3: Implementation of unified diagnostic services (UDS on CAN), ISO, 2004
Part 4: Requirements for emissions-related systems, ISO, 2005
ISO 16845 Road vehicles—Controller area network (CAN)—Conformance test plan, ICS 43.040.15, ISO, 2004
ISO 17025 General requirements for the competence of testing and calibration laboratories, ICS 03.120.20, ISO, 2005
ISO 22901 Road vehicles—Open diagnostic data exchange (ODX), ICS 43.180, ISO, 2008

ISO 26262 Road vehicles—Functional safety, ICS 43.040.10
ISO 27145 Road vehicles—Implementation of emissions-related WWH-OBD communication requirements, wird langfristig ISO 15031-5/SAE J1979 ersetzen, ICS 43.180, ICS 43.040.10
Part 2: Common emissions-related data dictionary, ISO, 2006
Part 3: Common message dictionary, ISO, 2006
Part 4: Connection between vehicle and test equipment, ISO, 2006

RTCA

RTCA/DO-178B Software Considerations in Airborne Systems and Equipment Certification, Radio Technical Commission for Aeronautics (RTCA), Washington D. C., 1. Dezember 1992

SAE

SAE J1587 Electronic Data Interchange Between Microcomputer Systems in Heavy-Duty Vehicle Applications, SAE, Juli 2008
SAE J1699 OBD II Related SAE Specification Verification Test Procedures, SAE, Januar 1998
J1699/3: Vehicle OBD II Compliance Test Cases, SAE, Dezember 2009
SAE J1708 Serial Data Communications Between Microcomputer Systems in Heavy-Duty Vehicle Applications, SAE, Oktober 2008
SAE J1850 Class B Data Communications Network Interface, SAE, Juni 2006
SAE J1930 J1930DA: Electrical/Electronic Systems Diagnostic Terms, Definitions, Abbreviations, and Acronyms, SAE, Mai 2010
SAE J1939 Recommended Practice for a Serial Control and Communications Vehicle Network, SAE, Februar 2010
J1939/11: Physical Layer, 250K bits/s, Twisted Shielded Pair, SAE, August 2009
J1939/15: Reduced Physical Layer, 250K bits/sec, UN-Shielded Twisted Pair (UTP), SAE, August 2008
J1939/21: Data Link Layer, SAE, Dezember 2006
J1939/31: Network Layer, SAE, Mai 2010
J1939/71: Vehicle Application Layer, SAE, Februar 2010
J1939/73: Application Layer—Diagnostics, SAE (Projektstart: Mai 2010)
J1939/84: OBD Communications Compliance Test Cases for Heavy Duty Components and Vehicles, SAE, März 2009
SAE J1962 Diagnostic Connector Equivalent to ISO/DIS 15031-3:December 14, 2001, SAE, April 2002
SAE J1979 E/E Diagnostic Test Modes, technisch gleichwertig zu ISO 15031-5, SAE, in Entwicklung (Projektstart: Juli 2008)
SAE J2012 Diagnostic Trouble Code Definitions, SAE, Dezember 2007
SAE J2284 High Speed CAN (HSC) For Passenger Vehicle Applications, SAE, zurückgezogen Mai 2001, siehe Teile 1 bis 3
J2284/1: High Speed CAN (HSC) for Vehicle Applications at 125 Kbps, SAE, 7. März 2002
J2284/2: High Speed CAN (HSC) for Vehicle Applications at 250 Kbps, SAE, 7. März 2002
J2284/3: High-Speed CAN (HSC) for Vehicle Applications at 500 Kbps, SAE, 2. März 2010
SAE J2411 Single Wire Can Network for Vehicle Applications, SAE, 14. Februar 2000, in Überarbeitung seit Mai 2007

Index

11-bit-Identifier 4
29-bit-Identifier 4

A
ACK 6
Acknowledgement-error 11
Advanced CAN 135
Analyzer-modus 156
Application tests 337
Arbitration 8
AUTOSAR 131, 193, 258

B
Babbling idiot 25
BACnet 280
Basic CAN 7, 25
Basic cyclesafety critical systems 32
Baud Rate Prescaler (BRP) 18
Behavior 2, 252
Bit coding 4
Bit error 10
Bit stuffing 4
Bit timing 18, 60, 115
Bit-timing-logic (BTL) 18
Bulk current injection (BCI) 100, 101, 268
Burst error 6
Bus off 12

C
Cable length, maximum 71, 73
CAN-Cores 135
CAN-IP 138
CAN-IP cores 138
CANoe 341
CAN-Software-Stack 319
Carbon dioxide (CO2) 112
Carrier Sense Multiple Access with Collision Detection and Arbitration on Message Priority (CSMA/CD+AMP) 8
Class A…D 256
Clocking 114
Component specification 284
Conformance 285
Conformance test 283, 285
Conformance test transceiver 296
Control field 5–7
Controller area network (CAN) 112, 113, 147, 155, 160, 319
Costs per node, CAN 280
CRC 6
CRC error 10
Cycle time 28, 204

D
Data Length Code (DLC) 6, 7
Direct Power Injection (DPI) 93
Disturbance 12
Dominant 8

E
Eco innovations 112
ECU test 334
EIB 280
Electromagnetic compatibility (EMC) 62, 63, 65, 66, 86, 87, 89
Electromagnetic interference 14
Electromagnetic noise 86
Electrostatic discharge (ESD) 42, 94
ELMOS Semiconductor AG 115
EMI 95
Emission 2, 45
Emission measurement 93
Encryption 334
End of frame 6
Error active 11, 12

W. Lawrenz (ed.), *CAN System Engineering,* DOI 10.1007/978-1-4471-5613-0,

© Springer-Verlag London 2013

Error correction 9
Error counter 12
Error detection 10, 34
Error flag 11
Error frame 10
Error handling 9, 10, 34
Error passive 11, 12
Event synchronization 33
Extended CAN frame 5

F
FACTS 333
Failure 27, 46, 206, 269, 286
Fault isolation 11
Filtering 137
First-in-first-out (FIFO) 149
FlexRay 257
Form error 10
FPGA 135
Full CAN 7, 25

G
Gateway 26, 158, 159
Generalized Interoperable Fault-tolerant (GIFT) 299
Geniax 279
GND shift 67
Go-to-sleep mode 45

H
Hard sync 19
High-Speed-CAN 257
Human body model (HBM) 94

I
Identifier extension 6
Identifier Extension Flag (IDE) 5
Idle 6
Immission 87
Immunity measurement 93
Implementation under test (IUT) 290
INHIBIT pin 114
Initialization phase 13
Inter Frame Space (IFS) 6
Interference 14
Interference suppression 102
Interframe space 6
International transceiver conformance test (ICT) 299
Interoperability 115, 288
Invariant tests 338
ISO 11519 4
ISO 11898 4
ISO norm 113
ISO 9646 289
ISO/OSI 4

J
J1939 132

L
Latency time 16
Lateral tester 325
Line losses 73
Line statement 66
Local disturbance 10, 11
LogiCORE™ 160
LON 280
Lower tester 305
Low-Speed-CAN 257

M
Masking 137
Model 1
Model based testing 330
Model check 329
Model testing 329
Modelling 81
MOST 257

N
Network propagation delay 54
Network simulation 340
Network topologies–design constraints 55, 56
Noise 14
Noise immunity 90
Normal mode 44

O
Observing tests 338
On-chip clocking 114
On-chip oscillator 115
Oscillator tolerance, calculation 22
OSEK/VDX 321
Overload 15, 16

P
Partial network mode 113
Partial networking 104, 112
Partial networking transceiver 115
Phase buffer segment 1/2 18
Phase error 20
Physical layer, high-speed 48
Power-modi 117
Propagation time segment 18
Protocol tests 335, 337

R
Receive error counter (REC) 12
Recessive 8
Redundant time transmitter 32
Remaining bus simulation 340
Remote transmission request (RTR) 5, 7
Residual system 325
Ringing 69

S
Selective Wake-up Interoperable Transceiver in CAN High-speed (SWITCH) group 113
Serial Linked IO (SLIO) 25
Signal quality 68
Simulation 56
Sleep mode 45
Soft sync 20
Software-test 319
SPICE 330
Spike, filtering 20
Standard CAN Frame 5
Standard-core 194
Stand-by Mode 44
Stand-by RAM 118
Stuff error 10
Stuff-bit 6
Substitute remote request 6
Supervisor 305
Synchronization 18–20
Synchronization Jump Width (SJW) 18

T
Target platform 322
Termination 51
Terminationstandard 51
Test 288, 289, 291–293
Time stamp 138
Time transmitter 32
Time triggered 4
Time-quantum 18
Time-Triggered CAN (TTCAN) 27, 28, 30, 32–34
Timing considerations 16
TLE 6250 G 125
TLE 6251-2G 128
Topology 53
Transceiver 43, 49, 51, 125, 128
Transmission line theory 56
Transmit error counter (TEC) 12
Trial and error 296
Trx_standby mode_swk 110
TSEG1, configuration 19
TSEG2, configuration 19
TwinCAN 148

U
Upper tester 305

V
Validation 80
Verification 115
VHDL-AMS 330

W
Wake-up 113, 115

Printed in the United States
By Bookmasters